SCHAUM'S OUTLINE OF

THEORY AND PROBLEMS

of

INTRODUCTORY SURVEYING

•

JAMES R. WIRSHING, B.S.
NCO-in-Charge, Drafting Department
Vandenberg Air Force Base

ROY. H. WIRSHING, B.I.E.
Registered Professional Engineer

SCHAUM'S OUTLINE SERIES
McGRAW-HILL

New York San Francisco Washington, D.C. Auckland Bogotá Caracas Lisbon
London Madrid Mexico City Milan Montreal New Delhi
San Juan Singapore Sydney Tokyo Toronto

To the Wirshing family who helped in the work:
Colette, Andree, and Susan

JAMES R. WIRSHING is currently a master sergeant in the United States Air Force. He is in a civil engineering squadron stationed at Vandenberg Air Force Base in California, where he is the noncommissioned officer in charge of the drafting department. He received his B.S. in business management and his A.M. (Associate in Management) from the University of Maryland. He is coauthor of *Civil Engineering Drafting*, published by McGraw-Hill, and is experienced in field engineering work as well as in the problems of the trade.

ROY H. WIRSHING, now retired from the civil engineering and teaching fields, is a professional engineer with a degree in industrial engineering from Ohio State University. His civil engineering experience includes problem solutions and highway design. He is coauthor of *Civil Engineering Drafting*.

Schaum's Outline of Theory and Problems of
INTRODUCTORY SURVEYING

17 CUS CUS 08

ISBN 0-07-071124-0

Sponsoring Editor, Jeff McCartney
Editing Supervisor, Nancy Warren
Production Manager, Nick Monti

Library of Congress Cataloging in Publication Data

Wirshing, James R.
  Theory and problems of introductory surveying.

  (Schaum's outline series)
  Includes index.
  1. Surveying. I. Wirshing, Roy H. II. Title.
TA545.W7555 1985   526.9   84-19427
ISBN 0-07-071124-0

McGraw-Hill
A Division of The **McGraw-Hill** Companies

Preface

We have attempted to write a book that will complement the five best-selling surveying textbooks used by students in junior colleges and technical schools. Eighty percent of the material covered in those texts is also covered in this one. In addition, this book should be a good supplement to any one of the leading texts used in elementary civil engineering courses.

The best way to learn how to solve civil engineering and surveying problems is to work a large number of problems. Most texts used in surveying and engineering courses give only a few completely solved examples, but this Outline contains a great many problems and examples paralleling the popular textbooks, and detailed solutions are given in simple and logical fashion.

Following through the different types of problems and concentrating on the method of solution will make elementary surveying far simpler than using a text which explains just a few types of problems. Studying the text and examples carefully and following through the solved problems should ensure success for the student in the course.

We wish to thank our sponsoring editor, Jeff McCartney, and our editing supervisor, Nancy Warren, for handling the details of the book.

We are grateful to the Florida Department of Transportation for their help in road design problems, and to the U.S. Army and U.S. Air Force for allowing us to use materials from their manuals.

Additional acknowledgments include David White Instruments, Charles Bruning Co., Dietzgen Corporation, Keuffel and Esser Co., and Faber-Castell Corporation.

JAMES R. WIRSHING
ROY H. WIRSHING

Contents

CONTENTS

CONTENTS

CONTENTS

Chapter 1

Trigonometry for Surveyors

1.1 TRIGONOMETRY

Trigonometry makes possible the computation of relationships between the lengths of the sides of a triangle and the sizes of its angles. It is a link between straight-line measurement and angular measurement. It has many uses and is essential in surveying computations.

1.2 ANGLES AND THEIR MEASUREMENT

An *angle* is the figure that is formed when a line, or *ray*, turns from some initial position *OC* to a terminal position *OD* (see Figs. 1-1 and 1-2). When the ray turns counterclockwise (CCW), as in Fig. 1-1, the angle is *positive*. Clockwise (CW) turning of the ray, as in Fig. 1-2, generates a *negative* angle.

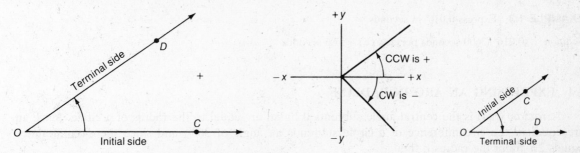

Fig. 1-1   A positive angle.   **Fig. 1-2**   A negative angle.

A *degree* (°) is a unit of angular measurement equal to 1/360 revolution (rev). When a circle is graduated, that is, divided into degrees,

$$1° = \frac{1}{360} = 0.00278 \text{ rev}$$

The fractional part of a degree can be expressed as a common fraction or as a decimal; angles written in decimal form are sometimes called *decimal degrees*.

EXAMPLE 1.1   Express the following in degrees or decimal degrees: 1/2 rev, 3/4 rev, 1/50 rev, 1/48 rev, 1/48 rev, 1/200 rev.

Solution:

$$\frac{1}{2}\text{ rev} = \frac{1}{2}(360°) = 180° \quad Ans. \qquad \frac{1}{48}\text{ rev} = \frac{1}{48}(360°) = 7.5° \quad Ans.$$

$$\frac{3}{4}\text{ rev} = \frac{3}{4}(360°) = 270° \quad Ans. \qquad \frac{1}{200}\text{ rev} = \frac{1}{200}(360°) = 1.8° \quad Ans.$$

$$\frac{1}{50}\text{ rev} = \frac{1}{50}(360°) = 7.2° \quad Ans.$$

1.3 EXPRESSING THE FRACTIONAL PART OF A DEGREE IN MINUTES AND SECONDS

Instead of using decimal degrees we may express the fractional part of a degree in minutes and seconds.

1

A *minute* (') is defined as one-sixtieth of a degree ($\frac{1}{60}°$). Thus

$$1° = 60 \text{ minutes}$$

A *second* (") is defined as one-sixtieth of a minute. Thus

$$1 \text{ minute} = 60 \text{ seconds}$$

$$1° = (60 \text{ minutes per degree}) (60 \text{ seconds per minute}) = 3600 \text{ seconds}$$

Minutes and seconds may be expressed in revolutions as follows:

$$1 \text{ minute} = \frac{1}{360(60)} = 0.0000462 \text{ rev}$$

$$1 \text{ second} = \frac{1}{360(60)(60)} = 0.000000772 \text{ rev}$$

EXAMPLE 1.2 How many minutes are in 0.6°?

Solution: (0.6°)(60 minutes per degree) = 36 minutes   *Ans.*

EXAMPLE 1.3 Express 0.016° in seconds.

Solution: (0.016°)(3600 seconds per degree) = 57.6 seconds   *Ans.*

1.4 EXPRESSING AN ARC IN RADIANS

A *radian* (rad) is the central angle subtended by an arc equal to the radius of a circle. Since an arc equal to the circumference of a circle subtends an angle of 360°, and since the circumference equals 2π times the radius *r*, then

$$2\pi \text{ rad} = 360° = 1 \text{ rev}$$

EXAMPLE 1.4 Convert 2.50 rad to decimal degrees.

Solution: Since 2π rad = 360°, π rad = 180°. So

$$(2.50 \text{ rad})\left(\frac{180°}{\pi \text{ rad}}\right) = 143.24°   \text{ \textit{Ans.}}$$

Degrees may also be expressed in radians. Use the following formula:

$$x°\left(\frac{\pi \text{ rad}}{180°}\right) = y \text{ rad}$$

EXAMPLE 1.5 Express 83.9° in radians.

Solution: $$83.9°\left(\frac{\pi \text{ rad}}{180°}\right) = 1.46 \text{ rad}   \text{ \textit{Ans.}}$$

Note: The following conversions will be useful in solving problems:

$$1° = 0.1745329 \text{ rad}$$
$$1 \text{ minute} = 0.00029088 \text{ rad} = 2.9088 \times 10^{-4} \text{ rad}$$
$$1 \text{ second} = 0.000004848 \text{ rad} = 4.848 \times 10^{-6} \text{ rad}$$
$$1 \text{ rad} = 57.29578°$$

All the examples and problems are easily solved by the use of a hand-held calculator.

1.5 ANGLE CONVERSIONS

Angles are converted in the same manner as are other units. The conversion factor is written as a fraction with the units to be canceled located in the denominator. Multiply the given angle by the fraction and round the result to the same number of significant figures as in the given angle.

Converting Revolutions to Degrees

EXAMPLE 1.6   Convert 0.5100 rev to degrees.

Solution: Use the conversion factor

$$1 \text{ rev} = 360°$$

So
$$(0.5100 \text{ rev})\left(\frac{360°}{1 \text{ rev}}\right) = 183.6°   \quad Ans.$$

Converting Revolutions to Radians

In converting revolutions to radians, the conversion factor to use is

$$1 \text{ rev} = 2\pi \text{ rad}$$

EXAMPLE 1.7   Convert 0.8450 rev to radians.

Solution: The conversion factor is

$$1 \text{ rev} = 2\pi \text{ rad}$$

So
$$(0.8450 \text{ rev})\left(\frac{2\pi \text{ rad}}{1 \text{ rev}}\right) = 5.309 \text{ rad}$$

Converting Degrees to Revolutions and Radians

EXAMPLE 1.8   Convert 151.3° to revolutions.

Solution: The conversion factor is

$$1 \text{ rev} = 360°$$

So
$$(151.3°)\left(\frac{1 \text{ rev}}{360°}\right) = 0.4203 \text{ rev}   \quad Ans.$$

EXAMPLE 1.9   Convert 92.572° to (*a*) revolutions and (*b*) radians.

Solution: (*a*) $(92.572°)\left(\dfrac{1 \text{ rev}}{360°}\right) = 0.25714 \text{ rev}   \quad Ans.$

(*b*) $(92.572°)\left(\dfrac{\pi \text{ rad}}{180°}\right) = 1.61569 \text{ rad}   \quad Ans.$

Converting Radians to Degrees and Revolutions

EXAMPLE 1.10   Convert 1.5362 rad to (*a*) degrees and (*b*) revolutions.

Solution: (*a*) $(1.5362 \text{ rad})\left(\dfrac{180°}{\pi \text{ rad}}\right) = 88.018°   \quad Ans.$

(*b*) $(1.5362 \text{ rad})\left(\dfrac{1 \text{ rev}}{2\pi \text{ rad}}\right) = 0.24449 \text{ rev}   \quad Ans.$

1.6 OPERATIONS WITH ANGLES EXPRESSED IN DEGREES, MINUTES, AND SECONDS

Conversions

An angle expressed in degrees, minutes, and seconds (DMS) that is to be converted to any other units must first be converted to decimal degrees. The decimal-degree angle then can be converted to an angle expressed in other units.

EXAMPLE 1.11   Convert 49°18′37″ to decimal degrees.

Solution: First convert the seconds to degrees:

$$(37 \text{ seconds})\left(\frac{1°}{3600 \text{ seconds}}\right) = 0.0103°$$

Next convert the minutes to degrees:

$$(18 \text{ minutes})\left(\frac{1°}{60 \text{ minutes}}\right) = 0.3000°$$

Express this answer to four places to match the four places contained in the conversion of seconds to degrees (0.0103).

Finally, add the degrees, minutes, and seconds:

$$49.0000 + 0.0103 + 0.3000 = 49.3103° \qquad Ans.$$

Instructional note: A hand calculator with a DMS key simplifies the work involved in solving this type of problem. Then the solution may be found as follows:

Enter degrees and depress decimal point.

Enter minutes and depress decimal point.

Enter seconds and depress DMS key.

The readout shows decimal degrees with extra significant figures. Round off to four decimal digits.

Note: Use the placeholder 0 for any single-digit entries. That is, enter 1° (or 1 minute or 1 second) as 01.

Converting Degrees, Minutes, and Seconds to Radians

EXAMPLE 1.12   Convert 5°46′12″ to radians.

Solution: First convert to decimal degrees:

$$(12 \text{ seconds})\left(\frac{1°}{3600 \text{ seconds}}\right) = 0.0033°$$

$$(46 \text{ minutes})\left(\frac{1°}{60 \text{ minutes}}\right) = 0.7667°$$

Add: $$0.0033° + 0.7667° + 5.0000° = 5.7700°$$

Next convert decimal degrees to radians:

$$(5.7700°)\left(\frac{\pi \text{ rad}}{180°}\right) = 0.10070 \text{ rad} \qquad Ans.$$

Converting Decimal Degrees to Degrees, Minutes, and Seconds

EXAMPLE 1.13   Convert 66.4941° to DMS.

Solution: Convert the decimal part to minutes:

$$(0.4941°)\left(\frac{60 \text{ minutes}}{1°}\right) = 29.65 \text{ minutes}$$

Convert the decimal part of 29.65 minutes to seconds:

$$(0.65 \text{ minute})\left(\frac{60 \text{ seconds}}{1 \text{ minute}}\right) = 39 \text{ seconds}$$

Putting it all together yields 66°29'39".   *Ans.*

Instructional note: If your hand-held calculator has a DMS key, you can use the inverse (INV) key with the DMS key. Then the solution may be obtained as follows:

Enter decimal degree: 66.4941

Press INV key

Press DMS key

Readout = 66.2939 = 66°29'39"

Addition and Subtraction

Add or subtract degrees, minutes, and seconds separately. Remove multiples of 60 from the sums of minutes and seconds and determine how many degrees are contained in the sum of minutes and how many minutes are contained in the sum of seconds.

EXAMPLE 1.14   Add 35°42'28" and 57°31'59".

Solution:	*Degrees*	*Minutes*	*Seconds*
	35	42	28
	+ 57	+ 31	+ 59
Add:	92	73	87
Remove multiples of 60		+ 1	− 60
and add to the next column:		74	27
	+ 1	− 60	
	93	14	27

Total = 93°14'27"   *Ans.*

EXAMPLE 1.15   Subtract 25°33'42" from 28°13'28".

Solution: Borrow from minutes and degrees so that subtraction can be performed.

	Degrees	*Minutes*	*Seconds*
	28	13	28
	− 1	+ 60	
	27	73	28
		− 1	+ 60
	27	72	88
	− 25	− 33	− 42
	2	39	46

So   28°13'28" − 25°33'42" = 2°39'46"   *Ans.*

Multiplication and Division

The method for multiplication and division of angles is as follows:

1. Convert the given angle to decimal degrees.
2. Perform the multiplication or division.
3. Convert back to DMS.

Multiplication of Angles

EXAMPLE 1.16   Multiply 13°28′35″ by 2.7354.

Solution: Convert to decimal degrees:

$$(28 \text{ minutes})\left(\frac{1°}{60 \text{ minutes}}\right) = 0.4667°$$

$$(35 \text{ seconds})\left(\frac{1°}{3600 \text{ seconds}}\right) = 0.0097°$$

$$13.0000° + 0.4667° + 0.0097° = 13.4764°$$

Multiply:

$$13.476(2.7354) = 36.862°$$

Convert back to DMS:

$$(0.862°)\left(\frac{60 \text{ minutes}}{1°}\right) = 51.7 \text{ minutes}$$

$$(0.7 \text{ minute})\left(\frac{60 \text{ seconds}}{1 \text{ minute}}\right) = 42 \text{ seconds}$$

So $(13°28′35″)(2.7354) = 36°52′42″$ *Ans.*

Division of Angles

EXAMPLE 1.17   Divide 72°32′ by 2.831.

Solution: Convert to decimal degrees:

$$(32 \text{ minutes})\left(\frac{1°}{60 \text{ minutes}}\right) = 0.533°$$

$$0.533° + 72.000° = 72.533°$$

Divide:

$$\frac{72.533°}{2.831} = 25.621°$$

Convert to DMS:

$$(0.621°)\left(\frac{60 \text{ minutes}}{1°}\right) = 37.26 \text{ minutes}$$

$$(0.26 \text{ minutes})\left(\frac{60 \text{ seconds}}{1 \text{ minute}}\right) = 16 \text{ seconds}$$

So $\dfrac{72°32′}{2.831} = 25°37′16″$ *Ans.*

1.7   AREAS OF TRIANGLES

The area of any triangle is equal to one-half the product of the base times the altitude to that base. See Fig. 1-3. The formula is

$$A = \tfrac{1}{2}bh$$

where A = area
          b = base
          h = height, or altitude

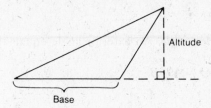

Fig. 1-3   Triangle showing base and
altitude.

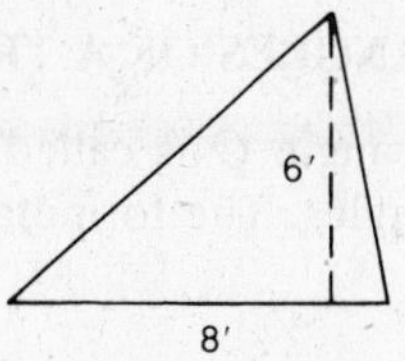

Fig. 1-4

EXAMPLE 1.18   Given: A triangle with a base of 8 ft and a height of 6 ft (see Fig. 1-4). Find the area of the triangle.

Solution: $A = \frac{1}{2}bh$

$A = \frac{1}{2}(6)(8) = 24$ square feet (ft^2)   *Ans.*

The formula for the area of a triangle having sides of lengths *a*, *b*, and *c* is as follows:

$$A = \sqrt{s(s-a)(s-b)(s-c)}$$

where $s = \dfrac{a+b+c}{2}$

EXAMPLE 1.19   Given: A triangle with sides $a = 81$ centimeters (cm), $b = 50$ cm, and $c = 60$ cm. (See Fig. 1-5.) Find the area in square centimeters (cm^2).

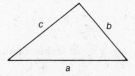

Fig. 1-5

Solution: Use the formula $s = (a + b + c)/2$ to find s:

$$s = \frac{81 + 50 + 60}{2} = 95.5 \text{ cm}$$

Use the area formula:

$$A = \sqrt{s(s-a)(s-b)(s-c)}$$
$$A = \sqrt{95.5(95.5 - 81)(95.5 - 50)(95.5 - 60)} = 1495.57 \text{ cm}^2 \qquad Ans.$$

1.8   INTERIOR ANGLES OF A TRIANGLE

The sum of the three interior angles of any triangle is 180°. The formula is

$$A + B + C = 180°$$

where *A*, *B*, and *C* are the three interior angles of the triangle.

EXAMPLE 1.20   Given: A triangle with angles $B = 80°$ and $C = 65°$. Find angle *A*.

Solution: $A + B + C = 180°$. So

$$A = 180° - B - C$$
$$A = 180° - 80° - 65° = 35° \qquad Ans.$$

1.9 EXTERIOR ANGLES OF A TRIANGLE

In Fig. 1-6 the angle Q is called an *exterior angle*. An exterior angle equals the sum of the two opposite interior angles. The formula is

$$Q = A + B$$

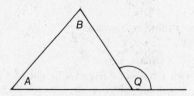

Fig. 1-6 Triangle showing an exterior angle.

EXAMPLE 1.21 See Fig. 1-6. Given: A triangle with angles $A = 30°$ and $B = 100°$ (see Fig. 1-6). Find angle Q.

Solution:
$$Q = A + B$$
$$Q = 30° + 100° = 130°$$

1.10 PYTHAGOREAN THEOREM

The pythagorean theorem applies to a right triangle (that is, any triangle having a right angle). The square of the hypotenuse (the side opposite the right angle) is equal to the sum of the squares of the other two sides, or legs. In equation form,

$$a^2 + b^2 = c^2$$

EXAMPLE 1.22 Given: A triangle with leg $a = 3.6$ inches (in) and leg $b = 4.7$ in. (See Fig. 1-7.) Find the hypotenuse c.

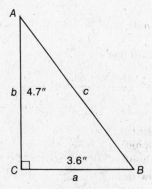

Fig. 1-7 Right triangle.

Solution: Use the pythagorean theorem:
$$a^2 + b^2 = c^2$$
$$c^2 = a^2 + b^2$$
$$c^2 = (3.6)^2 + (4.7)^2 = 12.96 + 22.09 = 35.05$$
So
$$c = \sqrt{35.05} = 5.92 \text{ in} \qquad Ans.$$

1.11 STANDARD POSITION FOR AN ANGLE

The standard position for an angle is on the coordinate axes with the vertex of the angle at the origin and the initial side of the angle along the positive x axis. See Fig. 1-8. Figure 1-8 shows the

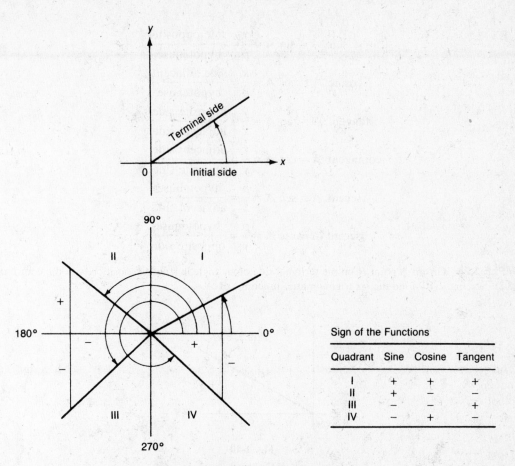

Fig. 1-8 Standard position for an angle.

Sign of the Functions			
Quadrant	Sine	Cosine	Tangent
I	+	+	+
II	+	−	−
III	−	−	+
IV	−	+	−

four quadrants of the cartesian coordinate system, and the signs of the trigonometric functions (see Sec. 1.12) according to quadrant.

1.12 TRIGONOMETRIC RATIOS

Let A be some angle in standard position (see Fig. 1-9). From any point R on the terminal side of the angle we draw a perpendicular to the x axis. This forms a right triangle ORS. Side OR is the hypotenuse of the triangle, and has a length p; RS is the side opposite angle A and has a length y; OS is adjacent to angle A and has a length x.

The six trigonometric ratios are now defined in terms of the coordinates of point R, as well as in terms of the sides of the right triangle ORS.

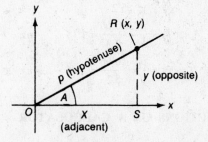

Fig. 1-9 Triangle showing trigonometric ratios.

$$\text{sine } A = \sin A = \frac{y}{p} = \frac{\text{side opposite}}{\text{hypotenuse}}$$

$$\text{cosine } A = \cos A = \frac{x}{p} = \frac{\text{side adjacent}}{\text{hypotenuse}}$$

$$\text{tangent } A = \tan A = \frac{y}{x} = \frac{\text{opposite side}}{\text{adjacent side}}$$

$$\text{cotangent } A = \cot A = \frac{x}{y} = \frac{\text{adjacent side}}{\text{opposite side}}$$

$$\text{secant } A = \sec A = \frac{p}{x} = \frac{\text{hypotenuse}}{\text{adjacent side}}$$

$$\text{cosecant } A = \csc A = \frac{p}{y} = \frac{\text{hypotenuse}}{\text{opposite side}}$$

EXAMPLE 1.23 Given: A point R on the terminal side of an angle in standard position with the coordinates of (9.3, 5.2). See Fig. 1-10. Find the six trigonometric functions of A.

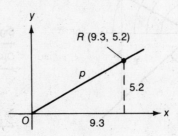

Fig. 1-10

Solution: Compute the distance OR by the pythagorean theorem.

$$c^2 = a^2 + b^2$$
$$(OR)^2 = (9.3)^2 + (5.2)^2 = 86.49 + 27.04 = 113.53$$

So

$$OR = \sqrt{113.53} = 10.66$$

Then, using the equations for the six trigonometric ratios, we find:

$$\sin A = \frac{y}{p} = \frac{5.2}{10.66} = 0.49 \quad Ans.$$

$$\cos A = \frac{x}{p} = \frac{9.3}{10.66} = 0.87 \quad Ans.$$

$$\tan A = \frac{y}{x} = \frac{5.2}{9.3} = 0.56 \quad Ans.$$

$$\cot A = \frac{x}{y} = \frac{9.3}{5.2} = 1.79 \quad Ans.$$

$$\sec A = \frac{p}{x} = \frac{10.66}{9.3} = 1.15 \quad Ans.$$

$$\csc A = \frac{p}{y} = \frac{10.66}{5.2} = 2.05 \quad Ans.$$

1.13 TRIGONOMETRIC FUNCTIONS ON A CALCULATOR

Sine, Cosine, and Tangent

Enter the angle and depress the SIN, COS, or TAN key.

EXAMPLE 1.24   Find sin 47.7° to four decimal places.

Solution: Be sure your calculator is in the degree, not radian, mode.

Enter 47.7.

Depress the SIN key.

Read 0.7396.

So
$$\sin 47.7° = 0.7396 \qquad Ans.$$

Cotangent, Secant, and Cosecant

Most calculators have no keys for these functions. We can use the reciprocal relationships:

$$\sin A = \frac{1}{\csc A} \qquad \cos A = \frac{1}{\sec A} \qquad \tan A = \frac{1}{\cot A}$$

We can obtain reciprocals on the calculator with the 1/X key.

EXAMPLE 1.25   Find cot 21.4° to four significant figures.

Solution: The calculator must be in the degree mode. Enter 21.4; depress the TAN key; depress 1/X key; read out 2.5517. So

$$\cot 21.4° = 2.5517 \qquad Ans.$$

Solved Problems

1.1   Express the following in degrees or decimal degrees: 1/3 rev, 2/3 rev, 1/8 rev, 1/4 rev.

Solution

$$\frac{1}{3}\,\text{rev} = \frac{1}{3}(360°) = 120° \qquad Ans.$$

$$\frac{2}{3}\,\text{rev} = \frac{2}{3}(360°) = 240° \qquad Ans.$$

$$\frac{1}{8}\,\text{rev} = \frac{1}{8}(360°) = 45° \qquad Ans.$$

$$\frac{1}{4}\,\text{rev} = \frac{1}{4}(360°) = 90° \qquad Ans.$$

1.2   Express the following fractional revolutions as decimal degrees: 1/52, 2/32, 1/32, 3/32, 1/16.

Solution

$$\frac{1}{52}\,\text{rev} = \frac{1}{52}(360°) = 6.92° \qquad Ans.$$

$$\frac{1}{32}\,\text{rev} = \frac{1}{32}(360°) = 11.25° \qquad Ans.$$

$$\frac{3}{32}\,\text{rev} = \frac{3}{32}(360°) = 33.75° \qquad Ans.$$

$$\frac{1}{16}\,\text{rev} = \frac{1}{16}(360°) = 22.5° \qquad Ans.$$

1.3   Express the following decimal degrees in minutes: 0.5°, 0.3°, 0.33°, 0.43°.

Solution

$$0.5°(60 \text{ minutes per degree}) = 30 \text{ minutes} \qquad Ans.$$
$$0.3°(60 \text{ minutes per degree}) = 18 \text{ minutes} \qquad Ans.$$
$$0.33°(60 \text{ minutes per degree}) = 19.8 \text{ minutes} \qquad Ans.$$
$$0.43°(60 \text{ minutes per degree}) = 25.8 \text{ minutes} \qquad Ans.$$

1.4   Express the following decimal degrees in minutes: 0.95°, 0.80°, 0.75°, 0.71°.

Solution

$$0.95°(60 \text{ minutes per degree}) = 57 \text{ minutes} \qquad Ans.$$
$$0.80°(60 \text{ minutes per degree}) = 48 \text{ minutes} \qquad Ans.$$
$$0.75°(60 \text{ minutes per degree}) = 45 \text{ minutes} \qquad Ans.$$
$$0.71°(60 \text{ minutes per degree}) = 42.6 \text{ minutes} \qquad Ans.$$

1.5   Convert 0.01° to seconds.

Solution

$$0.01°(3600 \text{ seconds per degree}) = 36 \text{ seconds} \qquad Ans.$$

1.6   Convert 0.006° to seconds.

Solution

$$0.006°(3600 \text{ seconds per degree}) = 21.6 \text{ seconds} \qquad Ans.$$

1.7   Express 2.32 rad in decimal degrees.

Solution

$$(2.32 \text{ rad})\left(\frac{180°}{\pi \text{ rad}}\right) = 132.93° \qquad Ans.$$

1.8   Express 1.34 rad in decimal degrees.

Solution

$$(1.34 \text{ rad})\left(\frac{180°}{\pi \text{ rad}}\right) = 76.78° \qquad Ans.$$

1.9   Express 65.2° in radians.

Solution

$$62.5°\left(\frac{\pi \text{ rad}}{180°}\right) = 1.09 \text{ rad} \qquad Ans.$$

1.10   Express 89.7° in radians.

Solution

$$89.7°\left(\frac{\pi \text{ rad}}{180°}\right) = 1.57 \text{ rad} \qquad Ans.$$

1.11   Convert 0.8240 rev to degrees.

Solution

Use conversion factor 1 rev = 360°.

$$(0.8240 \text{ rev})(360°/\text{rev}) = 296.6° \qquad Ans.$$

1.12   Convert 0.7351 rev to degrees.

Solution

Use conversion factor 1 rev = 360°.

$$(0.7351 \text{ rev})(360°/\text{rev}) = 264.6° \qquad Ans.$$

1.13   Convert 0.6932 rev to degrees.

Solution

Use conversion factor 1 rev = 360°.

$$(0.6932 \text{ rev})(360°/\text{rev}) = 249.6° \qquad Ans.$$

1.14   Convert 0.7215 rev to radians.

Solution

Conversion factor: $1 \text{ rev} = 2\pi \text{ rad}$
So: $(0.7215 \text{ rev})(2\pi \text{ rad}/\text{rev}) = 4.533 \text{ rad} \qquad Ans.$

1.15   Convert 0.6130 rev to radians.

Solution

Conversion factor: $1 \text{ rev} = 2\pi \text{ rad}$
So: $(0.6130 \text{ rev})(2\pi \text{ rad}/\text{rev}) = 3.852 \text{ rad} \qquad Ans.$

1.16   Convert 0.5125 rev to radians.

Solution

Conversion factor: $1 \text{ rev} = 2\pi \text{ rad}$
So: $(0.5125 \text{ rev})(2\pi \text{ rad}/\text{rev}) = 3.220 \text{ rad} \qquad Ans.$

1.17   Convert 85.253° to revolutions and radians.

Solution

$$85.253°\left(\frac{1 \text{ rev}}{360°}\right) = 0.23681 \text{ rev} \qquad Ans. \qquad 85.253°\left(\frac{\pi \text{ rad}}{180°}\right) = 1.48795 \text{ rad} \qquad Ans.$$

1.18   Convert 79.213° to revolutions and radians.

Solution

$$79.213°\left(\frac{1 \text{ rev}}{360°}\right) = 0.22004 \text{ rev} \qquad Ans. \qquad 79.213°\left(\frac{\pi \text{ rad}}{180°}\right) = 1.38253 \text{ rad} \qquad Ans.$$

1.19 Convert 61.111° to revolutions and radians.

Solution

$$61.111°\left(\frac{1\ \text{rev}}{360°}\right) = 0.16975\ \text{rev} \qquad \textit{Ans.}$$

$$61.111°\left(\frac{\pi\ \text{rad}}{180°}\right) = 1.06659\ \text{rad} \qquad \textit{Ans.}$$

1.20 Convert 1.4002 rad to (*a*) degrees and (*b*) revolutions.

Solution

(*a*) $$(1.4002\ \text{rad})\left(\frac{180°}{\pi\ \text{rad}}\right) = 80.226° \qquad \textit{Ans.}$$

(*b*) $$(1.4002\ \text{rad})\left(\frac{1\ \text{rev}}{2\pi\ \text{rad}}\right) = 0.22285\ \text{rev} \qquad \textit{Ans.}$$

1.21 Convert 1.3210 rad to (*a*) degrees and (*b*) revolutions.

Solution

(*a*) $$(1.3210\ \text{rad})\left(\frac{180°}{\pi\ \text{rad}}\right) = 75.688° \qquad \textit{Ans.}$$

(*b*) $$(1.3210\ \text{rad})\left(\frac{1\ \text{rev}}{2\pi\ \text{rad}}\right) = 0.21024\ \text{rev} \qquad \textit{Ans.}$$

1.22 Convert 1.2513 rad to (*a*) degrees and (*b*) revolutions.

Solution

(*a*) $$(1.2513\ \text{rad})\left(\frac{180°}{\pi\ \text{rad}}\right) = 71.694° \qquad \textit{Ans.}$$

(*b*) $$(1.2513\ \text{rad})\left(\frac{1\ \text{rev}}{2\pi\ \text{rad}}\right) = 0.19915\ \text{rev} \qquad \textit{Ans.}$$

1.23 Convert 26°27'28" to decimal degrees.

Solution

Convert seconds to degrees.

$$(28\ \text{seconds})\left(\frac{1°}{3600\ \text{seconds}}\right) = 0.0078°$$

Convert minutes to degrees.

$$(27\ \text{minutes})\left(\frac{1°}{60\ \text{minutes}}\right) = 0.4500°$$

Add.

$$0.0078° + 0.4500° + 26.0000° = 26.4578° \qquad \textit{Ans.}$$

1.24   Convert 10°15′39″ to decimal degrees.

Solution

Convert seconds to degrees.

$$(39 \text{ seconds})\left(\frac{1°}{3600 \text{ seconds}}\right) = 0.0108°$$

Convert minutes to degrees.

$$(15 \text{ minutes})\left(\frac{1°}{60 \text{ minutes}}\right) = 0.2500°$$

Add. $0.0108° + 0.2500° + 10.0000° = 10.2608°$ *Ans.*

1.25   Convert 11°50′30″ to radians.

Solution

First convert to decimal degrees.

$$(30 \text{ seconds})\left(\frac{1°}{3600 \text{ seconds}}\right) = 0.0083°$$

$$(50 \text{ minutes})\left(\frac{1°}{60 \text{ minutes}}\right) = 0.8333°$$

Add. $0.0083° + 0.8333° + 11.0000° = 11.8416°$

Convert decimal degrees to radians.

$$11.8416°\left(\frac{\pi \text{ rad}}{180°}\right) = 0.20667 \text{ rad}   \textit{Ans.}$$

1.26   Convert 23°46′16″ to radians.

Solution

First convert to decimal degrees.

$$(16 \text{ seconds})\left(\frac{1°}{3600 \text{ seconds}}\right) = 0.0044°$$

$$(46 \text{ minutes})\left(\frac{1°}{60 \text{ minutes}}\right) = 0.7667°$$

Add. $0.0044° + 0.7667° + 23.0000° = 23.7710°$

Convert decimal degrees to radians.

$$23.7710°\left(\frac{\pi \text{ rad}}{180°}\right) = 0.41488 \text{ rad}   \textit{Ans.}$$

1.27   Convert 77.5921° to DMS.

Solution

Convert the decimal part to minutes.

$$0.5921°\left(\frac{60 \text{ minutes}}{1°}\right) = 35.53 \text{ minutes}$$

Convert the decimal portion of 35.53 minutes to seconds.

$$(0.53 \text{ minute})\left(\frac{60 \text{ seconds}}{1 \text{ minute}}\right) = 31.8 \text{ seconds}$$

Add.   77°35'31.8"   *Ans.*

1.28   Convert 79.7979° to DMS.

Solution

Convert the decimal portion to minutes.

$$0.7979°\left(\frac{60 \text{ minutes}}{1°}\right) = 47.87 \text{ minutes}$$

Convert the decimal portion of 47.87 minutes to seconds.

$$(0.87 \text{ minute})\left(\frac{60 \text{ seconds}}{1 \text{ minute}}\right) = 52.2 \text{ seconds}$$

Add.   79°47'52.2"   *Ans.*

1.29   Add 37°46'26" to 61°21'10".

Solution	Degrees	Minutes	Seconds
	37	46	26
	+ 61	+ 21	+ 10
Add:	98	67	36
Remove multiples of 60	+ 01	− 60	
and carry forward:	99	07	36

Total = 99°07'36"   *Ans.*

1.30   Add 107°47'10" to 102°20'05".

Solution	Degrees	Minutes	Seconds
	107	47	10
	+ 102	+ 20	+ 05
Add:	209	67	15
	+ 1	− 60	
	210	07	15

Total = 210°07'15"   *Ans.*

1.31   Subtract 10°10'10" from 13°20'20".

Solution

Borrow from degrees and minutes (if needed).

Degrees	Minutes	Seconds
13	20	20
− 10	− 10	− 10
03	10	10

Answer = 3°10'10"   *Ans.*

In this case it was unnecessary to borrow from the degrees and minutes columns.

1.32   Multiply 10°10′10″ by 5.3216.

Solution

Convert to decimal degrees.

$$(10 \text{ minutes})\left(\frac{1°}{60 \text{ minutes}}\right) = 0.16667°$$

$$(10 \text{ seconds})\left(\frac{1°}{3600 \text{ seconds}}\right) = 0.00278°$$

Add: $0.1667° + 0.00278° + 10° = 10.16944°$

Multiply: $(10.170°)(5.3216) = 54.1177° = 54.121°$

Convert back to DMS.

$$(0.121°)(60 \text{ minutes per degree}) = 7.26 \text{ minutes}$$

$$(0.26 \text{ minute})(60 \text{ seconds per minute}) = 15.6 \text{ seconds}$$

So $(10°10′10″)(5.3216) = 54°7′15.6″$ *Ans.*

1.33   Divide 110°6′ by 4.500.

Solution

Convert to decimal degrees.

$$(6 \text{ minutes})\left(\frac{1°}{60 \text{ minutes}}\right) = 0.100°$$

Divide: $\dfrac{110.100°}{4.500} = 24.467°$

Convert to DMS.

$$(0.467°)(60 \text{ minutes per degree}) = 28.02 \text{ minutes}$$

$$(0.02 \text{ minute})(60 \text{ seconds per minute}) = 1.2 \text{ seconds}$$

So $\dfrac{110°6′}{4.500} = 24°28′1.2″$ *Ans.*

1.34   Given: Triangle of base 12.5 ft and height of 10.12 ft. Find the area of the triangle.

Solution

$$A = \tfrac{1}{2}bh = \tfrac{1}{2}(12.5 \times 10.12) = 63.25 \text{ ft}^2   \textit{Ans.}$$

1.35   Given: Triangle of base 32.16 ft and height of 22.49 ft (see Fig. 1-11). Find the area.

Solution

$$A = \tfrac{1}{2}bh = \tfrac{1}{2}(32.16)(22.49) = 361.64 \text{ ft}^2   \textit{Ans.}$$

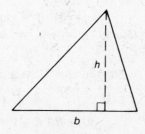

Fig. 1-11

1.36   Given: Triangle of height 7.69 ft and base of 2.79 ft (see Fig. 1-11). Find the area.

Solution

$$A = \tfrac{1}{2}bh = \tfrac{1}{2}(2.79)(7.69) = 10.73 \text{ ft}^2 \qquad Ans.$$

1.37   Given: Triangle with sides $a = 7.5$ in, $b = 8.9$ in, $c = 10.2$ in (refer to Fig. 1-5). Find the area.

Solution

Use the formula $s = (a + b + c)/2$ to find s.

$$s = \frac{7.5 + 8.9 + 10.2}{2} = 13.3 \text{ in}$$

Use the area formula:

$$A = \sqrt{s(s-a)(s-b)(s-c)} = \sqrt{13.3(13.3 - 7.5)(13.3 - 8.9)(13.3 - 10.2)} = \sqrt{1052.1896} = 32.44 \text{ in}^2 \qquad Ans.$$

1.38   Given: Triangle with sides $a = 11.32$ in, $b = 13.23$ in, $c = 14.92$ in (refer to Fig. 1-5). Find the area.

Solution

Use the formula $s = (a + b + c)/2$ to find s.

$$s = \frac{11.32 + 13.23 + 14.92}{2} = 19.74 \text{ in}$$

Use the area formula:

$$A = \sqrt{s(s-a)(s-b)(s-c)} = \sqrt{19.74(19.74 - 11.32)(19.74 - 13.23)(19.74 - 14.92)} = \sqrt{5215.3957}$$
$$= 72.22 \text{ in}^2 \qquad Ans.$$

1.39   Given: Triangle with sides $a = 93$ cm, $b = 72$ cm, $c = 23$ cm (refer to Fig. 1-5). Find the area.

Solution

Use the formula:

$$s = \frac{a + b + c}{2} = \frac{93 + 72 + 23}{2} = 94 \text{ cm}$$

Use the area formula:

$$A = \sqrt{s(s-a)(s-b)(s-c)} = \sqrt{94(94 - 93)(94 - 72)(94 - 23)} = \sqrt{146\,828} = 383.18 \text{ cm}^2 \qquad Ans.$$

1.40   Given: Triangle ABC with $B = 87°$ and $A = 38°$ (see Fig. 1-12). Find angle C.

Solution

$$A + B + C = 180°$$
$$C = 180° - A - B = 180° - 38° - 87° = 55° \qquad Ans.$$

1.41   Given: Triangle ABC with $C = 60°$ and $B = 71°$ (see Fig. 1-12). Find angle A.

Solution

$$A + B + C = 180°.$$
$$A = 180° - B - C = 180° - 71° - 60° = 49° \qquad Ans.$$

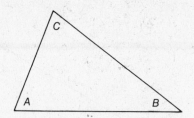

Fig. 1-12   Interior angles of a triangle.

1.42   Given: Triangle ABC with $A = 37°$ and $B = 56°$ (see Fig. 1-12). Find angle C.

Solution

$$A + B + C = 180°$$
$$C = 180° - 37° - 56° = 87° \qquad Ans.$$

1.43   See Fig. 1-6. Given: Angle $Q = 120°$, $A = 30°$. Find angle B.

Solution

$$Q = A + B$$
$$B = Q - A = 120° - 30° = 90° \qquad Ans.$$

1.44   See Fig. 1-6. Given: Angle $Q = 127°$, $B = 32°$. Find angle A.

Solution

$$Q = A + B$$
$$A = Q - B = 127° - 32° = 95° \qquad Ans.$$

1.45   See Fig. 1-6. Given: Angle $Q = 131°30'$, $A = 37°20'$. Find angle B.

Solution

$$Q = A + B$$
$$B = Q - A = 131°30' - 37°20' = 94°10' \qquad Ans.$$

1.46   See Fig. 1-6. Given: Angle $A = 40°$, $B = 30°$. Find angle Q.

Solution

$$Q = A + B$$
$$Q = 40° + 30° = 70° \qquad Ans.$$

1.47   See Fig. 1-13. Given: Side $a = 7.12$ in, side $b = 8.35$ in. Find side c.

Solution

$$c^2 = a^2 + b^2$$
$$c^2 = (7.12)^2 + (8.35)^2 = 50.69 + 69.72 = 120.47$$

So
$$c = \sqrt{120.41} = 10.97 \text{ in} \qquad Ans.$$

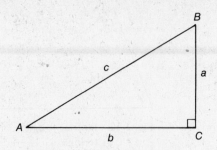

Fig. 1-13   Right triangle.

1.48   See Fig. 1-13. Given: Side $b = 10.27$ ft, side $c = 25.32$ ft. Find side a.

Solution

$$a^2 + b^2 = c^2$$
$$a^2 = c^2 - b^2 = (25.32)^2 - (10.27)^2 = 641.10 - 105.47 = 535.63$$

So
$$a = \sqrt{565.63} = 23.14 \text{ ft} \qquad \textit{Ans.}$$

1.49   See Fig. 1-13. Given: Side $a = 171.2$ ft, hypotenuse $c = 237.22$ ft. Find side b.

Solution

$$a^2 + b^2 = c^2$$
$$b^2 = c^2 - a^2 = (237.22)^2 - (171.2)^2 = 56\,273.33 - 29\,309.44 = 26\,963.89$$

So
$$b = \sqrt{26\,963.89} = 164.21 \text{ ft} \qquad \textit{Ans.}$$

1.50   See Fig. 1-13. Given: Side $a = 7.54$ ft, side $b = 9.81$ ft. Find side c.

Solution

$$a^2 + b^2 = c^2$$
$$c^2 = a^2 + b^2 = (7.54)^2 + (9.81)^2 = 56.85 + 96.24 = 153.09$$

So
$$c = \sqrt{153.09} = 12.37 \text{ ft} \qquad \textit{Ans.}$$

1.51   See Fig. 1-14. Given: A point on the terminal side of the angle with coordinates of (12.30, 6.10). Write the six trigonometric functions of A.

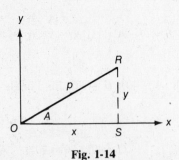

Fig. 1-14

Solution

Compute p by the pythagorean theorem.

$$a^2 + b^2 = c^2 \qquad \text{or} \qquad p^2 = x^2 + y^2$$
$$p^2 = (12.30)^2 + (6.10)^2 = 151.29 + 37.21 = 188.50$$

Taking the square root of both sides of the equation, we have

$$p = \sqrt{188.50} = 13.73$$

So we have $\quad\quad\quad\quad\quad x = 12.30 \quad\quad\quad y = 6.10 \quad\quad\quad p = 13.73$

Substituting into the six trigonometric equation ratios:

$$\sin A = \frac{y}{p} = \frac{6.10}{13.73} = 0.44 \quad\quad Ans.$$

$$\cos A = \frac{x}{p} = \frac{12.30}{13.73} = 0.90 \quad\quad Ans.$$

$$\tan A = \frac{y}{x} = \frac{6.10}{12.30} = 0.50 \quad\quad Ans.$$

$$\cot A = \frac{x}{y} = \frac{12.30}{6.10} = 2.02 \quad\quad Ans.$$

$$\sec A = \frac{p}{x} = \frac{13.73}{12.30} = 1.12 \quad\quad Ans.$$

$$\csc A = \frac{p}{y} = \frac{13.73}{6.10} = 2.25 \quad\quad Ans.$$

1.52 See Fig. 1-14. Given: Triangle in which $x = 7.0$, $y = 4.7$. Find the hypotenuse and the six trigonometric functions.

Solution

Compute p by the pythagorean theorem.

$$p^2 = x^2 + y^2 = (7.0)^2 + (4.7)^2 = 49.00 + 22.09 = 71.09$$

Taking the square root of both sides of the equation, we have

$$p = \sqrt{71.09} = 8.43 \quad\quad Ans.$$

Substituting into the six trigonometric equation ratios:

$$\sin A = \frac{y}{p} = \frac{4.7}{8.43} = 0.56 \quad\quad Ans.$$

$$\cos A = \frac{x}{p} = \frac{7.0}{8.43} = 0.83 \quad\quad Ans.$$

$$\tan A = \frac{y}{x} = \frac{4.7}{7.0} = 0.67 \quad\quad Ans.$$

$$\cot A = \frac{x}{y} = \frac{7.0}{4.7} = 1.49 \quad\quad Ans.$$

$$\sec A = \frac{p}{x} = \frac{8.43}{7.0} = 1.20 \quad\quad Ans.$$

$$\csc A = \frac{p}{y} = \frac{8.43}{4.7} = 1.79 \quad\quad Ans.$$

1.53 Using a calculator find the sine of 22.3°, 42.6°, 51.3°, 89.1°, 76.5°.

Solution

Enter 22.3 and depress SIN key = 0.3795   *Ans.*

Enter 42.6 and depress SIN key = 0.6769   *Ans.*

Enter 51.6 and depress SIN key = 0.7837   *Ans.*

Enter 89.1 and depress SIN key = 0.9999   *Ans.*

Enter 76.5 and depress SIN key = 0.9724   *Ans.*

1.54   Using a calculator find the cosine of 3.2°, 21.6°, 33.5°, 60°, 30°.

Solution

Enter 3.2 and depress COS key = 0.9984   *Ans.*
Enter 21.6 and depress COS key = 0.9298   *Ans.*
Enter 33.5 and depress COS key = 0.8339   *Ans.*
Enter 60 and depress COS key = 0.5000   *Ans.*
Enter 30 and depress COS key = 0.8660   *Ans.*

1.55   Using a calculator find the tangent of 9.4°, 17.2°, 21.1°, 35°, 88°.

Solution

Enter 9.4 and depress the TAN key = 0.1655   *Ans.*
Enter 17.2 and depress the TAN key = 0.3096   *Ans.*
Enter 21.1 and depress the TAN key = 0.3859   *Ans.*
Enter 35 and depress the TAN key = 0.7002   *Ans.*
Enter 88 and depress the TAN key = 28.6362   *Ans.*

Supplementary Problems

1.56   Express the following decimal degrees in minutes: (*a*) 0.69°, (*b*) 0.62°, (*c*) 0.51°, (*d*) 0.49°.
Ans.   (*a*) 41.4 minutes; (*b*) 37.2 minutes; (*c*) 30.6 minutes; (*d*) 29.4 minutes.

1.57   Express 3.72 rad in decimal degrees.   *Ans.*   213.14°

1.58   Express 4.73 rad in decimal degrees.   *Ans.*   271.01°

1.59   Express 5.03 rad in decimal degrees.   *Ans.*   288.20°

1.60   Express 6.12 rad in decimal degrees.   *Ans.*   350.65°

1.61   Express 97.8° in radians.   *Ans.*   1.71 rad

1.62   Express 201.6° in radians.   *Ans.*   3.52 rad

1.63   Convert 0.5871 rev to degrees.   *Ans.*   211.36°

1.64   Convert 0.4192 rev to degrees.   *Ans.*   150.9°

1.65   Convert 0.4159 rev to radians.   *Ans.*   2.613 rad

1.66   Convert 0.3250 rev to radians.   *Ans.*   2.042 rad

1.67   Convert 0.2222 rev to radians.   *Ans.*   1.396 rad

1.68   Convert 59.103° to revolutions and radians.   *Ans.*   0.16417 rev, 1.03154 rad

1.69   Convert 41.123° to revolutions and radians.   *Ans.*   0.11423 rev, 0.71773 rad

1.70   Convert 1.0021 rad to (*a*) degrees and (*b*) revolutions.   *Ans.*   (*a*) 57.416°; (*b*) 0.15949 rev

1.71   Convert 148.2° to revolutions.   *Ans.*   0.4117 rev

1.72   Convert 132.7° to revolutions.   ·*Ans.*   0.3686 rev

1.73   Convert 89°56'32" to decimal degrees.   *Ans.*   89.9422°

1.74   Convert 78°51'52" to decimal degrees.   *Ans.*   78.8644°

1.75   Convert 47°16'23" to radians.   *Ans.*   0.82507 rad

1.76   Convert 56°17'27" to radians.   *Ans.*   0.98246 rad

1.77   Convert 89.5960° to DMS.   *Ans.*   89°35'45.6"

1.78   Add 119°67'51" to 03°10'17".   *Ans.*   123°18'08"

1.79   Add 09°59'59" to 05°51'51".   *Ans.*   15°51'50"

1.80   Subtract 13°22'16" from 78°10'06".   *Ans.*   64°47'50"

1.81   Multiply 23°26'15" by 7.6521.   *Ans.*   179°20'31.2"

1.82   Divide 140°58' by 2.000.   *Ans.*   70°29'2.4"

1.83   Given: Triangle of base 10.42 ft and altitude of 10.42 ft (see Fig. 1-11). Find the area.   *Ans.*   54.29 ft^2

1.84   Given: Triangle of base 104.97 ft and altitude, or height, of 56.25 ft (see Fig. 1-11). Find the area.
            Ans.   2952 ft^2

1.85   Given: Triangle with sides *a* = 20 ft, *b* = 22 ft, *c* = 10 ft (refer to Fig. 1-5). Find the area.
            Ans.   49.92 ft^2

1.86   Given: Triangle sides of *a* = 21.2 in, *b* = 20.1 in, *c* = 30.2 in (refer to Fig. 1-5). Find the area.
            Ans.   212.56 in^2

1.87   Given: Triangle ABC with $B = 50°$ and $C = 55°$ (see Fig. 1-12). Find angle A.   *Ans.*   75°

1.88   Given: Triangle ABC with $A = 55°10'$ and $B = 60°20'$ (see Fig. 1-12). Find angle C.   *Ans.*   64°30'

1.89   See Fig. 1-6. Given: Angle $A = 29°30'$, $B = 36°20'$. Find angle Q.   *Ans.*   65°50'

1.90   See Fig. 1-13. Given: Side $a = 153.27$ in, side $b = 120.10$ in. Find side c.   *Ans.*   194.72 in

1.91   See Fig. 1-13. Given: Side $a = 19.21$ cm, side $c = 27.32$ cm. Find side b.   *Ans.*   19.43 cm

1.92   See Fig. 1-14. Given: Triangle ORS, y coordinate is 22.39, x coordinate is 45.21. Find p and the six trigonometric functions for this triangle.
       Ans. $p = 50.45$, sin $A = 0.44$, cos $A = 0.90$, tan $A = 0.50$, cot $A = 2.02$, sec $A = 1.16$, csc $A = 2.25$

1.93   See Fig. 1-14. Given: Triangle in which $x = 14.31$, $y = 14.31$. Find p and the six trigonometric functions.
       Ans. $p = 20.24$, sin $A = 0.71$, cos $A = 0.71$, tan $A = 1.00$, cot $A = 1.00$, sec $A = 1.41$, csc $A = 1.41$

1.94   See Fig. 1-14. Given: Triangle in which $x = 10.00$, $y = 20.00$. Find p and the six trigonometric functions.
       Ans. $p = 22.36$, sin $A = 0.89$, cos $A = 0.45$, tan $A = 2.00$, cot $A = 0.50$, sec $A = 2.24$, csc $A = 1.12$

1.95   Using a calculator find the sine of the following angles: (*a*) 10.7°, (*b*) 21.6°, (*c*) 42.7°, (*d*) 70°, (*e*) 82°.
       Ans.   (*a*) 0.1857; (*b*) 0.3681; (*c*) 0.6782; (*d*) 0.9397; (*e*) 0.9903

1.96   Using a calculator find the cosine of the following angles: (*a*) 33.1°, (*b*) 45°, (*c*) 63.2°, (*d*) 80°, (*e*) 85.1°.
       Ans.   (*a*) 0.8377; (*b*) 0.7071; (*c*) 0.4509; (*d*) 0.1736; (*e*) 0.0854

1.97   Using a calculator find the cotangent, secant, and cosecant of (*a*) 51°, (*b*) 22.7°, and (*c*) 18.3°.
       Ans.   (*a*) 0.8098, 1.5890, 1.2868; (*b*) 2.3906, 1.0840, 2.5913; (*c*) 3.0237, 1.0533, 3.1848

Chapter 2

Field Notes

2.1 REASON FOR FIELD NOTES

The field notes made by the surveyors in any given construction project are the only permanent record of the work done in the field. If they are incorrect or incomplete, most of the time spent making the accurate measurements will be lost. A field book with the information gathered over a period of weeks will be worth thousands of dollars. The field notes must contain a complete record of all measurements made during the survey, along with any necessary sketches, diagrams, or narrations which help clarify the notes.

The office personnel normally use the data in the field book to make their drawings and computations. Therefore, the notes must be intelligible without any further explanation to anyone looking at the field notebook. The gothic style or Reinhardt system of slope lettering is generally employed for clarity and speed. Notes are all freehand, and clarity and speed are the most essential qualities.

Field notebooks are legal documents and are often used in court to review boundaries of property. They are therefore an important factor in litigation. They may be used as references in land transactions for generations, so they must be indexed and properly preserved. Cash receipts in a surveyor's office may be kept in an unlocked desk drawer, but the survey field books are always stored in a fireproof safe.

Original notes are the ones taken at the time measurements are being made. Anything done later is a "copy" and must be marked as such. Copied notes are not acceptable in court. There might be a question concerning possible omissions to the notes or even an addition to them.

The value of distances or angles placed in the field book from memory 5 or 10 minutes after the observation was made is unreliable. The observation must be entered at the same time it is made.

No one should scribble notes on sheets of scrap paper for later transference in neater form to the regular field book. This is sometimes done, even by veteran survey crews, but is definitely bad practice.

It should be remembered that the best field survey is of little value if the notes are not complete and clear. The field notes are the only record that is left after the survey party departs from the field survey site.

2.2 REQUIREMENTS FOR GOOD NOTES

The following points are most important in valuing a set of field notes:

1. Accuracy. This is number one for all surveying operations. If the angles, distances, and entering of these facts are incorrect, the whole survey is meaningless.
2. Legibility. Notes that are illegible are valueless.
3. Integrity. All measurements must be entered at the time of observation. One single missed detail will nullify the entire survey. Never "fudge" the notes to improve closures. [*Closure*, which will be discussed fully in Chap. 7, refers to the closing of a geometric polygon made up of a series of consecutive lines that have known lengths and directions (a traverse). Closure provides a check on the measured angles and distances; thus field notes should never be altered to show perfect closing or the checking effect would be nullified.]
4. Clarity. Plan the survey so that the notes will not be crowded or have any omissions in detail.
5. Arrangement. Use forms for the notebook that are appropriate for the survey undertaken. This will help the legibility, accuracy, and integrity of the field book notes.

2.3 TYPES OF FIELD BOOKS

Several kinds of field books may be purchased. Bound and loose-leaf are the most commonly used. Poor-quality books should never be used; the valuable data contained in the field book must be permanent and it would be economically unfeasible to allow the material to be lost through deterioration of the field book.

Bound books have sewed binding and stiff covers of imitation leather or impregnated canvas; they usually contain 80 leaves.

There are also bound books which can be used for duplicating notes through carbon paper. Every other page is perforated in these books so the alternate page can be easily removed.

Loose-leaf books offer a number of advantages:

1. They allow for easy filing of individual project notes.

2. They allow for easy transfer of partial sets of notes between office and field.

3. They allow the use of different ruling spaces in the same book.

4. They cost less than a bound book.

5. There is less waste than there is with a bound book; you need use only the number of sheets required for the individual survey.

2.4 KINDS OF NOTES

There are three general types of notes; a combination of these three types is generally used in practice. The three types are as follows:

1. *Tabulations.* The numerical measurements are recorded in columns according to a prescribed plan depending on the instrument used, order of accuracy of the survey, and the type of measurement. See Fig. 2-1.

2. *Sketches.* Sketches clarify field notes and should be used liberally. They may be drawn to scale or approximate scale, or exaggerated for clarity. A plane table sheet is an example of a sketch drawn to scale. The measurements should be added directly on the sketch or keyed in some way to the tabular data. Legibility is a very important requirement of any sketch. See Fig. 2-2.

3. *Descriptions.* Tabulations with or without added sketches can also be supplemented with descriptions. A description may be only one or two words to clarify the recorded measurements, or it may be a lengthy narration if it is to be used at some future date, possibly years later, to locate a survey monument. When in doubt about the need for information, include it and make a sketch. Too much information is far better than too little. See Fig. 2-3.

2.5 ARRANGEMENT OF FIELD NOTES

Left- and right-hand pages of the field book are always used in pairs and carry the same number. The title of the survey should be lettered across the top of the left-hand page and often extends onto the right-hand page. Titles may be abbreviated on following pages for the same survey project.

Location and type of work are placed under the title. Note forms may be quite flexible as long as the resulting notes are clear to the reader. To permit easy location of desired data, the field book must have a table of contents which is kept current daily.

The upper part of the left- or right-hand page must contain four items:

1. Party: First initials and last names of party members, and their duties. Jobs may be shown with a symbol of a transit for an instrumentperson, Greek lowercase letter phi (ϕ) for rodperson, N for notekeeper, HT for head taper.

2. Date, time (morning or afternoon), start and finish time.

STATION	TABULATIONS			
Hub	Sta.	Dist.		
A	0+00			
		220.10'		
B	2+51.01			
		266.07'		
C	5+87.21			
		100.51'		
D	6+98.01			
		316.88'		
E	10+06.64			
		390.34		
A	12+94.10			
Σ	1294.10	1293.90		

Ratio of error $= \dfrac{0.20}{1294.00} = \dfrac{1}{6470}$

Fig. 2-1 Legible description on field book page.

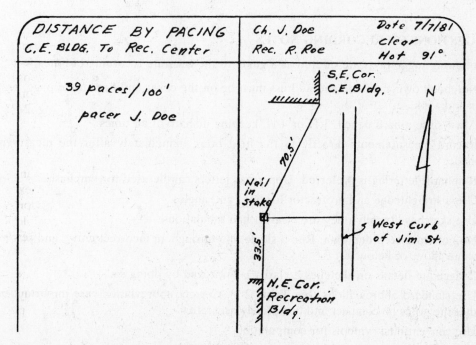

DISTANCE BY PACING
C.E. BLDG. To Rec. Center

Ch. J. Doe
Rec. R. Roe

Date 7/7/81
clear
Hot 91°

39 paces/100'

pacer J. Doe

S.E. Cor.
C.E. Bldg.

N

70.5'

Nail
in
Stake

West Curb
of Jim St.

33.5'

N.E. Cor.
Recreation
Bldg.

Fig. 2-2 Sketch in field book.

DESIGNATION			DATE		19—		10 July 1970 through 30 July 1970			
Tapes and Calibration Data:								Remarks		
Tape #16 20# Tension-K = 0.009							Tape #16 accidently broken on			
Supported throughout (T/o) = 100.028							15 July 1970. Returned to Co. Hqs.			
" at 0, 50, & 100 = 100.026							for repair. Tape #5 received as			
" at 0, & 100 = 100.019							a replacement on 16 July, 1970.			
Tape #2 20 lbs. Tension-K = 0.011										
Supported throughout (T/o) = 100.002										
" at 0, 50, & 100 = 99.999										
" at 0, & 100 = 99.991										
Tape #5 20 lbs. Tension-K = 0.012										
Supported throughout (T/o) = 100.034										
" at 0, 50, & 100 = 100.031										
" at 0, & 100 = 100.022										
						①				

Fig. 2-3 Tabulation of data. (*Courtesy of U.S. Air Force*)

3. Weather. Wind velocity and temperature are important. Rain, snow, sunshine, and fog all have an effect on surveying operations. Weather details are necessary when field notes are reviewed. They are also needed for applying tape corrections due to temperature variations.

4. Instrument type and number. Identification of the instrument will be of help in finding errors made in the survey.

2.6 SUGGESTIONS ON RECORDING NOTES

The following suggestions are offered as a guide to the accurate recording of notes:

1. Notebook owner's name and address must be on the cover and first inside page, preferably in India ink.

2. Always use a hard pencil, 3-H or 4-H, keeping it sharp at all times.

3. Record measurements directly in the field book immediately after the measurement is made.

4. Reinhardt lettering is preferred. Uppercase letters can be used for emphasis.

5. Use a straightedge and protractor for lines and angles.

6. Use sketches liberally; they are clearer than tabulations.

7. Do not erase recorded data. Run a single line through an incorrect entry, and place correct value above or below it.

8. Exaggerate details on sketches if clarity is improved by doing so.

9. Use standard abbreviations (see Table 2-1). Correct abbreviations are important in order that the notes be compact and completely accurate.

10. Use conventional symbols for computations.

11. Use a north arrow on all sketches.

Table 2-1   Field Notebook Standard Notations*

Abbreviation	Meaning	Abbreviation	Meaning	Abbreviation	Meaning
A.M.	Morning	Ho.	House	Rd.	Road
∤, &, or ⊄ *	And	Hor.	Horizontal	Ret. wall	Retaining wall
⊄	Angle	Ht.	Height	RH	Redhead
Asph.	Asphalt	IFS	Intermediate foresight	RR	Railroad
Ave.	Avenue			Rt.	Right
Avg.	Average	in	Inch	R/W or	Right of way
BC	Beginning of curve	Inst.	Instrument	ROW	
		IP	Iron pipe	φ	Rod
Bldg.	Building	Lat.	Latitude	S	South or sewer
Blvd.	Boulevard	LH	Lamp hole	San.	Sanitary
BM	Bench mark	Lin ft	Lineal feet	SD	Storm drain
BS	Backsight	Long.	Longitudinal	Sect.	Section
BVC	Beginning of vertical curve	Lt.	Left	S/O	South of
		LW	Low water	Spk.	Spike
CB	Catch basin	Max.	Maximum	ft²	Square foot
Cem.	Cement	MGD	Million gallons per day	SS	Sanitary sewer
CIP	Cast iron pipe			St.	Street
₵	Center line	MH	Manhole	₷	Street line
Conc.	Concrete	MB	Mailbox	Sta.	Station
Cor.	Corner	MHW	Mean high water	Stk.	Stake
Ctr.	Center	MHT	Mean high tide	TBM	Temporary bench mark
Cu.	Cubic	MLW	Mean low water		
Culv.	Culvert	MSL	Mean sea level	Tel.	Telephone (pole)
Decl.	Declination	N	North	Temp.	Temperature
°, deg.	Degree	No.	Number	TH	Test hole
DH	Drill hole	N/O	North of	Tk.	Tack
Dia.	Diameter	OD	Outside diameter	TL	Traverse line
DMH	Drop manhole	Opp.	Opposite	Topo.	Topography
E	East	Pg.	Page	⋏	Transit
EC	End of curve	Pav.	Pavement	TP	Turning point
El. or elev.	Elevation	PC	Party chief or point of curvature	TS	Traffic signal
E/O	East of			TW	Traveled way
Etc.	Et cetera			USGS	United States Geological Survey
Ex. or exist.	Existing	PI	Point of intersection		
FH	Fire hydrant	PL or ₽	Property line	VC	Vertical curve
FL	Flow line	P.M.	Afternoon	Vert.	Vertical
FS	Foresight	POB	Point of beginning	W	West
ft	Foot	Prof.	Profile	WL	Water line
gpm	Gallons per minute	Pt.	Point	W/O	West of (or without)
		PT	Point of tangent		
Gr.	Grade	RC	Rear chainperson	W/	With
HC	Head chainperson	RCP	Reinforced concrete pipe	X-sect.	Cross section
HI	Height of instrument				

* If a term or phrase is used only once or twice, it is better practice not to abbreviate it. Abbreviations used but not listed should be defined within the notes.

12. Repeat aloud values to be recorded before entering them in the field book. They can then be verified by the taper or instrumentperson.

13. Use a zero before the decimal point for numbers less than 1.

14. Use significant figures. Record 7.90 instead of 7.9 only if the reading has been obtained in hundredths.

15. Complete all closures and ratios of error while in the field.

16. Sign surname and initials in the lower right-hand corner of the right-hand page of the original notes.

EXAMPLE 2.1 Given: Southeast (SE) corner of house 105 on Bay Avenue and northeast (NE) corner of neighboring house 103 on Bay Avenue are 29.3 paces apart. In this case J. Doe, who is acting party chief, has found his pace to equal 39 paces in 100 ft of taped length. The two houses have a setback of 25 ft from Bay Avenue. Set up the field book sketch to show these dimensions in paces. Do not forget the north arrow.

Solution: See Fig. 2-4.

$$100 \text{ ft} = 39 \text{ paces} \qquad \text{so} \qquad \frac{100}{39} = 2.56 \text{ ft per pace}$$

The setback is 25 ft, so $25/2.56 = 9.77$ paces (round off to 10 paces because of the inability of judging less than 1 pace) from the front of the houses to the west edge of Bay Avenue. The distance between houses was given in the data as 29.3 of J. Doe's paces.

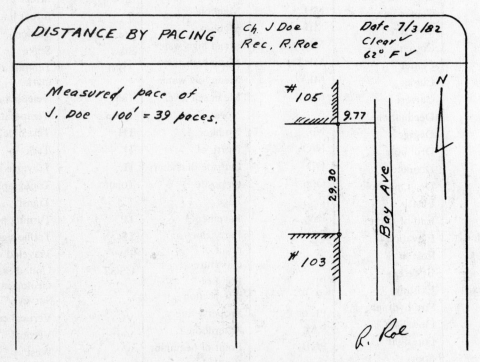

Fig. 2-4 Field book sketch to show pacing.

2.7 PROPER METHODS

The following information is taken directly from the standards book of a large consulting engineering firm, and Tables 2-1 to 2-6 are the instructions and diagrams given to their employees to ensure that they follow correct procedures in the field. Table 2-2 gives the list of equipment taken on field trips, and Tables 2-3 to 2-6 present a basic format for preparing a field notebook.

Field notes must be carefully prepared since they are very important for the successful completion of the survey. The party chief or notekeeper must keep in mind that the notes will be interpreted by others.

One objective of this section is to ensure that survey data are recorded within certain accepted limits of form, arrangement, neatness, etc., and that each set of notes will have a strong similarity to other notes so prepared.

The survey situations under which a party chief is required to exercise personal judgment are varied. Some aspect of every survey is different. However, the notekeeper should follow the guidelines set forth as closely as possible.

A good set of field notes must be:
- Neat and legible (use sharp pencil)
- Complete (review the notes)

Table 2-2   Survey Party Equipment List

Job _____
Job. No. _____
Date (Today) _____

Quantity
- _____ Camera (film)
- _____ Compass
- _____ Crow bar
- _____ Cup tacks
- _____ Field book & pages
- _____ Flagging
- _____ Flashlight
- _____ Fluorescent dye
- _____ Fluorescent paint
- _____ Hammer
- _____ Hatchet
- _____ Hand level
- _____ Iron pipe
- _____ Keel or chalk
- _____ Level
- _____ Machete
- _____ Magnifying glass
- _____ Measuring wheel
- _____ Nails (concrete penetration)
- _____ Pencil (hard/straightedge, etc.)
- _____ Pencil sharpener (knife, sandpaper)
- _____ Pickaxe
- _____ Plumb bob
- _____ Plumb bob string
- _____ Range pole

Quantity
- _____ Rod
- _____ Rule (6 ft)
- _____ Sample bags (soil)
- _____ Sample bottles
- _____ Saw
- _____ Shovel
- _____ Spikes
- _____ Star drill
- _____ String
- _____ Tables (mathematical)
- _____ Tape
- _____ Taping pins
- _____ Target
- _____ Transit
- _____ Transit legs
- _____ Walkie-talkie
- _____ Wooden hubs
- _____ Wooden wedge stakes
- _____ Other _____

Personal Equipment
- _____ Rubber boots
- _____ Gloves
- _____ Insect repellent
- _____ Rain gear
- _____ Hat

Comments: _____
_____
_____

Table 2-3   Titles and Listings for the Start of Each Job

SURVEY NOTEBOOK DATA

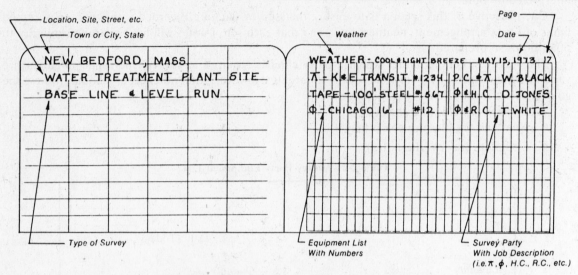

The above titles and listings must be written at the start of each job and must be repeated for each day on the job or with a change of job type, location, crew, etc. Every page of loose leaf notebooks should repeat client and date.

Table 2-4   Example of Level Run

EXAMPLE LEVEL RUN

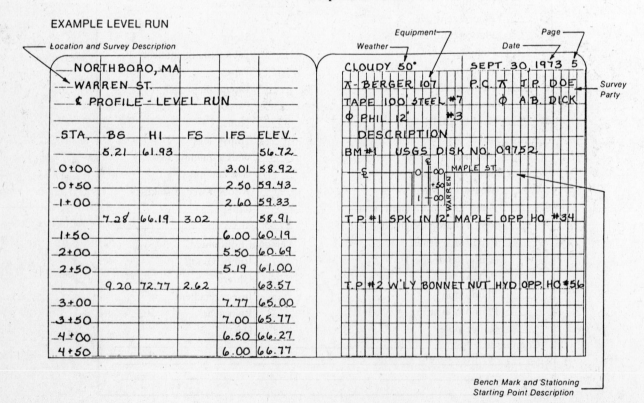

Table 2-5   Example of a Topo Survey

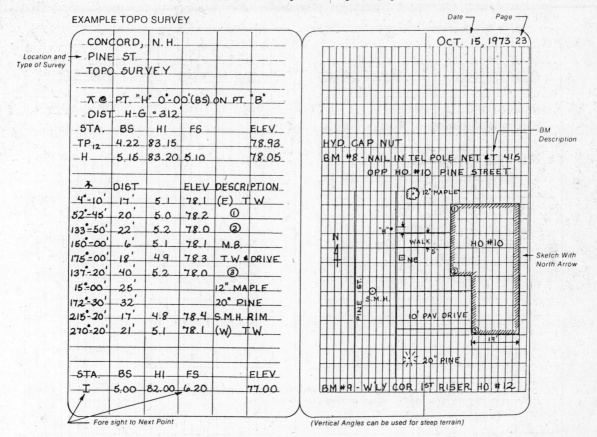

- Self-explanatory (enlarge details and draw sketches)
- Accurate (record actual measurements, not estimates)
- Self-checking
- Indexed for logical filing

The quality of a survey depends on the quality of the field notes. Field notes must stand alone. Interpretation should not require recourse to the one who prepared them, who may or may not be in the office or even with the company.

Often the party chief will be handed a plan or similar material, not signed, with the title block incomplete, and be told to use it. The party chief should protect him- or herself by recording the name of the person who provided the information, the nature of the material, and any information used from it.

Improper recording of data can result in lost time and money, repeated trips to the job site, and design errors. Field data may sometimes represent the individual or the firm in personnel or legal actions. For these reasons the data must be precise.

Sketches or written descriptions of special situations should be used to complete the notes. Any discussions with the client, municipal representatives, or others should be noted.

Often residents talk to survey crews and more often than not volunteer information which is important. For example, a party chief might be told of someone moving or destroying survey stakes. Items of this nature should be recorded.

If traverse lines are required, the traverse points should be marked so that they can be found and identified later in the field. Distances between traverse points should be measured twice and angles

Table 2-6 Indexing

INDEXING

DESCRIPTION	PAGES		JOB NUMBER	DATE	PARTY CHIEF
1. NEW BEDFORD, MASS.			309-15-SU	Nov. 10, 1970	A. SMITH
Sludge Lagoons Topo	5-12				
2. BILLERICA, MASS.			52-54-6U	Nov. 15, 1970	B. JONES
Pond St. Easement	13-17				
3. EVERETT, MASS.					
Green St. Level Run	18-25		580-9-SU	Nov. 20, 1970	A. SMITH
Green St. & Hope St. Topo	25-27		580-9-SU	Nov. 25, 1970	A. SMITH
Green St. Detail	28-40		580-9-SU	Dec. 1, 1970	A. SMITH
Shipyard-Control Traverse	41-50		580-9-SU	Dec. 15, 1970	A. SMITH

NOTE:
All survey work shall be indexed using the above format and submitted
to the   library for filing after plotting.

should be doubled. Levels run to set bench marks (BMs) for sewer construction should be "double-rodded," or "looped." Similarly, all bench marks and turning points must be checked and described in the field notes.

All leaves in a field notebook must be numbered. In other words, only the right-hand page of an open field book is numbered consecutively in the upper right-hand corner.

Proper indexing and subdivision, using title pages, page headings, and adequate references and cross-references, improve the continuity of field notes. It is the responsibility of the party chief to index the work upon completion of each job or job step. Indexes are usually arranged chronologically; as each increment of work is completed, it is indexed. The existence of a master index in the library does not relieve the party chief from indexing within the field notes.

The notekeeper should use standard abbreviations, symbols, and codes to save time, space, and ensure clarity. The list of standard field notebook notations in Table 2-1 should be used to facilitate notekeeping.

2.8 SIGNIFICANT FIGURES

In notekeeping an important feature that indicates the accuracy attained in recording measurements is the number of *significant figures* that are recorded. The number of significant figures in any value is made up of the known digits plus only one estimated digit.

For example, a distance recorded as 375.63 ft is said to have five significant figures; the first four digits are certain and the last digit may be estimated. All data should be recorded with the correct number of significant figures.

Dropping a significant figure when recording values wastes the time used in making the measurement precise. Likewise, data recorded with more significant figures than those reflected by the measurement give a false accuracy to the survey.

Do not confuse significant figures with the number of decimal places. Decimal places are used to maintain the correct number of significant figures, but do not themselves indicate significant figures.

Examples of significant figures follow:

Two significant figures: 4.4; 0.44; 0.0044; 0.40

Three significant figures: 56.4; 0.000564; 0.0440; 0.404

Four significant figures: 87.65; 0.0008765; 44.00

2.9 ROUNDING OFF NUMBERS

When a number is rounded, one or more digits are dropped so the answer will contain only the number of significant digits necessary to subsequent computations.

If the digit to be dropped is less than 5, the preceding digit remains the same. Thus 85.493 becomes 85.49.

When the digit to be dropped is 5, the nearest even number is used for the preceding digit. Thus 85.375 becomes 85.38, and 85.385 is also rounded to 85.38.

When the digit to be dropped is greater than 5, the preceding digit is increased by 1. Thus 85.376 becomes 85.38.

Solved Problems

The following problems use several surveying terms that will be discussed more thoroughly in upcoming chapters. To facilitate problem solving here, the terms are defined briefly as follows: A *backsight* is a sight taken backward, as opposed to the forward direction of a survey, toward a rod held on a point of known or assumed elevation. Another name for backsight is *plus* (+) *sight*. A *foresight* is a sighting in the forward direction of the survey toward a rod whose elevation is to be determined; a foresight is often called a *minus* (−) *sight*. A *traverse* is a series of angles and distances, bearings and distances, or azimuths and distances which connect successive points.

2.1　　Given: A building which must have both the length and width dimensions determined by pacing. Averaging the party chief's paces, we find 155 paces are made in a distance of 400 ft. To get this average, J. Doe paced a 400-ft distance north and south twice, giving him the average of 155 paces. The building is 208 paces in a north-south direction, and 100 paces in an east-west direction. Set up the field notes required for the problem. Be sure to draw the sketch, add a north arrow, show the original pacing of J. Doe to establish his length of pace, and convert paces to feet to give the length of the building in feet.

Solution

Part A of Fig. 2-5 shows by sketch the pacing of 400 ft in the north-south direction to establish J. Doe's pace length:

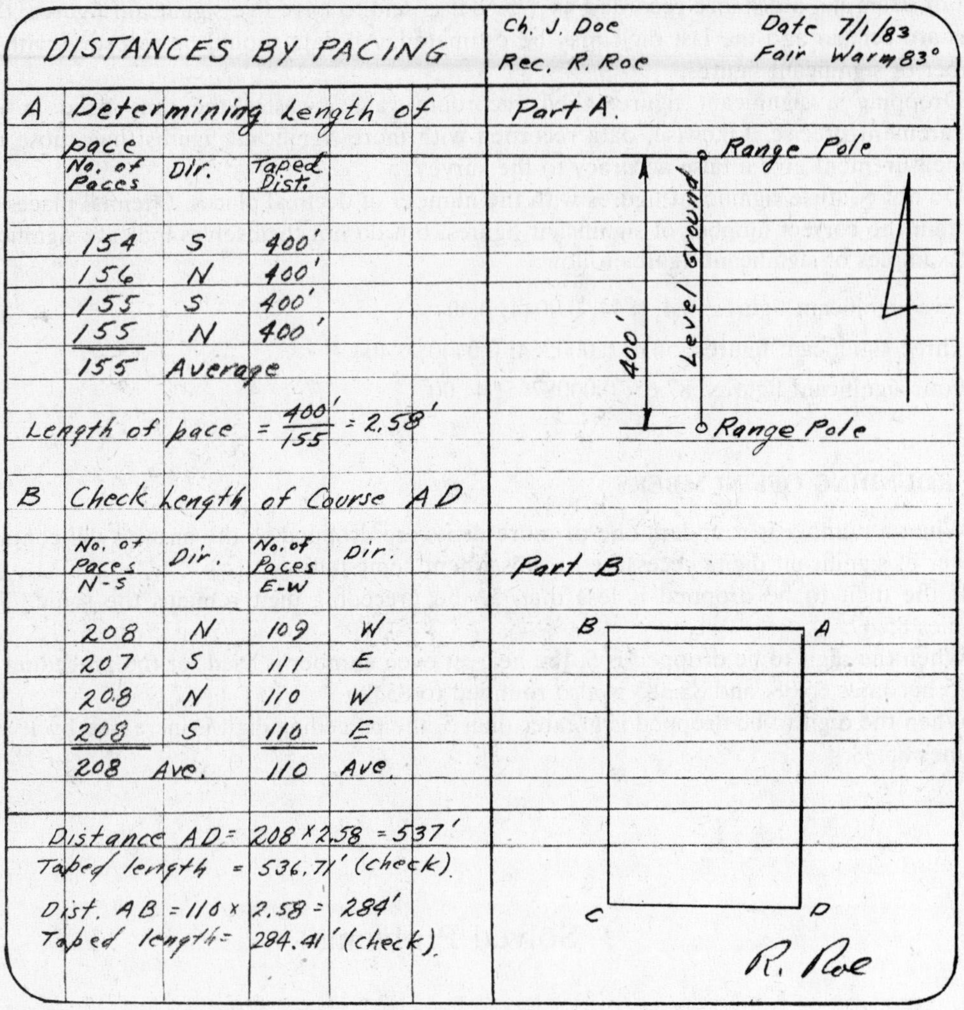

Fig. 2-5 Determining building size by pacing.

$$\text{Average number of paces for 400 ft} = 155$$

$$\text{Length of pace} = \frac{400 \text{ ft}}{155 \text{ paces}} = 2.58 \text{ ft per pace}$$

Part B on the left-hand page describes the paces in two directions to average the number of paces for the two sides of the building:

$$\text{North-south} = 208 \text{ paces}$$
$$\text{East-west} = 110 \text{ paces}$$

Finally, convert the paces to feet by multiplying the average number of paces times the length of the pace:

$$208(2.58) = 537 \text{ ft for } AD \text{ distance}$$
$$110(2.58) = 284 \text{ ft for } AB \text{ distance}$$

Enter the information in the field book as shown in Fig. 2-5.

2.2   Given: Bench marks are available on two adjacent buildings on the Ohio State University campus. These buildings are the Military Science building and the women's dormitory. Bench mark 1 has an elevation of 1053.182. It is located on the bottom step of the main entrance to the Military Science building. It is a chiseled multi in the concrete step. A backsight from a turning point where the level is stationed gives a rod reading of 1.605. Three forward sightings are made to BM 2. Three are made in order to average the sightings; the weather is clear but cold, making it difficult to attain accurate readings. The three foresights on BM 2 read 11.304, 11.302, and 11.291. BM 2 is a cross in the first step of the women's dormitory entrance.

The survey is then reversed and the level repositioned to another turning point between the buildings. From this point a backsight to BM 2 reads 12.203, and three separate foresights on BM 1 are 2.517, 2.526, and 2.523.

Without using a sketch set up the field notes for the reciprocal leveling survey between these two buildings.

Solution

See Fig. 2-6. Backsights are labeled with a plus (+) sign since they are added in surveying computations; foresights are labeled with a minus (−) sign since they are subtracted in surveying computations.

$$
\begin{array}{lr}
\text{BM 1 elevation} = & 1053.182 \\
\text{Add the backsight} & +\quad 1.605 \\
\text{Elevation} = & 1054.787 \\
\text{Subtract average foresight} & -\quad 11.299 \\
\text{Elevation BM 2} = & 1043.488
\end{array}
$$

To find the elevation difference subtract the back- and foresights:

$$
\begin{array}{lr}
\text{Average foresight} = & 11.299 \\
\text{Subtract backsight} & -\ 1.605 \\
\text{Elevation difference} = & 9.694
\end{array}
$$

The elevation of BM 2 is 1043.488, so we have:

$$
\begin{array}{lr}
\text{BM 2 elevation} = & 1043.488 \\
\text{Add the backsight} & +\quad 12.203 \\
\text{Elevation} = & 1055.691 \\
\text{Subtract average foresight} & -\quad 2.522 \\
\text{Elevation} = & 1053.169
\end{array}
$$

Find the elevation difference by subtracting fore- and backsights:

$$
\begin{array}{lr}
 & 12.203 \\
 & -\ 2.522 \\
\text{Elevation difference} = & 9.681
\end{array}
$$

To find closure difference, find sum of backsights (+) and subtract sum of foresights (−):

$$
\begin{array}{lr}
\text{Backsights } (+) & 13.808 \\
\text{Foresights } (-) & -\ 13.821 \\
\text{Closure difference} = & -\ 0.013
\end{array}
$$

$$
\text{Mean of elevation differences} = \frac{9.694 + 9.681}{2} = 9.688
$$

Divide the closure difference of −0.013 by 2 to give average closure difference:

$$
\frac{-0.013}{2} = -0.006
$$

Subtract average closure difference from BM 2 elevation to obtain mean elevation:

$$
\text{Mean elevation BM 2} = 1043.488 - (-0.006) = 1043.488 + 0.006 = 1043.494
$$

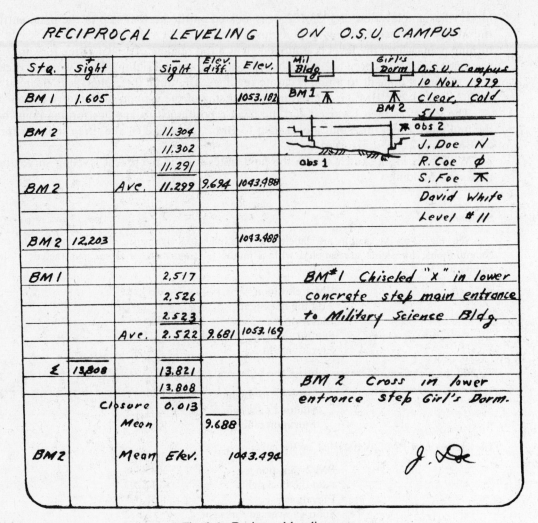

Sta.	+Sight		−Sight	Elev. diff.	Elev.	Mil Bldg	Girl's Dorm	O.S.U. Campus
BM 1	1.605				1053.182	BM 1		10 Nov. 1979
							BM 2	clear, cold
BM 2			11.304				obs 2	51°
			11.302					J. Doe N
			11.291			obs 1		R. Coe ∅
BM 2		Ave.	11.299	9.694	1043.488			S. Foe
								David White
								Level #11
BM 2	12.203				1043.488			
BM 1			2.517					BM #1 Chiseled "x" in lower
			2.526					concrete step main entrance
			2.523					to Military Science Bldg.
		Ave.	2.522	9.681	1053.169			
	Σ 13.808		13.821					BM 2 Cross in lower
			13.808					entrance step Girl's Dorm.
		Closure	0.013					
		Mean		9.688				
BM 2		Mean Elev.			1043.494			J. Doe

Fig. 2-6 Reciprocal leveling notes.

2.3 Given: A monument numbered 1072 set in 1945 by the U.S. Geological Survey 50 ft east of the NE corner of the women's gymnasium is numbered BM 3 and its elevation is 1046.122. Backsight to BM 3 from an instrument turning point (TP) is 2.011, and three foresights taken from the turning point are 10.671, 10.675, and 10.670.

Survey re-sets up and returns to BM 3: Backsight on BM 4 from TP 2 is 10.992; foresights taken on BM 3 are 2.327, 2.330, and 2.332. BM 4 is located 50 ft north of NW corner of swimming pool. Without using a sketch, work out the page of the field book for this problem.

Solution

See Fig. 2-7.

BM 3 elevation =	1046.122
Add backsight	+ 2.011
	1048.133
Subtract average foresight	− 10.672
Elevation BM 4 =	1037.461
Add backsight	+ 10.992
	1048.453
Minus foresight	− 2.330
Elevation BM 3 =	1046.123

Sta.	+ sight		− sight	Elev. Diff.	Elev.	RECIPROCAL LEVELING / OSU CAMPUS
						O.S.U. Campus
						12 Dec. 1980
BM3	2.011				1046.122	Cold, Foggy
						32°
BM4			10.671			
			10.675			J. Doe N
			10.670			R. Coe Ø
BM 4		Ave.	10.672	8.661	1037.461	S. Foe
						David White
BM 4	10.992				1037.461	Level #8
BM3			2.327			BM3 Monument 1072 set
			2.330			in 1945 by U.S.C.&G.S. 50'
			2.332			east of N.E. corner of the
		Ave.	2.330	8.662	1046.123	girl's gymnasium.
Σ	13.003		13.002			BM 4 Monument 50' north
			13.003			of N.W corner of swimming
	Closure		0.001			pool.
	mean			8.661		
BM4		mean Elev.			1037.461	J. Doe

Fig. 2-7 Leveling notes.

2.4   Take the data of Prob. 2.3 and draw a sketch of the problem.

Solution

See Fig. 2-8.

2.5   Given: A traverse taken on Pershing Field. Data: Magnetic north arrow to be 10° west of true north. Point A is at the intersection of the radius of a circle of 27 ft from a 10-in oak and the radius of a circle of 31 ft from the fire hydrant located close by. Point B is located at the intersection of a circle of 15-ft radius with center at a 15-in fir tree, and a circle of 29-ft radius whose center is the SW corner of a concrete catch basin.

Point C is at the intersection of a 16.8-ft-long perpendicular line from a concrete sidewalk. This line is intersected by a circle of 30-ft radius which has its center at an 8-in pine tree. Point D is developed from the intersection of two circles: one has a center at a 6-in cedar tree and a 26-ft radius; the other has a streetlight pole as a center and a 41-ft radius.

The final point E is located off a 33.6-ft radius from the SE corner of Buler Hall and a circle that has as a radius 21 ft struck with a 12-in oak tree as its center. This traverse is to be laid out through pacing. A distance of 400 ft was paced in 160 paces. This establishes the length of a single pace as 2.50 ft. Paces on the various courses of the traverse are shown in the accompanying table.

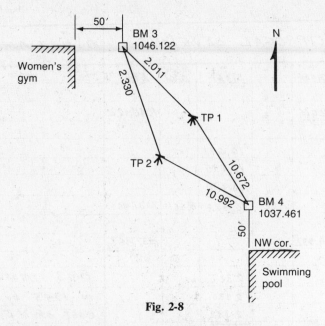

Fig. 2-8

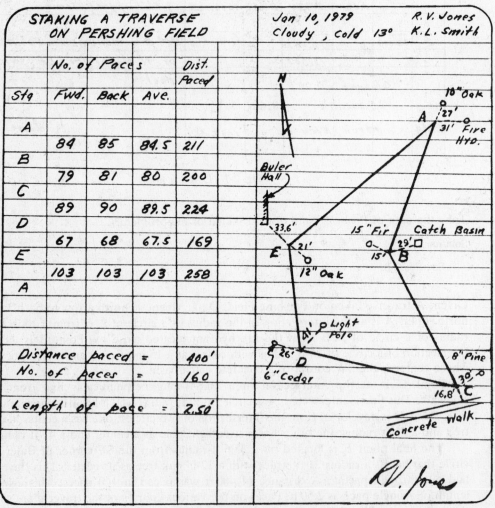

Fig. 2-9 Pershing field traverse.

Station	Number of Paces			Distance Paced, ft
	No. Forward	No. Back	Average	
A	84	85	84.5	211
B	79	81	80	200
C	89	90	89.5	244
D	67	68	67.5	169
E	103	103	103	258

Set up this problem as it would appear in a field book.

Solution

See Fig. 2-9.

Many other types of notekeeping will be explained throughout the text so no extra solved problems will be presented here. For further examples, see Figs. 14-3, 14-5, 14-6, 14-19, and 14-21 to 14-23.

Horizontal Distances

3.1 METHODS OF MEASUREMENT

Several methods used in surveying to obtain linear measurements follow:

1. Pacing

2. Odometer readings

3. Stadia

4. Electronic distance measuring equipment

5. Taping

Measuring the distance from one point to another is a fundamental part of surveying. With today's modern equipment, one can look at the readout of an electronic distance measuring (EDM) machine and read the exact distance correctly, but since these devices are expensive, they are not always available. In this chapter you will learn the conventional steel tape and pins method.

Distance measuring equipment in use today includes steel tapes, microwave instruments, electro-optical instruments for use in light-wave measuring systems, and stadia boards and calibrated instruments used with the stadia method. When used properly, the microwave and light-wave methods (see Figs. 3-1 and 3-2) yield extremely accurate results over distances ranging from less than 5 ft [1.5 meters (m)], to over 1 mile [1.6 kilometers (km)]. Unfortunately, this capability is also reflected in the initial cost of the equipment. Thus most schools will only be concerned with teaching the taping and stadia methods. However, knowledge of these two basic methods is very valuable in understanding the other distance measuring techniques and other aspects of surveying.

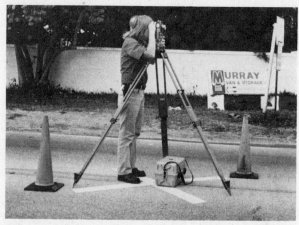

Fig. 3-1 An electronic distance measuring (EDM) device in use on State Road A-1-A. (*Midsouth Engineering Co.*)

3.2 PACING

Accuracy attained by pacing is sufficient for a great many purposes in surveying, geology, agriculture, and military field sketching, to name a few. It can also be used efficiently for detecting blunders that might have occurred in either taping or stadia reading.

Pacing consists of counting the paces in a distance you are measuring. First the person who will do the measurement must have the length of his or her pace determined. The method of doing this is to walk in natural steps back and forth over a measured level course about 400 ft long. The average is

Fig. 3-2   Rodperson holding an optical prism which reflects the signal directly back to the electronic distance meter. (*Midsouth Engineering Co.*)

taken by dividing the total distance by the number of steps taken. Figures 2-1, 2-4, 2-5, and 2-9 give the notes required in a field book for several pacing problems.

A 3-ft step is possible, but very tiring for a person of average height, so the method of averaging the pace over a longer distance is preferable to using a 3-ft pace. For long distances, a pocket instrument called a *pedometer* may be carried; it registers the distance traveled on foot.

Pacing is a valuable tool. It requires no equipment, and experienced pacers can measure distances of 100 ft or longer to a precision of 1:100 if terrain is level.

3.3   ODOMETER READINGS

Figure 3-3 shows a surveyor using a roller marker, or odometer, on a paved street in a new subdivision. An *odometer* is a wheel which rolls on the surface of the ground and converts the number of revolutions into a slope-distance measurement. The odometer is easy to use and very fast, although not as accurate as properly used tape. It is a good method of rough-checking measurements made by other methods. A precision of 1:200 may be expected, with greater precision on smooth road surfaces. The odometer can be used from a vehicle for some preliminary surveys in route-location projects.

3.4   HORIZONTAL SIGHTS BY STADIA

An optics discussion of transit telescopes will be neglected so that the theory of stadia can be simplified. Assume that externally focusing telescopes are used. These instruments have a focusing screw that causes the objective lens to move (see Fig. 3-4), where it is desired to determine the distance D from the center of the instrument to the rod. This distance equals c (distance from the center of the instrument to the center of the objective lens) plus f (the focal distance) plus d (the distance from the focal point to the rod).

Fig. 3-3 Odometer, or roller marker, in use.

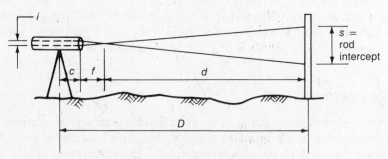

Fig. 3-4 Stadia principle.

If the distance from the top stadia hair to the bottom stadia hair is i, from similar triangles we can write the following equation and from it determine the distance D.

$$\frac{i}{f} = \frac{s}{d} \quad \text{where } s = \text{rod intercept}$$

$$d = \frac{f}{i} s$$

$$D = \frac{f}{i} s + c + f$$

In most transits the value of f/i, or K, as it is usually designated, is 100, and the value of $c + f$, called the *stadia constant*, varies from approximately 0.8 to 1.2 ft with an average value of 1 ft, most often used in the formula. If it is not 1, substitute the *actual* stadia constant shown on the instrument box.

Manufacturers show the exact value of the stadia constant on the instrument box. Generally, with the telescope in a horizontal position, the horizontal distance H from center of the rod to center of the instrument is

$$H = Ks + 1 \qquad (3\text{-}1)$$

Unequal refraction and unintentional inclination of the rod by the rodperson make stadia intervals larger than they actually are. In order that this feature of stadia measurement errors may be offset, the stadia constant is usually neglected.

Many transits of recent manufacture have internally focusing telescopes. These instruments have a stadia constant of only a few tenths of a foot, and it is even more reasonable to neglect it.

EXAMPLE 3.1   Given: Transit with $f/i = K = 101.5$ and the stadia constant $c + f = 1.0$ ft. The horizontal sight rod intercept is 4.62 ft. Find the stadia distance.

Solution:

Horizontal distance = H

$K = \dfrac{f}{i} = 101.5$

Stadia constant = $c + f = 1.0$ ft

s = sight rod intercept = 4.62 ft

Formula: $\qquad\qquad H = Ks + 1$

$\qquad\qquad\qquad\quad H = 101.5(4.62) + 1 = 468.93 + 1 = 469.93 = 470$ ft   *Ans*.

3.5   ELECTRONIC DISTANCE MEASUREMENTS

Discoveries in the last few decades make possible the use of light waves, electromagnetic waves, infrared waves, or lasers to measure distances precisely. Some of these waves are affected by changes in temperature, pressure, and humidity, but the effects are small and can be accurately corrected. Normally the corrections amount to less than a few centimeters in several miles. Several portable electronic devices using these wave types have been developed. These EDM machines permit measurement of distance with extreme precision.

These devices have not replaced chaining or taping, but are now commonly used by private and government survey crews. The prices for the EDM machines, although still prohibitive for most schools, is coming to a level that is affordable by private survey companies.

Electronic distance measuring devices (EDMs) have some advantages over other measurement methods. They easily measure lines that are difficult of access, such as across lakes and rivers, highways, standing farm crops, etc. For distances of several miles, the time required to get the distance accurately by EDM is in minutes. A taping party would require hours to perform the same job. Two trained workers can do the work better and faster than a regular four-person crew.

Disadvantages of EDM measurements are the cost, size, and weight of the equipment. EDMs measure slope distance from the EDM unit to the reflector. Provision must be made to obtain the vertical angle or difference in elevation between EDM unit and reflector. Taping is still the usual method of measuring short distances even though EDMs may be used there also.

3.6   PRECISION OF VARIOUS MEASURING METHODS

We will take up taping as a method of measuring distance in the next section. First a few of the measuring methods will be evaluated for their precision. Pacing has a precision of 1:50 to 1:200 and is usually used for reconnaissance and rough planning. The odometer has a precision of 1:200, and it is also used mostly in rough planning. The stadia method has a precision of 1:250 to 1:1000 and it is very common in mapping, rough surveys, and checking of other types of measurement. Ordinary taping has a precision of 1:1000 to 1:5000 and is used for regular land surveys and building construction. Precision taping has a precision of 1:10000 1:30000 and is employed for excellent land surveys, precise construction work, and city surveys. Electronic distance measuring equipment has a precision of ±0.04 ft (instrument constant), ±1:300 000 of the length measured. It has been used in the past for precise government surveys and is now becoming more common for land development and very precise construction work.

3.7 TAPING

There are two basic methods for measuring distance with a tape. These methods are *slope taping* and *horizontal taping*. In the slope taping method, the tape is held as required by the slope of the ground, the slope of the tape is measured, and the horizontal distance is computed. In the horizontal taping method, the tape is held horizontally and the required graduation is projected to the ground with a plumb bob. Under certain conditions, each of the methods of taping has its advantages. The slope taping method is the most precise and is always used for taping baselines and second- and higher-order traverse distances. (*Note*: A first-order survey is the most accurate—see Sec. 5.4.)

Third-order traverse measurements are usually made using slope taping; horizontal taping can be used, but it is not recommended. The horizontal taping method is generally used for lower-order tape traverses in mapping projects and construction surveys.

Party Organization

Figure 3-5 shows a four-person taping party. The minimum number of people in a taping party for horizontal taping is two people. This is not recommended except for very low order surveys. The standard taping party should consist of at least four people: a recorder, a stretcher, and a rear and front taper, as shown in Fig. 3-5.

Fig. 3-5 Four-person taping party.

Level Party

For precise tape measurements, the slope of the tape is usually determined by direct (differential) leveling. This type of leveling requires an instrumentperson (see Fig. 3-6), a rodperson, and a recorder.

For third- and lower-order taping, the slope of the tape is usually determined using an Abney topographic hand level. Where using the hand level, the tapers or the stretchers are responsible for the slope measurement.

Fig. 3-6 Instrumentperson using a universal level-transit.
(*David White Instruments*)

3.8 SLOPE TAPING PROCEDURE

In the slope taping method, the tape may lie on smooth ground, on a paved road, or on some other fairly smooth and uniformly sloping surface, or its ends may be supported by taping stools or stakes. The following procedures are generally used when taping third- or lower-order traverse distances. Normally, the equipment consists of a range pole, one set of taping pins, one steel tape, one tape thermometer, and one Abney hand level.

The range pole is set along the line to be measured and slightly behind the point toward which the taping will proceed.

The rear taper, with 1 of the 11 taping arrows (pins), is stationed at the point from which the taping will start. The head taper, with the zero end of the tape and 10 taping arrows, moves toward the end point. When the head taper has gone nearly the full tape length, the rear taper gives a signal to halt by calling out "tape." The rear taper holds the last graduation at the initial point and directs the head taper into alignment with the previously set range pole. The head taper then pulls the tape taut, bringing it onto line. The tapers exert and maintain the required tension on the tape (assuming there are no stretchers). When the exact graduation is on the initial point, the rear taper calls "mark" or "stick," and the head taper marks the distance by sticking an arrow into the ground at an angle of 45° with the ground and perpendicular to the tape at the zero graduation of the tape. The head taper then calls "marked" or "stuck" and both tapers release the tension.

It is considered good practice to immediately check this measurement. When ready to repeat the measurement, the head taper calls "check," and the measurement is repeated as a check. The distance is now recorded along with the tape temperature, and the Abney hand level reading is now taken and recorded along the line coincident with or parallel to the slope of the tape.

An Abney hand level and clinometer have a limited application in measuring vertical angles and slopes, and for direct leveling. Included are an arc graduated in degrees up to 90°, a vernier reading to 10 minutes, and several scales for slopes ranging from ratios of 1:1 to 1:10.

Before moving forward, the rear taper pulls out the rear taping pin and carries it along. The front taper leaves the newly placed pin in its position. Thus, one taping pin always remains in the ground, and the number of pins held by the rear taper indicates the number of full tape lengths from the initial point to the remaining arrow.

Both tapers then move forward and make another complete measurement starting from the taping pin that marks the end of the previous measurement.

When the taped distance is more than 10 tape lengths, the head taper signals the rear taper to come forward at the end of the tenth length and to bring the 10 taping pins from the rear. Both tapers check the number of pins to see that none have been lost. The head taper checks the recording book to see that 10 taped distances, temperature readings, and slope measurements have been recorded. The rear taper gives 10 pins to the head taper, and the measurement is then continued.

At the end of the line, the head taper holds the zero graduation on the terminal point. The rear taper pulls the tape taut and holds the appropriate full graduation at the taping pin. The fractional unit is now read by the head taper, the procedure depending on the type of tape and its end graduation.

At this point, two important items should be noted. First, if less than a full tape length is being measured, then the tension applied to the tape must be prorated. This is done by dividing the amount of tension applied to a full tape length by the length of a full tape, and then multiplying the result by the fractional length to be measured.

EXAMPLE 3.2 20 pounds (1b) [89 newtons (N)] of tension is normally applied to a certain 300-ft (91-m) tape. The fractional length at the end of a line is approximately 45 ft (14 m) (within 1 ft). How much tension should be applied to the tape for the final section?

Solution:
$$\frac{20 \text{ lb}}{300 \text{ ft}} = 0.0666 \text{ lb/ft}$$

$$(0.0666 \text{ lb/ft})(45 \text{ ft}) = 2.997 \text{ lb}$$

Therefore, 3 lb (13 N) of tension should be applied to the final 45-ft (14-m) section.   *Ans.*

Tape Alignment

One of the requirements of taping is that the taping be done on a straight line between stations. The order of the work being done will define the permissible limits of misalignment between stations. For precise taping the allowable misalignment is very small, and a theodolite is used to keep the alignment of the tape within limits. For third- and lower-order taping, the tape can be aligned by having the rear taper "eye in" the tape.

The error in a 6-in misalignment is shown in Fig. 3-7 and is computed as follows: Distance A to B is measured in three taped lengths of 100 ft each. The first measurement of 100 ft is off line by 6 in. We can therefore see that the straight-line distance we are supposed to be measuring will be shorter than 100 ft. Using the pythagorean theorem, we can find the off-line error:

$$a^2 + b^2 = c^2$$
$$a^2 = c^2 - b^2$$
$$a^2 = 100^2 - (0.5)^2$$
$$a = 99.999 \text{ ft}$$

Thus the off-line error is +0.001 ft.

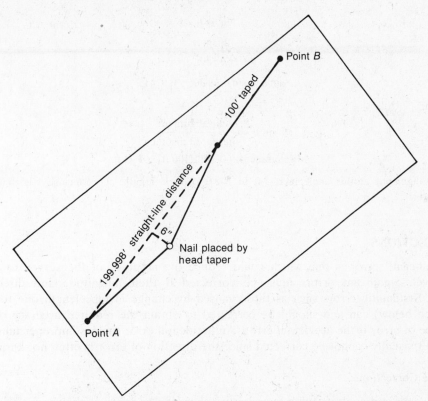

Fig. 3-7   Error in 6-in misalignment. (*From C. A. Herubin, Principles of Surveying, Reston Publishing, Reston, Va., 1982*)

Continuing the measurement gives the second measurement on line in the forward direction, but since it started from the 6-in off-line point, we have again the same right triangle that we had on the first measurement. This means the error due to misalignment is twice the +0.001-ft error, or +0.002 ft in the 300-ft measurement (see Fig. 3-7). This is an accuracy of 1:150000. The error due to misalignment of 6 in in 100 ft is not significant because accidental errors in measuring the distance are usually of greater magnitude.

EXAMPLE 3.3   Given: A misalignment of 0.01 ft at the end of a 100-ft measurement. Determine whether this causes a measurable error in taping. (A measurable error in alignment would be around 1 ft off line.)

Solution: See Fig. 3-8.

$$\cos a = \frac{\text{true distance}}{\text{measured (off-line) distance}}$$

$$\cos a = \frac{99.99 \text{ ft}}{100.00 \text{ ft}} \qquad \text{(an error of 0.01 ft)}$$

This is a measurable error when measurements are read to the second decimal place.

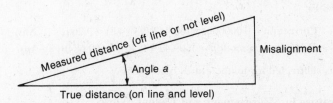

Fig. 3-8   Significant error caused by tape misalignment. (*From C. A. Herubin, Principles of Surveying, Reston Publishing, Reston, Va., 1982*)

$$\cos a = \frac{99.99 \text{ ft}}{100.00 \text{ ft}} = 0.9999$$

Thus
$$a = 0°48'30''$$

$$\sin a = \sin 0°48'30'' = 0.01411$$

From Fig. 3-8,

$$\sin a = \frac{\text{misalignment}}{\text{measured distance}}$$

So
$$\text{Misalignment} = 0.1411(100 \text{ ft}) = 1.41 \text{ ft}$$

Thus the misalignment at this angle is 1.41 ft in 100 ft of tape length; the conclusion is that this causes a measurable error.

3.9 CORRECTIONS

A "significant" error is one which could reduce the accuracy of the survey to less than the required level. Significant errors must be corrected if their magnitude and direction can be determined. Systematic errors such as those caused by change in tape length due to temperature variation (see below) can and should be corrected to obtain the required accuracy of the survey. Another type of error is the accidental error. An accidental error such as improper taping technique (see Sec. 3.13) usually cannot be corrected since the direction of error is often not known.

Temperature Corrections

Steel expands and contracts with temperature changes. An index of the change in size that steel undergoes because of temperature changes is the steel's coefficient of expansion. The *coefficient of expansion* is the percentage that any length will expand or contract with a change of temperature of 1°. A coefficient of 0.01 indicates a 1 percent change in length for a 1° change in temperature.

For example, 1 ft would expand to 1.01 ft with a 1° temperature rise and to 1.05 ft with a 5° rise. A length of 100 ft would increase to 101 ft with a 1° rise in temperature, and decrease to 99 ft with a 1° drop in temperature.

Steel tapes have a coefficient of expansion of 0.00000645. This is an important number in surveying. A steel tape lengthens 0.00000645 ft for each foot of its original length with each 1° increase in the surrounding temperature.

The tapes are standardized for 68°F (20°C). At 68°F the tape should be the correct length.

EXAMPLE 3.4 Given: A 100.00-ft tape accurate at 68°F. It is a hot summer day; the temperature is 93°F. A survey is taken where a measured distance at 93°F is 1609.42 ft. Find the correction that must be applied to the tape and the measured distance plus the correction.

Solution: The true distance is greater than the distance read on the tape since the tape increases in length and the numbers on the tape became farther apart. The error thus is negative, making the correction positive.

The correction that must be applied is the coefficient of expansion (0.00000645) times the measured distance (1609.42 ft) times the temperature difference between outside temperature (93°F) and the temperature where readings are accurate (68°F):

$$\text{Correction} = 0.00000645(1609.42)(93 - 68) = 0.26 \text{ ft} \quad Ans.$$
$$\text{Distance} + \text{correction} = 1609.42 + 0.26 = 1609.68 \text{ ft} \quad Ans.$$

Note: The correction is subtracted for temperatures below 68°F.

Adjusting for Temperature in Measurement and Layout Work

There are two taping operations: Measurement work, where a measure is taken between fixed points (already set in the ground), and layout work, where a second point is to be set with respect to a first point. Depending on the outside temperature, a tape either expands (is longer) or contracts (is shorter) from its true length at 68°F (20°C).

For measurement work if the tape is longer (temperature greater than 68°F), it does not sufficiently cover the fixed distance. Therefore the recorded distance is smaller than the actual distance and a correction must be *added*. If the tape is shorter, the opposite is true.

For layout work if the tape is longer (temperature greater than 68°F), it covers too great a distance on the ground and the correction must be *subtracted*. If the tape is shorter, the opposite is true.

The adjustments are summarized in the accompanying table.

Sign of the Correction

Tape	Measurement	Layout
Longer than actual length	+	−
Shorter than actual length	−	+

EXAMPLE 3.5 We are to set a point 1409.39 ft from a known point, using the same tape from which the original measurement was made. The temperature is 96°F. What distance should be measured on the tape?

Solution: The true distance (1409.39 ft) is *greater* than the distance shown on the tape because the temperature on this day is 96°F instead of the 68°F at which the tape is at its true length. Therefore, the length read on the tape must be less than 1409.39 ft.

The distance to be laid out (that is, measured on the tape) is the true distance minus the correction:

$$\text{Layout distance} = 1409.39 \text{ ft} - 0.00000645(1409.39 \text{ ft})(96 - 68)$$
$$= 1409.39 - 0.25 = 1409.14 \text{ ft} \qquad Ans.$$

3.10 SLOPE COMPUTATIONS

In measuring the distance between two points on a steep slope, it is often desirable to tape along the slope and determine the angle of inclination a or the difference in elevation d (see Fig. 3-9) rather than break tape every few feet. It is a real advantage to use long tapes of 200 to 500 ft for measuring along slopes. In Fig. 3-9, if angle a is determined, the horizontal distance between points A and B can be computed from the relation

$$H = L \cos a \tag{3-2}$$

where H = horizontal distance between the points
          L = the slope distance between the points
          a = the vertical angle from the horizontal, usually obtained with the Abney hand level, clinometer, or transit

EXAMPLE 3.6 Given: A slope measurement L of 581.25 ft and a slope angle a of 4°. Find the horizontal distance H.

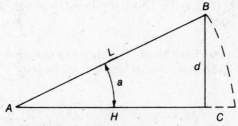

Fig. 3-9 Slope measurement.

Solution: See Fig. 3-9. Use Eq. (*3-2*).

$$H = L \cos a = 581.25 \cos 4° = 581.25(0.9975641) = 579.83 \text{ ft} \qquad Ans.$$

Approximate Slope Formula for Baseline Work

The difference in elevation, d (see Fig. 3-9), between the ends of the tape is found by leveling, and the horizontal projection computed. From Fig. 3-9:

$$C = L - H \qquad \text{where } C = \text{slope correction}$$
$$d^2 = L^2 - H^2 = (L - H)(L + H) = C(L + H)$$

So

$$C = \frac{d^2}{L + H}$$

The approximate formula from the above is:

$$C = \frac{d^2}{2L} \tag{3-3}$$

The error in the approximate formula for a 100-ft length gets larger with increasing slope. The answers for slopes for inclinations less than 10° are correct to the nearest 0.001 ft.

EXAMPLE 3.7 Compute the slope corrections for difference in elevation of 5 ft and a slope length of 500.00 ft; then compute the horizontal distance H by the pythagorean theorem and compare results.

Solution: See Fig. 3-9. Use Eq. (*3-3*).

$$C = \frac{d^2}{2L} = \frac{25}{1000} = 0.025$$
$$H = L - C = 500 - 0.025 = 499.975 \text{ ft} \qquad Ans.$$

By the pythagorean theorem,

$$H^2 + d^2 = L^2$$
$$H^2 = L^2 - d^2$$
$$H^2 = 250\,000 - 25 = 249\,975$$
$$H = \sqrt{249\,975} = 499.975 \text{ ft} \qquad Ans.$$

The results agree.

3.11 STATIONING

In route surveying, the centerline is stationed from a starting point designated as station $0 + 00$ or $10 + 00$ or some other point of beginning. The term *full station* is applied to each 100 ft of length, where a stake is normally set. The position of any other point is given by its total distance from the point of beginning. For example, in Fig. 3-10 station $145 + 60$ is a unique point 2560.00 ft from the begin project station of $120 + 00$ (14 560 ft − 12 000 ft). The partial length beyond a full station, in this example, 60.00 ft, is termed a *plus*.

EXAMPLE 3.8 Given: A project near Lake Mead. The beginning of the project is station $120 + 00.00$. If the end of the project is station $186 + 73.12$, find the total length of the project.

Solution: See Fig. 3-10.

$$
\begin{aligned}
\text{End project} &= 186 + 73.12 = 18\,673.12 \text{ ft} \\
\text{Begin project} &= 120 + 00.00 = 12\,000.00 \text{ ft} \\
\text{Total length of project} &= \overline{6\,673.12 \text{ ft}}
\end{aligned}
$$

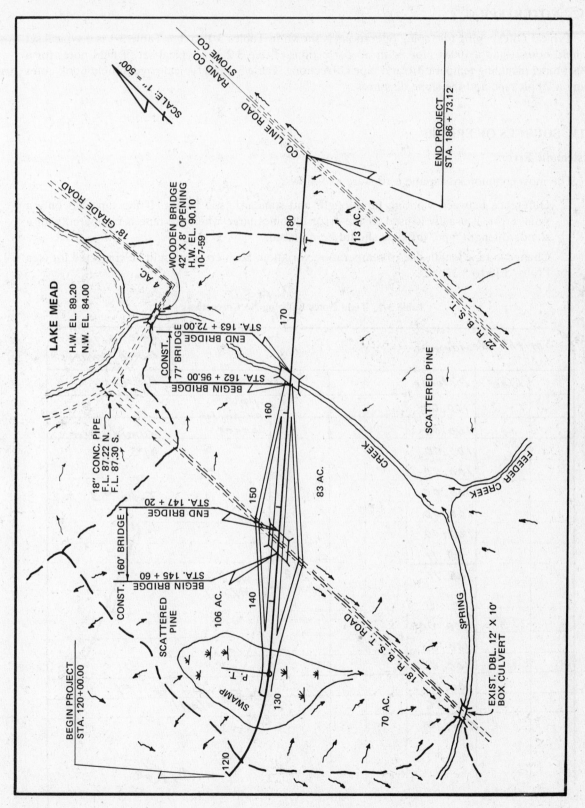

Fig. 3-10 Stationing project near Lake Mead.

3.12 NOTEKEEPING

Typical notekeeping for taping operations is shown in Tables 3-1 to 3-3. Table 3-1 is a typical set of field notes using a 100-ft tape, with no corrections. Table 3-2 is a typical set of field notes for a 200-ft tape, including temperature and tape corrections. Table 3-3 is a typical page of field book notes using a 200-ft tape and the slope distances.

3.13 SOURCES OF ERROR

Systematic Errors

The most common systematic errors are as follows:

1. Difference between working tape length and standard tape length. If the tape has only a small error, it usually is used. If the error becomes large when the tape is compared with a standard-length tape, the tape should be replaced.

2. Change in tape length due to temperature variation. Such errors should be corrected for (see Probs. 3.15 to 3.22).

Table 3-1   Field Notes Showing No Corrections

Smith's Residence						Date: 6/7/82 D. Roe - chief
Line	Distance					B. Jones
1 - 2	100.00				#1 • I.P. Property Corner	
	100.00					Humid - Hot
	100.00					91°
	100.00					
	100.00					Tape #3
	100.00					
	100.00					N
	88.71					
	788.71				#2 ▫ Wood stk. W/nail	
2 - 3	100.00					
	100.00					
	100.00					
	100.00					
	0.36				#3 ▫ Wood stk. W/nail	
	400.36					
						D. Roe

Accidental Errors

The most common sources of accidental errors are as follows:

1. Incorrectly estimating the second decimal place on the tape.
2. Incorrectly plumbing the tape over a point.
3. Using improper tension on the tape.
4. Misalignment, either vertical or horizontal.
5. Wind blowing the tape, making it curve to one side. This may be controlled by applying greater tension and taping shorter distances.

3.14 COMMON MISTAKES

The most common mistakes made in surveying operations are as follows:

1. Misreading a number. To avoid this problem the numbers should be read on each side of the number to be recorded. In addition, viewing of the distance measured will prevent mistakes such as recording 96 ft for 69 ft.
2. Recording a number incorrectly. To avoid this problem the number should be said aloud while it is being recorded. The other taper should listen and verify the number.

Table 3-2   Field Notes Showing Corrections for Temperature and Tape

Southern Bell Telephone Line, Atlanta, Ga.   Date: 8/12/83

Line	Meas.	Tape Corr.	Temp. Corr.	Corr. Dist.		
						W. Jones, chief
						S. Smith, Tape
1-2	31.60				I.P.@ Base of pole	No. 22.
	28.50					Clear - 56°F
	75.12					
	46.21					N
	181.43	-0.02	-0.14	181.38		
2-3	131.12					
	162.28				I.P.@ Fence	
	293.40	-0.03	-0.02	293.32		

3 • I.P.

W. Jones

Table 3-3 Slope Distance Using 200-ft Tape

Line	Meas. Dist.	V ∡	Corr(-)	Corr. Dist.		D. Roe, chief
						H. Doe
1-2	196.34	4°-30'	0.61	195.73'		Fair cool
						Tape A-6
2-3	141.61	2°-15'	0.11	141.50'		

Township Line Road, Montgomery County   Date 6/6/83

N

1 Nail & Tab in ⅊ spring st.

2 Nail in Wood stk.

3 Nail in Wood stk

D. Roe.

3. Using an incorrect location for zero of the tape. Check to see if zero is at the end of the ring or on the tape.

4. Omitting a tape length. This can happen easily, so check carefully to eliminate this problem.

5. Allowing the tape to touch brush or other objects so it is not hanging freely between the two taping points.

Solved Problems

3.1 Given:

K value of transit = 101.0

Stadia constant = 1.0 ft

Sight intercept $s = 3.96$ ft

Find horizontal distance H.

Solution

Use Eq. (*3-1*).

$$H = Ks + 1 = 101.0(3.96) + 1 = 399.96 + 1 = 401 \text{ ft} \qquad Ans.$$

3.2   Given:

K value of transit = 100.3
Stadia constant = 1.1 ft
Sight intercept s = 3.74 ft
Find horizontal distance H.

Solution

$$H = Ks + 1.1 = 100.3(3.74) + 1.1 = 375.12 + 1.1 = 376 \text{ ft} \qquad Ans.$$

3.3   Given: K = 101.5; s = 3.12 ft; stadia constant = 0.8 ft. Find horizontal distance H.

Solution

$$H = Ks + 0.8 = 101.5(3.12) + 0.8 = 317 \text{ ft} \qquad Ans.$$

3.4   Given: K = 101.2; s = 3.01 ft; stadia constant = 1 ft. Find horizontal distance H.

Solution

$$H = Ks + 1 = 101.2(3.01) + 1 = 306 \text{ ft} \qquad Ans.$$

3.5   Given: K = 99.7; s = 2.71 ft; stadia constant = 1 ft. Find horizontal distance H.

Solution

$$H = Ks + 1 = 99.7(2.71) + 1 = 271 \text{ ft} \qquad Ans.$$

3.6   Given: K = 101.3; s = 3.98 ft; stadia constant = 0.9 ft. Find horizontal distance H.

Solution

$$H + Ks + 0.9 = 101.3(3.98) + 0.9 = 404 \text{ ft} \qquad Ans.$$

3.7   Given: A normal tape tension of 20 lb (89 N) is applied to a 300-ft (91-m) tape. The fractional length of the tape at the end of the line is 96 ft (30 m) (within 1 ft). How much tension should be applied to the tape for the final section?

Solution

$$\frac{20 \text{ lb}}{300 \text{ ft}} = 0.0666 \text{ lb/ft}$$

$$0.0666(96) = 6.394 \text{ lb} \qquad Ans.$$

Therefore, 6.4 lb (28 N) of tension should be applied to the final 96-ft (30-m) section.

3.8   Given: Normal tape tension of 20 lb (89 N) for a 100-ft tape. The final measurement in the line of tapes is 77 ft (24 m). How much tension should be applied to the spring balance on the end of the tape to give the proper tension to the 77-ft increment?

Solution

$$\frac{20 \text{ lb}}{100 \text{ ft}} = 0.200 \text{ lb/ft}$$

$$0.200(77) = 15 \text{ lb} \qquad Ans.$$

Therefore, 15 lb (67 N) of tension should be applied to the 77-ft increment of tape.

3.9   Given: Hot summer day temperature of 96°F. The measurement on this day is 1782.51 ft. Find the correction and the measured distance plus the correction for this measurement.

Solution

$$\text{Correction} = 0.00000645(1782.51)(96 - 68) = 0.32 \text{ ft} \qquad Ans.$$

$$\text{Distance} + \text{correction} = 1782.51 + 0.32 = 1782.83 \text{ ft} \qquad Ans.$$

3.10   Given: On a fall day at a temperature of 78°F a measurement of 2132.61 ft was made. Find the correction and the measured distance plus correction for this measurement.

Solution

$$\text{Correction} = 0.00000645(2132.61)(78 - 68) = 0.14 \text{ ft} \qquad Ans.$$

$$\text{Distance} + \text{correction} = 2132.61 + 0.14 = 2132.75 \text{ ft} \qquad Ans.$$

3.11   Given: Measurement of 1372.13 ft at temperature of 13°F. Find the measured distance adjusted for the correction.

Solution

$$\text{Corrected distance} = 1372.13 - 0.00000645(1372.13)(68 - 13) = 1372.13 - 0.49 = 1371.64 \text{ ft} \qquad Ans.$$

3.12   Given: Measurement of 697.13 ft at temperature of 72°F. Find the corrected measurement at this temperature.

Solution

$$\text{Corrected distance} = 697.13 + 0.00000645(697.13)(72 - 68) = 697.13 + 0.02 = 697.15 \text{ ft} \qquad Ans.$$

3.13   Set a point 1721.71 ft from a known point with the same tape at the same temperature. The temperature on this day is 79°F. What distance should be laid out on the tape?

Solution

The true distance is greater than the amount to be measured, or laid out, on the tape because the temperature is 79° instead of the 68° at which the tape is standardized as its true length. Thus the distance to be laid out is the true distance minus the correction.

$$\text{Layout distance} = 1721.71 - 0.00000645(1721.71)(79 - 68) = 1721.71 - 0.12 = 1721.59 \text{ ft} \qquad Ans.$$

3.14   Set a point 6791.19 ft from a known point with the same tape at the same temperature—87°F.

Solution

The distance to be laid out is the true distance minus the correction:

$$\text{Layout distance} = 6791.19 - 0.00000645(6791.19)(87 - 68) = 6791.19 - 0.83 = 6790.36 \text{ ft} \qquad Ans.$$

3.15   Given: A slope measurement of 791.32 ft. The angle measured with an Abney hand level is 5°30′. Find the horizontal distance H for this slope distance.

Solution

See Fig. 3-9. Use the formula $H = L \cos a$, where $a = 5°30′ = 5.5°$.

$$H = L \cos 5.5 = 791.32(0.9953962) = 787.68 \text{ ft} \qquad Ans.$$

3.16   Given: A slope measurement of 900.00 ft and a slope angle of 7°30′. Find the horizontal distance H.

Solution

See Fig. 3-9. Use the formula $H = L \cos a$, where $a = 7°30′ = 7.5°$.

$$H = L \cos 7.5 = 900.00(0.9914449) = 892.30 \text{ ft} \quad \textit{Ans.}$$

3.17   Given: A slope distance of 400 ft with a difference in elevation of 6 ft. Compute the horizontal distance H by the approximate formula $C = d^2/2L$ and compare the result with that obtained by the pythagorean theorem.

Solution

See Fig. 3-9.

$$C = \frac{d^2}{2L} = \frac{6^2}{2(400)} = \frac{36}{800} = 0.045$$

$$H = L - C = 400 - 0.045 = 399.96 \text{ ft} \quad \textit{Ans.}$$

By the pythagorean theorem,

$$H^2 + d^2 = L^2$$
$$H^2 = L^2 - d^2$$
$$H^2 = 400^2 - 6^2 = 160\,000 - 36 = 159\,964$$
$$H = \sqrt{159\,964} = 399.96 \text{ ft} \quad \textit{Ans.}$$

The results agree.

3.18   Referring to Fig. 3-10, determine the first bridge length if the bridge starts at station $145 + 60$ and ends at station $147 + 20$.

Solution

See Fig. 3-10.

$$
\begin{aligned}
\text{End of bridge station} &= 147 + 20 = 14\,720 \text{ ft} \\
\text{Start bridge station} &= 145 + 60 = \underline{14\,560 \text{ ft}} \\
\text{Length of bridge} &= 160 \text{ ft} \quad \textit{Ans.}
\end{aligned}
$$

3.19   Referring to Fig. 3-10, determine how far it is in feet from the end of the second bridge at station $163 + 72.00$ to the end of the project.

Solution

See Fig. 3-10.

$$
\begin{aligned}
\text{End of project} &= 186 + 73.12 = 18\,673.12 \text{ ft} \\
\text{End of bridge} &= 163 + 72.00 = \underline{16\,372.00 \text{ ft}} \\
\text{Distance} &= 2\,301.12 \text{ ft} \quad \textit{Ans}
\end{aligned}
$$

Supplementary Problems

3.20 Given: K value of transit $= 100.5$, stadia constant $= 1.0$ ft, sight intercept $s = 3.89$ ft. Find horizontal distance H.   *Ans.*   392 ft

3.21 Given: $K = 100.5$; stadia constant $= 1.0$ ft; $s = 3.27$ ft. Find horizontal distance H.   *Ans.*   330 ft

3.22 Given: $K = 101.5$; stadia constant $= 1$ ft; $s = 2.96$ ft. Find horizontal distance H.   *Ans.*   301 ft

3.23 Given: $K = 101.5$; stadia constant $= 1$ ft; $s = 4.07$ ft. Find horizontal distance H.   *Ans.*   414 ft.

3.24 Given: Normal tape tension of 20 lb (89 N) for a 100-ft length of tape suspended at the two ends. Vary the tension in proportion to a length of 45 ft (14 m), which is the final distance measured in the line of stations. How much tension should be applied to the final section?
Ans.   9 lb (40 N) should be applied to the 45-ft segment of tape.

3.25 Given: A taped distance of 100 ft. At the end of the 100-ft tape there is a misalignment of 0.02 ft. Determine whether this is a measurable error. See Fig. 3-8.
Ans.   Misalignment $= 2.01$ ft in 100 ft. This is a measurable error and would be unacceptable for the survey.

3.26 Given: Temperature at survey time of 31°F; measurement of 782.13 ft. Find the correction and the corrected measured distance for this measurement.   *Ans.*   Correction $= 0.19$ ft, Distance $= 781.94$ ft

3.27 Set a point 3212.69 ft from a known point with a 100-ft tape at a constant temperature of 59°F. What distance should be laid out on the tape?   *Ans.*   3212.88 ft

3.28 Given: A slope measurement of 879.21 ft with an angle of slope of 6°15'. Find the horizontal distance H. See Fig. 3-9.   *Ans.*   873.98 ft

3.29 Given: A slope measurement of 1000.06 ft and a slope angle of 5°30'. Find the horizontal distance H. See Fig. 3-9.   *Ans.*   995.46 ft

3.30 Given: A slope distance of 325 ft with a difference in elevation of 3.5 ft. Compute the horizontal distance H with the approximate formula $C = d^2/2L$ and compare the result with that obtained by the pythagorean theorem.
Ans.   Using the approximate formula, $H = 324.98$ ft. Using the pythagorean theorem, $H = 324.98$ ft. The results agree.

3.31 Given: A slope distance of 267 ft with a difference in elevation of 4.27 ft. Compute the horizontal distance H by the approximate formula $C = d^2/2L$ and compare the result with that obtained by the pythagorean theorem.
Ans.   Using the approximate formula, $H = 266.97$ ft. Using the pythagorean theorem, $H = 266.97$ ft. The results agree.

Chapter 4

Transits

4.1 GENERAL INFORMATION

The engineer's or surveyor's transit (see Fig. 4-1) is often called the universal surveying instrument because of its many uses. It can be used for observing horizontal angles and/or directions, for observing vertical angles and differences in elevation, for prolonging straight lines, and for measuring distances by stadia. Although transits of various manufacturers differ in appearance, the parts and mode of operation are essentially the same. Figure 4-1 shows the parts of a common transit, which are listed and identified here:

1. Plate level vials. In addition to the telescope level vial (located directly below the telescope; not shown in Fig. 4-1), the transit also has two plate level vials which are used to level the instrument within the horizontal plane.

2. Compass. The transit has a built-in surveying compass for jobs requiring magnetic directional readings. The compass is graduated to 1° and numbered in quadrants. The W and E on the compass are reversed from the normal map position because the dial surface is attached to the instrument and revolves with it. The needle remains the fixed line and indicates the direction the telescope is facing.

3. Compass locking screw. The compass locking screw disconnects the needle to reduce wear on the needle bearing of the compass when not in use.

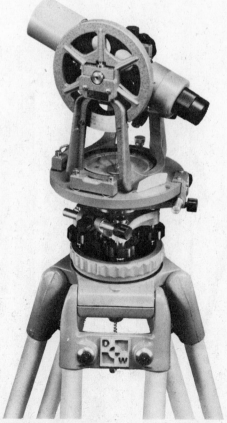

Fig. 4-1 The engineer's transit. (*David White Instruments*)

4. Vertical clamp screw. The telescope can be locked to the approximate vertical angle with the vertical clamp screw.

5. Vertical tangent screw. Fine vertical settings can be made with the vertical tangent screw. The clamp must be hand-tightened firmly before the tangent screw will function.

6. Lower horizontal clamp screw. The upper horizontal clamp screw secures the horizontal circle to the standard. The lower horizontal clamp screw secures the circle to the leveling head.

7. Tripod. The tripod (see Fig. 4-2) is the base or foundation which supports the survey instrument and keeps it stably attached to the ground during observations. It consists of a tripod head to which the instrument is attached, three wooden or metal legs which are hinged at the head, and metal pointed shoes on each leg to press or anchor into the ground to achieve a firm setup.

Figure 4-3 is a good view of the engineer's transit; the tripod can be seen firmly supporting the instrument.

Two types of tripods are available to the surveyor: the fixed-leg tripod and the extension-leg tripod. The legs of the fixed-leg tripod are made of a single piece of material, and, for this reason, they must be swung either in or out in varying amounts in order to level the head and to control the

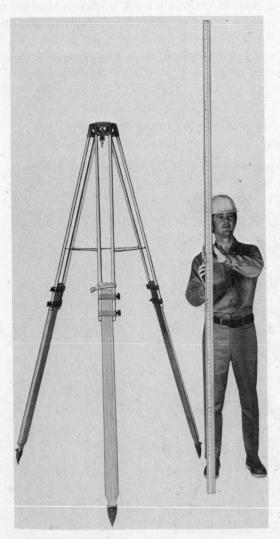

Fig. 4-2   Tripod. (*David White Instruments*)   **Fig. 4-3**   Engineer's transit mounted on a tripod.

instrument's height. The legs of the extension-leg tripod are made in two sections which slide up and down. This feature is particularly useful for setting up over rough terrain.

4.2 DIFFERENCES BETWEEN TRANSITS AND THEODOLITES

There is no universally accepted understanding about the difference between the terms *transit* and *theodolite*. *Transiting* means the plunging or reversing, or inverting, of the telescope—hence the term *transit*. Thus a transit is a universal level with the telescope mounted so that it can be transited, or reversed.

Divergent basic characteristics between the transit and theodolite determine which instrument is which. Transits have metal circles read by means of verniers (sliding graduated scales). Theodolites have glass circles from which readings are taken from finely graduated glass scales or micrometers viewed through internal microscopic optical systems. Generally speaking, theodolites are capable of greater accuracy in angle measurements than are transits, and are gaining acceptance in the United States. The engineer's transit is designed for every option required in surveying and has been the type of transit used most extensively in the United States in the past; it is still widely used in schools for instruction. Properly operated the engineer's transit will measure horizontal angles to the most frequently required accuracy—vertical angles can be measured to ±10 seconds, and elevations to third-order accuracy.

All problems in this text are for the engineer's transit. If a surveyor masters this instrument he or she can use any transit or theodolite with ease, because of the similarity in function between these two fundamental instruments.

4.3 RELATIONSHIPS OF ANGLES AND DISTANCES

In making measurements of angles with the transit the following relationships between angles and distances are important (see Fig. 4-4):

sin 1 minute of arc = tan 1 minute = 0.0002909

sin 1° = tan 1° = 0.01745

1 minute of arc = 0.03 ft at 100 ft, or 3 cm at 100 m

1 minute of arc = 0.09 ft or approximately 1 in at 300 ft

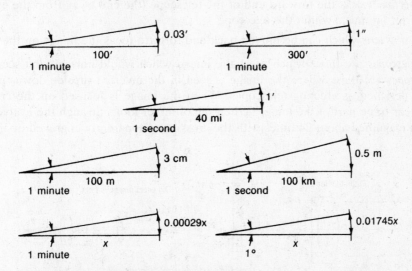

Fig. 4-4 Relationship of angles and distances. (*From R. C. Brinker and P. R. Wolf, Elementary Surveying, IEP Publishing, Dun-Donnelley, New York, 1977*)

1 second of arc = 1 ft at 40 mi, or 0.5 m at 100 km

All of the above are approximate figures.

Note: Only laboratory instruments such as Wild T-4 can read, or are graduated to, 0.1 second of arc.

In accordance with these relationships, using a theodolite, which is slightly more accurate than a transit, theoretically you could read to the nearest 0.1 second and measure the angle between two points 1 ft apart at a 40-mi distance.

Example 4.1   Given: A distance of 500 ft and an angle of 1 minute. Find the perpendicular offset at 500 ft for a 1-minute angle.

Solution: Draw a figure with sides of 500 ft and a 1-minute angle (exaggerate the angle). See Fig. 4-5. From relationships of angles and distances (Fig. 4-4):

$$\text{Offset} = 0.00029x \qquad \text{where} \qquad x = 500 \text{ ft}$$
$$\text{Offset} = 0.00029(500) = 0.145 \text{ ft} \qquad Ans.$$

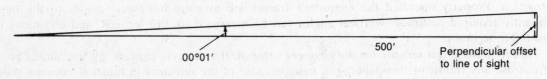

00°01′   500′   Perpendicular offset
                                                        to line of sight

Fig. 4-5

4.4 TELESCOPIC SIGHT

Common to both surveying level and transit is the telescopic sight. The modern telescopic sight (see Fig. 4-6), consists of the following:

1. A *reticle* providing the *cross hairs* near the rear of the telescope tube.

2. An *eyepiece* which magnifies the cross hairs and must be focused on them according to each surveyor's eyesight.

3. An *objective lens* at the forward end of the telescope (the end away from the eyepiece). This lens forms an image within the telescope.

4. A *focusing lens* which can be moved back and forth to focus the image on the cross hairs.

Erecting telescopes are those which erect the image which would normally be seen as inverted. *Inverting telescopes* are those where the image is seen in the inverted (upside-down) position. Most telescopes are designed as erecting telescopes. When the image is focused on the cross hairs, the cross hairs appear to be part of the image. When the observer looks through the eyepiece, he or she sees the object magnified about 24 times with the cross hairs apparently engraved upon it.

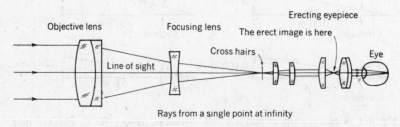

Objective lens   Focusing lens   Erecting eyepiece
                                                The erect image is here

                        Line of sight   Cross hairs   Eye

Rays from a single point at infinity

Fig. 4-6   The telescopic sight.

4.5 LINE OF SIGHT

When viewed through the telescope a point on an object will be on a straight line through the optical center of the objective lens. A straight line from the cross hairs through the optical center of the lens will strike the point on the object where the observer sees the cross hairs apparently located. The *line of sight* of a telescopic lens therefore is defined by the cross hairs and the optical center of the objective.

A properly focused telescopic sight will allow the observer to move his or her eye slightly without changing the position of the cross hairs on the object. In a rifle sight the eye must be accurately aligned with the sight to determine where it is pointing. This is not the case with the telescopic sight on the transit. Since the telescopic sight magnifies the object 24 times, the diameter of the field of view is small, about 1°, or 1.75 ft at 100 ft.

4.6 VERNIERS

Verniers are short auxiliary scales set parallel with and adjacent to a primary scale. The vernier is so constructed that when it is placed so that both primary and vernier scales have a line in coincident position, the fractional part of the smallest division of the primary scale can be obtained without interpolation.

As illustrated in Fig. 4-7, the vernier has n divisions in a space covered by $n - 1$ of the smallest divisions on the scale. Thus

$$(n - 1)d = nv \qquad \text{or} \qquad v = \frac{(n - 1)d}{n}$$

where d is the length of a main scale division and v is the length of a vernier division.

EXAMPLE 4.2   Given: In Fig. 4-7, $n = 10$, $d = 0.01$ ft, and $v = 0.09/10 = 0.009$ ft; three settings of the vernier are shown. What are the vernier readings for Fig. 4-7a, b, and c? *Note*: The numbers on the main scale represent tenths of a foot.

Solution: Reading Fig. 4-7a gives 0.400 ft.   *Ans.*

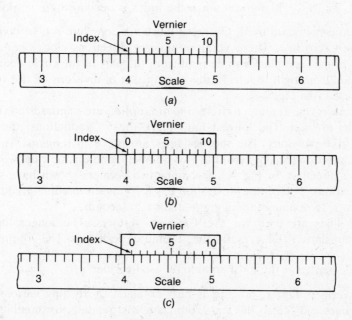

Fig. 4-7 Verniers.

In Fig. 4-7b the vernier is moved so that its first graduation from zero coincides with the first graduation of the scale beyond 0.400. Thus the vernier index has moved a distance equal to

$$d - v = 0.01 - 0.009 = 0.001 \text{ ft}$$

The reading of Fig. 4-7b is therefore

$$0.4 + 0.0001 = 0.401 \text{ ft} \qquad Ans.$$

If the vernier is moved so that the second graduation on the vernier coincides with the representation of 0.42 on the main scale, the movement from the position in Fig. 4-7a is $2(d - v) = 0.002$ ft.

To determine the reading of Fig. 4-7c, note the position where the vernier and the primary scales are coincident; the vernier scale is at a position of 8. Since each vernier graduation = 0.001 ft, a reading of 8 indicates 0.008 ft. To this add the primary reading of 0.4 ft to get

$$0.4 + 0.008 = 0.408 \text{ ft} \qquad Ans.$$

When a vernier is used, $d - v$ is the smallest reading obtainable without interpolating. It is called the *least count*, or *least reading*, of the vernier. It is expressed as follows:

$$\text{Least count} = \frac{\text{value of the smallest division on the main scale}}{\text{number of divisions on the vernier}} = \frac{d}{n}$$

To be certain the scale and vernier are being read correctly, an observer must determine the least count.

In selecting the vernier line which is coincident with a scale division, an observer should take a position directly opposite the lines, or over them, to avoid parallax. The second graduation on each side of the coincident lines must be checked to see that a symmetrical pattern is formed about them. For example, in 4-7c, vernier graduations 6 and 10 fall inside (toward division 8) the main scale graduations by equal distances; therefore, 8 (that is, 0.008) is the correct reading.

Typical mistakes in reading the vernier for minutes and seconds follow:

1. Not using a magnifying glass

2. Failing to determine the least count correctly

3. Reading in the wrong direction from zero

4. Omitting 10, 15, 20, or 30 minutes when the index is beyond these marks.

Ordinarily *double verniers* are used. On these there are complete sets of divisions running both ways from a common zero line. These verniers indicate directions clockwise or counterclockwise whenever needed. Verniers are placed to be read from the part of the circle nearest the observer; a clockwise angle is read from right to left. In this case the set of divisions on the vernier to the left of the central mark is used (see Fig. 4-8).

The patterns of lines on the graduated circles and the verniers are standardized. The top circle in Fig. 4-8 has three lengths of lines. The longest lines mark the 5° graduations; the next longest lines are used for the degree positions. The shortest lines are the $\frac{1}{2}$° increments. The 10° positions are numbered. Each 10° position (except zero) has two numbers, one for the clockwise and the other for the counterclockwise direction. In Fig. 4-8 the other two verniers shown have similar features but different scale graduations. The center illustration has a 20-minute reading graduated to 30 seconds; the lower vernier has a 15-minute reading graduated to 20 seconds.

Two lengths of lines are used for the 1-minute verniers. The longer lines mark 5-minute graduations, and the shorter lines mark the 1-minute positions. The 10-minute positions are numbered.

Note: The double vernier is the most commonly used vernier.

To understand verniers better, practice is best obtained by reading various types of verniers. Then you may calculate and sketch the least count, or least reading, for combinations of scale and vernier divisions.

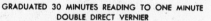

GRADUATED 30 MINUTES READING TO ONE MINUTE
DOUBLE DIRECT VERNIER

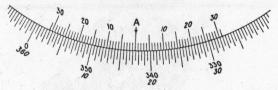

GRADUATED 20 MINUTES READING TO 30 SECONDS
DOUBLE DIRECT VERNIER

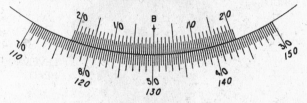

GRADUATED TO 15 MINUTES READING TO 20 SECONDS
DOUBLE DIRECT VERNIER

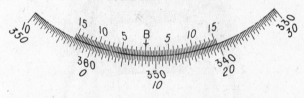

Fig. 4-8 Verniers. (*From P. Kissam, Surveying Practice, McGraw-Hill, New York, 1978. Courtesy of Keuffel and Esser Co.*)

4.7 TRANSIT GEOMETRY

The following are geometrically required in any transit (Fig. 4-9):

1. The inside and outside tapered surfaces of the outer center must be concentric. Without this, the alidade will not remain level when the circle is turned.

2. The vertical and horizontal axis must be perpendicular.

3. The inner and outer spindles and the bearings of the horizontal axis must be perfectly fitted. Thus the instrument will turn about geometric lines.

4. The vertical axis, horizontal axis, and line of sight must meet at the instrument center and be mutually perpendicular (see Fig. 4-10).

5. The line of sight must be perpendicular to the horizontal axis. The lens used for focusing must be guided so that, when it moves, it does not change the direction of the line of sight. (See Fig. 4-11.)

6. The vertical circle must read zero when the line of sight is perpendicular to the vertical axis.

7. There must be concentricity between the graduations on the vertical circle and the horizontal axis.

8. The vertical circle must be concentric with and perpendicular to the horizontal axis.

9. The graduations on the horizontal circle must be concentric with the vertical axis.

10. The horizontal graduated circle must be concentric with and perpendicular to the vertical axis.

Fig. 4-9 Transit showing alidade, leveling
head, and circle.

11.   The telescope bubble must center when the line of sight is horizontal.

12.   The plate bubbles must center when the vertical axis is vertical.

In actual daily usage by the survey crew, the instrument is operated so that if any of these geometric requirements are not met exactly, the errors introduced are neutralized and their effects are eliminated. Random human errors and systematic errors such as loose bearings cannot be eliminated.

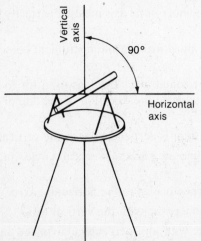

Fig. 4-10 Horizontal   and   vertical
axis of transit.

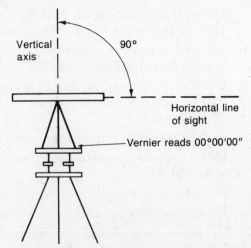

Fig. 4-11 Vernier circle reading zero.

4.8   THE COMPASS

Engineer's transits are equipped with a compass. Most transits have a compass box mounted on the upper plate between the standards. The compass needle has at the center a conical jewel bearing which rests on a hardened steel pivot. There is a lifting screw and lifting lever built into the compass so that when the compass is not being used the needle can be lifted off the pivot by the lever and screw.

When the needle is in place, a slight jar might cause the jewel to damage the fine point of the pivot so that the needle becomes slow-moving. Loss of magnetism is seldom a problem in the compass needle. Usually a slow needle is due to a dull pivot. On the needle is a small coil of brass wire (brass is not affected by magnetism) that may be moved along the needle to balance the effect of the dip of the earth's magnetic field. The needle aligns itself with the horizontal component of the earth's magnetic field. This component is called the *magnetic meridian*. The circle which surrounds the needle is graduated in degrees and half degrees and numbered to show the bearing of the line of sight of the telescope. Since the circle turns with the alidade and not with the compass needle, the east and west indications must be shown on the face of the compass dial in reversed positions so that bearings can be read directly.

4.9   MAGNETIC DECLINATION

The horizontal angle between the magnetic meridian and the true geographic meridian is called the *magnetic declination*. The armed forces term this as *deviation*, whereas it is called *variation of the compass* by navigators.

Declination east is obtained if the magnetic meridian is east of true north. *West declination* is obtained if the magnetic meridian is west of true north. The declination value at a particular location is determined by establishing the true meridian by astronomical observations and then reading the compass while sighting along the true meridian.

Figure 4-12 is a United States Geological Survey map for the year 1975 showing the distribution of magnetic declination in the United States. An *agonic line* is a line joining locations having zero declination. Along this line the magnetic needle points to true as well as magnetic north. An *isogonic line* is a line connecting locations having the same declination.

4.10   VARIATIONS IN MAGNETIC DECLINATION

Variations in the magnetic declination of any point occur with the passage of time. There are four variations, which are listed as follows:

1. Secular variation. Secular variation is the most important variation because it is the largest and changes over long periods of time. At various times in the past variation values were obtained by referring to tables and charts compiled through observation over time. Retracing old property lines determined by compass makes it necessary to allow for the difference in magnetic declination at the time of the original survey from that of the present date. Secular variation is usually the greatest cause of the difference.

2. Daily variation. Every day the variation of the magnetic needle's declination is an average arc of 8 minutes in the United States. The needle reaches an extreme eastern position at 0800 hours (h). The most westerly reading occurs at 1330 h. Daily variation may be neglected as it is within the range of error expected from compass readings. Another term for daily variation is *diurnal variation*.

3. Annual variation. Annual variation is less than 1 minute of arc and can be neglected.

4. Irregular variation. Magnetic disturbances and electrical storms can cause short-term variations of as much as 1°.

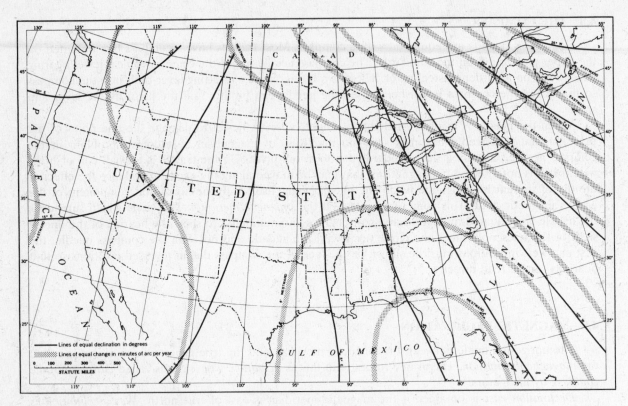

Fig. 4-12   United States Geological Survey map for the year 1975.

4.11   LOCAL ATTRACTION

Metallic objects (ferrous metals) and direct current electricity cause local attraction of the compass needle. If the source of local attraction is fixed, all bearings from that station will be in error by the same amount. Angles calculated from bearings taken at that station will, however, be correct. You can tell if local attraction is present when forward and back bearings of a line differ by more than normal observational error.

EXAMPLE 4.3   (*a*) Given: A compass bearing for line *AB* of N23°15′W was observed; then a back bearing of S23°10′E was noted. (*b*) Given: Bearing *CD* of N62°00′E and back bearing *DC* of S63°30′W. In each case determine whether a local attraction was present, and, if so, the location of the attraction.

Solution: (*a*)   Subtracting 23°10′ from 23°15′ leaves 00°05′. This difference is so small that you can assume it is just normal observational error. Thus it indicates local attraction is not present.

   (*b*)   Subtract *CD* from *DC* to get a difference of 1°30′. This is a large difference, more than should occur through normal observational error. Thus you have a local attraction occurring at point *D* or *C*. If the local attraction is at *D*, the compass needle is being deflected 1°30′ to the west of north.

4.12   SOURCES OF ERROR IN COMPASS WORK

Some sources of error in using the transit compass are as follows:

1.   Magnetic variation

2.   Local attraction

3.   Weak magnetism of needle

4. Compass not level

5. Pivot, needle, or sight vanes bent

Possible sources of local attraction are (1) chaining pins, (2) metal range poles, (3) axes, (4) loose-leaf field books, (5) a penknife, (6) metal in a shirt pocket, (7) nearby power lines, (8) a parked car. Any of these objects can create a local attraction for your compass needle.

4.13   MISTAKES IN COMPASS WORK

Typical mistakes encountered while doing compass work follow:

1. Failing to check both forward and back bearings when possible.

2. Reading the wrong end of the needle.

3. Setting the declination to the wrong side of north when attempting to observe true bearings.

4. Not making a sketch showing known and desired items.

5. Parallax (reading while looking from the side of the needle rather than along it).

4.14   DETERMINING AN ACCURATE COMPASS BEARING

The procedure for determining an accurate compass bearing follows:

1. Make sure the north point (zero) on the compass circle is at zero of the declination arc.

2. Set up at a spot where there is no local attraction. Aim at a well-defined point.

3. Read the angle at each end of the needle. Aim at the point with the instrument reversed, and again read the angle at each end of the needle. Average the four angles and add the proper letters for the direction of the point used.

4. In reading the needle, place the eye in line with the needle and estimate each reading to the nearest 5 minutes.

5. The value obtained may be corrected for the time of day and time of year by values obtained from the National Geodetic Survey.

EXAMPLE 4.4   Given: A line *AB* with a magnetic bearing of N26°E. It is known that there is an east declination of 20°. Find the true bearing of *AB*.

Solution: Draw a sketch (see Fig. 4-13). Draw true north vertically. The compass needle is pointing to north but it has a deviation at this particular location of 20° to the east. Thus the east deviation is shown 20° east of true north, and then the magnetic bearing of the line *AB*, observed, is shown in place 26° clockwise from the compass needle line.

Calculations:

$$\begin{array}{ll} \text{Magnetic bearing of } AB = & \text{N26°E} \\ \text{Declination} = & \underline{20°E} \\ \text{True bearing of } AB = & \overline{\text{N46°E}} \end{array}$$

EXAMPLE 4.5   Given: Magnetic bearing *AB*, taken in the year 1868, of N48°30′E. In 1983 the same line was observed to have a bearing of N10°27′E and is represented by line *AD*. Find the change in magnetic bearing over the years.

Solution: Draw a sketch (see Fig. 4-14). Subtract *AD* from *AB*:

$$\begin{array}{l} AB = 48°30′ \\ AD = \underline{10°27′} \\ \overline{38°03′} \end{array}$$

The change in magnetic bearing is 38°03′ and it has been progressing east.

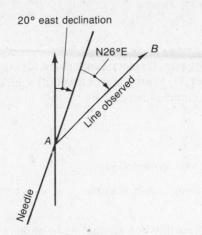

Fig. 4-13 Bearings.   **Fig. 4-14** Bearings.

4.15 MEASURING A HORIZONTAL ANGLE

To measure a horizontal angle, use the following procedure:

1. Free both motions and set the *A* vernier at zero with the upper motion (read the *B* vernier if necessary); see clamp screw information at the beginning of this chapter.

2. Point at the initial (left-hand) point, using the lower motion, clamp with lower clamp, come on the point using clockwise turn of the lower tangent screw.

3. Loosen the upper clamp point at the second point, using the upper motion and follow as above in step 2.

4. Free the lower motion.

5. Read the *A* vernier clockwise (and the *B* vernier, if necessary).

Note: If the *A* vernier is used throughout, the reading of the *A* vernier is taken as the value of the angle. If the *B* vernier is also used, the value of the angle is the average of the *A* and *B* readings minus the initial reading, which is the average of the *A* and *B* readings when the vernier is set at zero.

4.16 FIELD RECORDS

Usually only the *A* vernier is used. In this case almost any form of record of the angle is satisfactory. When the *B* vernier is also read, the standard form of the record is shown in Fig. 4-15. The explanation of the record for the two angles in the figure is shown in Table 4-1.

Note: The *A* vernier readings are recorded in full. Only the minutes and seconds in the *B* reading are recorded. When a *B* reading is 1 minute greater than the number of minutes in the *A* record, 60 seconds is added to the number of seconds in the *B* record. When the *B* reading is 1 minute less than the *A* record, a line, or overbar, is placed over the number of seconds in the *B* record.

EXAMPLE 4.6 Given: *A* record and *B* reading shown in the first two columns of the accompanying table. Determine the *B* record and the average for the field notes.

A record	B reading	B record, seconds	Average
87°10′45″	267°11′15″	75	87°11′00″
17°20′15″	197°19′30″	$\overline{30}$	17°19′52.5″

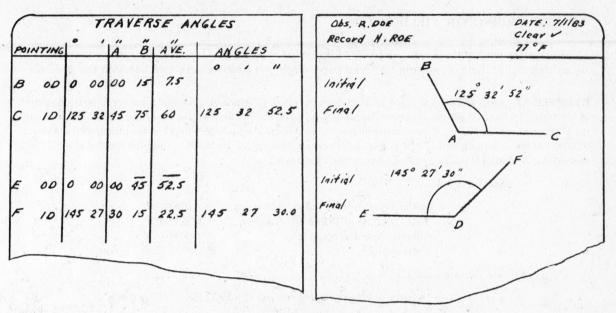

Fig. 4-15 Field book pages.

Table 4-1 Angles for Direct Reading of Verniers

Angle *EDF*: Telescope direct (D) throughout			
Pointing	*A* Vernier	*B* Vernier	Average
B (initial) (D)	0°00′00″	180°00′15″	0°00′07.5″
C (D)	125°32′45″	305°33′15″	125°33′00.0″
To compute angle, subtract initial average: 125°32′52.5″			
Angle *EDF*: Telescope direct (D) throughout			
Pointing	*A* Vernier	*B* Vernier	Average
E (initial) (D)	0°00′00″	179°59′45″	359°59′52.5″
F (D)	145°27′30″	325°27′15″	145°27′22.5″
To compute angle, subtract initial average: 145°27′30.0″			

Solution: See the last two columns of the accompanying table. Note that 60 seconds is added to the number of seconds in the *B* record because the number of minutes in the *B* reading is 1 greater than the number of minutes in the *A* record. Note also that when the *B* reading is less than the *A* record by 1 minute, a bar is placed over the number of seconds in the *B* record.

The initial average is obtained by averaging the number of seconds of the *A* and *B* records when the vernier is set at zero. For the angle in the first row of the table:

$$\frac{45 \text{ seconds} + 75 \text{ seconds}}{2} = \frac{120 \text{ seconds}}{2} = 60 \text{ seconds} = 1 \text{ minute}$$

The 1 minute is added to the number of minutes in the average angle, and the number of seconds becomes 0; the average is thus 87°11′00″. For the angle in the second row of the table, the average is obtained as follows:

$$\frac{15 \text{ seconds} + (30 + 60 \text{ seconds})}{2} = \frac{105 \text{ seconds}}{2} = 52.5 \text{ seconds}$$

Thus the average angle is 17°19′52.5″.

4.17 RECORDING THE FIELD NOTES

Recording field notes is not complicated but it takes practice. The computations required should be written out in their usual form on scrap paper until you become very familiar with the system.

EXAMPLE 4.7   See Table 4-1 and Fig. 4-15. (a) Given: Angle BAC which has an initial transit reading from vernier A of $0°00'00''$ for line AB. On the B vernier the reading is $180°00'15''$. For line AC the A vernier reads $125°32'45''$, and the B vernier reads $305°33'15''$. Compute the final angle. (b) Given: Angle EDF which has an initial transit reading of $0°00'00''$ for the A vernier, and $179°59'45''$ for the B vernier. On line DF on the A vernier the reading is $145°27'30''$, and on the B vernier it is $325°27'15''$. Compute the final angle.

Solution: (a)   See Table 4-1.

Initial average of A and B verniers =	$0°00'07.5''$
BAC average of A and B verniers =	$125°33'00.0''$
Subtract initial average	$- \quad 0°00'07.5''$
Final angle =	$125°32'52.5''$ *Ans.*

(b)   See Table 4-1 and Fig. 4-15.

Initial average pointing toward $E = 59'52.5'' = -0.75$ seconds

Pointing to F the average is $145°27'22.5''$. To compute the final angle subtract the initial average from the average bearing:

Final angle $= 145°27'22.5'' - (-07.5'') = 145°27'30''$ *Ans.*

Note: The bar over the 52.5 seconds in fig. 4-15 means there is a double negative—a negative quantity is subtracted from the DF quantity; a double negative results in a positive, so add the 07.5 seconds to get the final angle.

4.18 SOURCES OF ERROR IN TRANSIT WORK

Errors can usually be classified in three categories: instrument errors, natural errors, and personal errors.

Instrument errors are as follows:

1. Plate bubbles out of adjustment
2. Centers on verniers slightly eccentric
3. Axis of telescope bubble not parallel to axis of sight
4. Horizontal and vertical axes not mutually perpendicular
5. Axis of sight not perpendicular to the horizontal axis

Natural sources of error are as follows:

1. Wind
2. Tripod settling
3. Refraction
4. Temperature changes

Common personal errors follow:

1. Poor focus of telescope
2. Unsteady tripod
3. Vernier misinterpreted

4. Level bubbles not correctly centered

5. Instrument not correctly set over point

6. Improper use of tangent screws

7. Careless plumbing over the observing station

8. Careless plumbing of the target

In addition to the personal errors listed above, the following are common mistakes made in transit work:

1. Sighting on or setting up over the wrong point

2. Turning the wrong tangent screw

3. Reading the wrong circle

4. Calling out or recording the incorrect value

Solved Problems

4.1 Given: A distance of 1000 ft and an angle of 1 minute. Find the offset for this angle.

Solution

From the relationships shown in Fig. 4-4,

If $x = 1000$ ft: $\qquad$ Offset $= 0.00029x = 0.00029(1000) = 0.29$ ft $\qquad$ *Ans.*

4.2 Given: 1 second of arc and a 100-ft length for the side of the angle. What is the offset at this distance?

Solution

See Fig. 4-4. If offset for 1 minute $= 0.00029x$, then offset for 1 second is one-sixtieth of that for 1 minute, or

$$\text{Offset} = \frac{0.00029}{60} x = 0.0000048(100) = 0.00048 \text{ ft} \qquad \textit{Ans.}$$

4.3 Given: Radius of an arc $= 50$ mi. Find the offset for a 1-second angle.

Solution

$$\text{Offset at 1 minute} = 0.00029x$$

So $\qquad$ $$\text{Offset at 1 second} = \frac{0.00029}{60}x = 0.0000048x$$

Since 1 mi $= 5280$ ft, then

$$\text{Offset} = 0.0000048(50)(5280) = 1.27 \text{ ft} \qquad \textit{Ans.}$$

4.4   Given: The vernier shown in Fig. 4-16*a*. Find the correct angular reading.

Solution

Reading clockwise on the circle you get 58° first (graduation progressing to the left). Next you are beyond 30 minutes on the main scale, so find the coincident line on your vernier. It is 17. Add 30 minutes because the vernier index is past the 30-minute mark: $17 + 30 = 47$ minutes. So the reading for Fig. 4-16*a* is 58°47′.

Reading counterclockwise on Fig. 4-16*a* you have 301° on the main scale. This index is before the 30-minute mark on the main scale so reading the coincident lines on the vernier and main scale of the transit you find coincidence at the 13-minute mark. The correct reading of Fig. 4-16*a* is thus $301°00′ + 13′ = 301°13′$.

Note that the vernier is always read in the same direction from the index as the direction of numbering of the scale, that is, on the side of the double vernier in the direction of the increasing angle. Sum of the two angles = 360°00′.

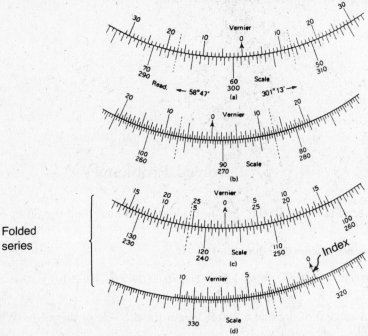

Fig. 4-16   Different types of verniers. (*From R. C. Brinker and P. R. Wolf, Elementary Surveying, IEP Publishing, Dun-Donnelley, New York, 1977. Courtesy of Keuffel and Esser Co.*)

4.5   Given: The vernier shown in Fig. 4-16*b*. Find the correct angular reading.

Solution

Reading the inner set of numbers from the double vernier gives 91°20′. The coincident line on the vernier is at 7 minutes. So add minutes:

$$20 \text{ minutes} + 07 \text{ minutes} = 27 \text{ minutes}$$

So the correct angle is 91°27′.

Reading the outer set of figures gives 268°20′, and

$$20 \text{ minutes} + 13 \text{ minutes} = 33 \text{ minutes}$$

The correct angle is 268°33′.

4.6   Given: The vernier shown in Fig. 4-16*c*. Find the correct angular reading.

Solution

The reading at the index of the vernier is 117° on the clockwise scale. Next the match line on the vernier scale is at 5′30″, so the reading is 117°05′30″.

Instructional note: This is a folded vernier so that when reading clockwise from zero you read zero to 15 to the left of zero and 15 to 30 to the right of zero.

Read the outer numbers (progression to right) in Fig. 4-16c. First read 242°30′ on the scale; coincidence is at 24′30″ on the counterclockwise vernier folding scale, so you have 242°30′ + 24′30″ = 242°54′30″.

4.7 Read Fig. 4-16d.

Solution

The scale reads 321°10′. Vernier reads 3′20″. So

$$321°10′ + 3′20″ = 321°13′20″ \qquad \text{for a clockwise angle} \qquad Ans.$$

4.8 Given: Fig. 4-8 (top)—double direct vernier graduated to 30 minutes reading to 1 minute. What is the correct angular reading?

Solution

Read inside scale: 342°30′. Coincident line is at 5 minutes on the vernier. So

$$342°30′ + 5 \text{ minutes} = 342°35′ \qquad Ans.$$

Read outside scale on same double direct vernier: 17° + 25 minutes on the vernier. So we have

$$17°25′ \qquad Ans.$$

4.9 Given: Fig. 4-8 (middle)—double direct vernier graduated to 20 minutes reading to 30 seconds. What is the correct angular reading?

Solution

Read inside scale: 49°40′. Add 10 minutes on the vernier. So

$$49°40′ + 10 \text{ minutes} = 49°50′ \qquad Ans.$$

Read outside scale on the same vernier scale as in the first part of the problem: 130°. Vernier coincides at 10 minutes. So

$$130° + 10 \text{ minutes} = 130°10′ \qquad Ans.$$

4.10 Given: Fig. 4-8 (bottom)—double direct vernier graduated to 15 minutes reading to 20 seconds. What is the correct angular reading?

Solution

Read the inside scale: 8°15′. Vernier reads 10′00″. So

$$8°15′ + 10′00″ = 8°25′00″ \qquad Ans.$$

Read the outside scale: 351°30′. Add vernier reading of 6 minutes. So

$$351°30′ + 6 \text{ minutes} = 351°36′ \qquad Ans.$$

4.11 Given: Bearing BC of N77°40′W and a back bearing CB of S77°40′E. Determine whether there is any variation or local attraction.

Solution

Subtract *CB* from *BC*. You get 0°. Thus there is no local attraction at these stations and no observational error.

4.12   Given: Magnetic bearing *AB* of N35°15′E. Declination is 15° east. Find the true bearing of *AB*.

Solution

Draw a sketch (see Fig. 4-17). Computations:

$$
\begin{aligned}
\text{Magnetic bearing of } AB &= \quad \text{N35°15′E} \\
\text{Declination east} &= \quad +\ \text{15°00′E} \\
\text{True bearing of } AB &= \quad \text{N50°15′E} \qquad Ans.
\end{aligned}
$$

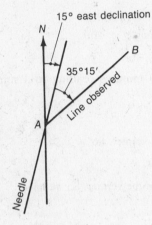

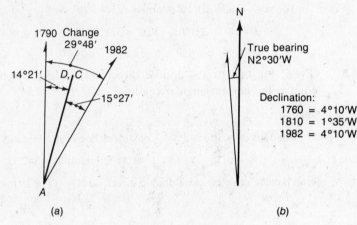

Fig. 4-17   Bearings.   Fig. 4-18   Bearings.

4.13   Given: Magnetic bearing *AD* of N14°21′E taken in the year 1790. The same line shown as *AC* has a bearing taken in 1982 of N15°27′W. Find the change in the magnetic declination from 1790 to 1982.

Solution

From Fig. 4-18*a* it can be seen from the quadrants occupied by the two bearings that one bearing is east and the other is west, so the bearings will have to be added to get the answer.

$$
\begin{aligned}
& 14°21′ \\
+\ & 15°27′ \\
\hline
& 29°48′ \qquad Ans.
\end{aligned}
$$

The later bearing is in the western quadrant, so the change in magnetic declination is N29°48′W, and the magnetic meridian has been moving easterly since the year 1790.

4.14   At a certain location the declination was 4°10′W in 1760, 1°35′W in 1810, and 4°10′W in 1982 (see Fig. 4-18*b*). If the line has a true bearing of N2°30′W, what was its magnetic bearing in (*a*) 1760, (*b*) 1810, (*c*) 1982?

Solution

(*a*)
$$
\begin{aligned}
& 4°10′W \\
-\ & 2°30′W \\
\hline
& 1°40′W
\end{aligned}
$$

The magnetic bearing in 1760 was N1°40′W.   *Ans.*

(b)   2°30'W
                                                                  − 1°35'W
                                                                 ──────────
                                                                   00°55'W

The magnetic bearing in 1810 was N00°55'W.   *Ans.*

(c)   4°10'W
                                                                  − 2°30'W
                                                                 ──────────
                                                                    1°40'W

The magnetic bearing in 1982 was N1°40'W.   *Ans.*

4.15 Convert the following magnetic bearings to true bearings: (a) N74°39'E, declination 10°10'W, (b) S10°30'E, declination 10°45'W.

Solution

(a) From a sketch (see Fig. 4-19a), compute the following:

$$
\begin{array}{r}
74°39' \\
-\,10°10' \\
\hline
\text{N64°29'E}
\end{array}
$$

True bearing = N64°29'E   *Ans.*

(b) See Fig. 4-19b.

$$
\begin{array}{r}
10°30' \\
+\,10°45' \\
\hline
\text{S20°75'}
\end{array}
$$

Since 20°75' = 21°15', the true bearing is S21°15'E.   *Ans.*

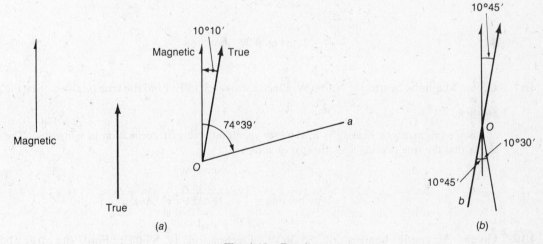

Fig. 4-19 Bearings.

4.16 Convert the following magnetic bearings to true bearings: (a) N2°30'W, declination of 5°40'E; (b) S4°35'W, declination 5°40'E.

Solution

Use the following rule to solve this problem.

Rule: Magnetic bearings may be changed to true bearings by *adding east* declinations in the northeast and southwest quadrants and *subtracting* east declinations in the southeast and northwest quadrants.

(a) By the sketch (Fig. 4-20), we see that the magnetic bearing is in the NW quadrant. Therefore, the declination is subtracted, and, as the sketch shows, the true bearing is *east*.

$$\begin{array}{r} 5°40'E \\ -\ 2°30'W \\ \hline \end{array}$$

True bearing =   N3°10'E   *Ans.*

(b) By the given rule, we add if the east declination is in the SW quadrant. As the sketch shows, the true bearing is *west*.

$$\begin{array}{r} S4°35'W \\ +\ 5°40'E \\ \hline 9°75' \end{array}$$

9°75' = 10°15'   so   True bearing = S10°15'W   *Ans.*

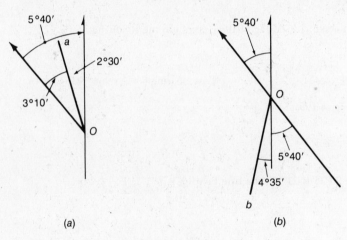

(a)   (b)

Fig. 4-20   Bearings.

4.17   Given: Magnetic bearing of N3°50'W. Declination is 3°15'E. Find the true bearing. (See Fig. 4-21.)

Solution

Since the magnetic bearing is in the NW quadrant, the east declination is subtracted. The sketch shows that the true bearing is in the east quadrant.

$$\begin{array}{r} N3°50'W \\ -\ 3°15'E \\ \hline \end{array}$$

True bearing =   N0°35'E   *Ans.*

4.18   Given: Magnetic bearing of S5°30'W. Declination is 5°30'E. Find the true bearing. (See Fig. 4-22.)

Solution

$$\begin{array}{r} S5°30'W \\ +\ 5°30'E \\ \hline 10°60' \end{array}$$

10°60' = 11°   so   True bearing = S11°00'W   *Ans.*

4.19   The magnetic bearing of a line of an old survey is recorded as N10°30'E. It is now N2°45'W. What has been the change in magnetic declination and its direction? (See Fig. 4-23.)

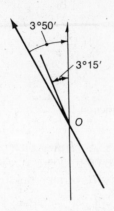

Fig. 4-21   Bearings.

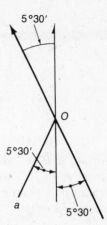

Fig. 4-22   Bearings.

Solution

$$
\begin{array}{r}
\text{N10°30'E} \\
+ \quad 2°45'\text{W} \\
\hline
12°75' \quad = 13°15'
\end{array}
$$

The magnetic declination has changed 13°15' to the east.   *Ans.*

4.20 The magnetic bearing of a line of an old survey is recorded as N17°30'E. At the present date the magnetic bearing shows as N3°15'E. What has been the change in magnetic declination and its direction? (See Fig. 4-24.)

Solution

$$
\begin{array}{r}
\text{N17°30'E} \\
- \quad 3°15'\text{E} \\
\hline
14°15'
\end{array}
$$

Change in magnetic declination was 14°15' to the east.   *Ans.*

4.21 The observed bearing of a line is N87°35'E, and the true bearing is S89°25'E. Calculate the value and direction of the local attraction. (See Fig. 4-25.)

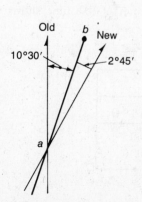

Fig. 4-23   Bearings.

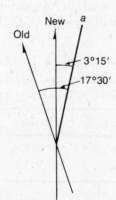

Fig. 4-24   Bearings.

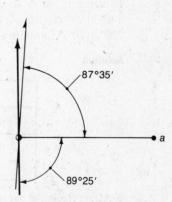

Fig. 4-25   Bearings.

Solution

$$S89°25'E$$
$$+ N87°35'E$$
$$\overline{177°00'}$$

$$180°00'$$
$$- 177°00'$$
$$\overline{3°00'E \text{ local attraction}}$$

4.22 Given: An angle *BAC* which has a reading on the *A* vernier of 0°00′00″ with the transit pointing from *A* to *B*. The *B* vernier now reads 180°00′10″. The *A* vernier reads 130°30′10″ when the transit is pointing toward *C* from *A*. The *B* vernier this time shows 210°31′10″. Compute the final angle.

Solution

See Table 4-1 and Example 4.7 to refresh your memory on how to handle this type of problem.

Angle *BAC*: Telescope direct			
Point	*A* Vernier	*B* Vernier	Average
B (init.)	0°00′00″	180°00′10″	0°00′05″
C (D)	130°30′10″	210°31′10″	130°30′40″

Subtract the initial average:

$$130°30'40'' - 0°00'05'' = 130°30'35''\quad Ans.$$

The angle would be written in the field book as follows (see Fig. 4-15).

Traverse angles			
Point	° ′ ″	″ avg.	Angle
B (0 D)	0 00 00	05	
C (1 D)	130 30 10	30	130°30′35″

4.23 Given: An angle *BAC* which has a reading on the *A* vernier of 0°00′00″ with the transit pointing from *A* to *B*. The *B* vernier reads at this pointing 180°00′30″. The transit reads 140°30′00″ on the *A* vernier when pointed from *A* to *C*. The *B* vernier at this time shows 220°30′20″. Compute the final angle.

Solution

Angle *BAC*: Telescope direct			
Point	*A* Vernier	*B* Vernier	Average
B (init.)	0°00′00″	180°00′30″	0°00′15″
C (D)	140°30′00″	220°30′20″	140°30′10″

Subtract the initial average:

$$140°30'10'' - 0°00'15'' = 140°29'55''.\quad Ans.$$

4.24   Given: An angle *BAC* which has a reading on the *A* vernier of 0°00′00″ with the transit
pointing from *A* to *B*. The *B* vernier reads at this pointing 180°00′50″. The *A* vernier reads
156°15′00″ when the transit is pointed from *A* to *C*. The *B* vernier at this time shows
336°14′00″. Compute the final angle.

Solution

Angle *BAC*: Telescope direct			
Point	*A* Vernier	*B* Vernier	Average
B (init.)	0°00′00″	180°00′50″	0°00′25″
C(D)	156°15′00″	336°14′00″	156°14′30″

Subtract the initial average:

$$156°14′30″ - 0°00′25″ = 156°14′05″ \qquad \textit{Ans.}$$

Supplementary Problems

4.25   Given: Side of an arc of 300 ft and central angle of 1°. Find the offset.   *Ans.*   5.235 ft

4.26   Given: Radius of an arc = 400 km. Find the offset for a 1-second angle. See Fig. 4-4.   *Ans.*   2 m

4.27   Given: Bearing *DE* of N89°36′E and *ED* of 288°26′W. Determine whether there is local attraction.
Ans.   Subtract *ED* from *DE* to find a difference of 1°10′. This is more than an observational error, so there
is a local attraction at *D* or *E*.

Chapter 5

Leveling

5.1 INTRODUCTION

Measuring the difference in elevation between points on the earth's surface is a fundamental part of surveying. These differences in elevation can be determined by various methods of leveling.

Leveling is the operation of determining the difference in elevation between points on the earth's surface. A level reference surface, or datum, is established and an elevation assigned to it. Differences in the determined elevations are subtracted from or added to this assigned value and result in the elevations of the points.

A *level surface* is one in which every point is perpendicular to the direction of the plumb line. It differs from a plane surface, which is flat and is perpendicular to a plumb line at only one point. A body of still water will assume a level surface. If the changes in the surface of the ocean caused by such influences as tides, currents, winds, atmospheric pressure, and the rotation of the earth could be eliminated, the resulting surface would be level.

The ocean's level surface is determined by averaging a series of tidal height observations over a Metonic cycle (approximately 19 calendar years). This average, called *mean sea level*, is the most common datum for leveling and is usually assigned an elevation of zero. This datum remains in effect until continuing observations show a significant difference and it becomes worthwhile to change to the new datum. In the United States, the mean sea level datum of 1929 is still in effect.

5.2 TYPES OF LEVELING

Leveling operations are divided into two major categories. *Direct leveling* is usually referred to as *differential*, or *spirit*, *leveling*. In this method the difference in elevation between a known elevation and the height of the instrument, and then the difference in elevation from the height of the instrument to an unknown point, are determined by measuring the vertical distance with a precise or semiprecise level and leveling rods. This is the only method that will yield accuracies of the third or a higher order.

The second method of leveling, *indirect leveling*, is further subdivided into two separate methods, *trigonometric* and *barometric*. The trigonometric method applies the principles of trigonometry to determine differences in elevation; a vertical angle (above or below a horizontal plane) and a horizontal distance or slope distance (measured or computed) are used to compute the vertical distance between two points. This method is generally used for lower-order leveling where the terrain is prohibitive to direct leveling.

Barometric leveling uses the differences in atmospheric pressure as observed with a barometer or altimeter to determine differences in elevation. This is the least used and least accurate method of determining differences in elevation. It should only be used in surveys where one of the other methods is unfeasible or would involve large amounts of time or money. Because of its limited use, barometric leveling will not be discussed further.

5.3 LEVELING EQUIPMENT

Levels

The engineer's level can be compared to a carpenter's level. The difference is that the engineer's level is used by mounting it on a tripod (to hold it steady) and sighting through a telescope in order to transfer the level line to another point. While the carpenter's level can be used to determine if two points a few inches apart are on a level surface, the engineer's level can tell if two points a few hundred feet apart are on a level line.

Just as there are many different types of carpenter's levels—for example, water levels and line levels—so too there are many different types of engineer's levels. Although some of these levels have certain features that others do not have, they all share certain basic parts.

Figure 5-1 identifies the important parts of an engineer's level.

1. Eyepiece. The adjustable lens through which the observer looks. The eyepiece is rotated to bring the cross hairs into focus.
2. Telescope. The tube which holds all the lenses and focusing gears in their proper positions.
3. Sunshade. A metal or plastic extension which can be placed over the objective lens to protect the lens from damage and to reduce glare when the level is in use.
4. Focusing knob. An adjustment knob which internally focuses the level on the desired target.
5. Horizontal circle.
6. Leveling screws. Adjustable screws used to level the instrument.
7. Base. A $3\frac{1}{2}$ by 8 in (89 by 203 mm) threaded base which secures the instrument to the tripod.
8. Plumb bob, hook, and chain. A hook and chain, centered under the level, to which the plumb bob is attached if angles will be turned. These items are not illustrated.
9. Shifting center. A design feature which permits exact placement over a given point.
10. Name and serial number plate.
11. Horizontal tangent screw. An adjustment screw which allows exact alignment of the cross hairs and the target within the horizontal plane.

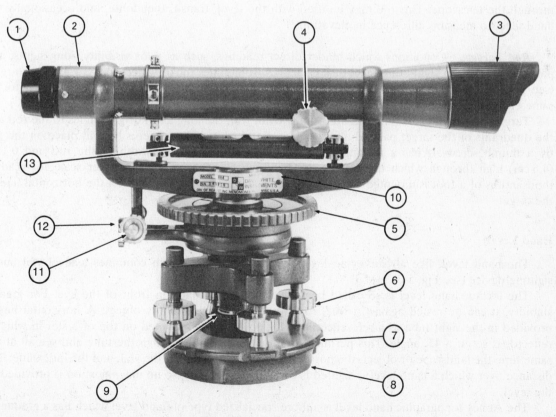

Fig. 5-1 Universal level. (*David White Instruments*)

12. Horizontal clamp screw. An adjustment screw which allows approximate alignment of the cross hairs and the target within the horizontal plane.

13. Level vial. A graduated, liquid-filled glass vial which is parallel to the line of sight of the telescope.

Tripods

The tripod supports the level base and keeps it stable during the observations (see Fig. 4-2). For a complete explanation of the tripod see Sec. 4.1.

Leveling Rods

Refer to Fig. 4-2 for a good picture of a leveling rod. A leveling rod is essentially a tape supported vertically and used to measure the vertical distance (difference in elevation) between a line of sight and a specific point above or below the line of sight. The point may be a permanent station, such as a bench mark, or it may be some natural or constructed surface.

Leveling rods are available in several different styles. The Florida rod, the California rod, and the Detroit rod are but a few of the possible variations, but the Philadelphia rod is by far the most commonly used type. The standard Philadelphia rod is a graduated wooden rod made of two sections. It can be extended from 7.1 to 13.1 ft (2.2 to 4.0 m). The graduations on the rod are feet, tenths of feet, and hundredths of feet. Instead of using a small line or tick to mark hundredths, the spaces between alternate pairs of graduations are painted black on a white background. Thus, the mark for each hundredth is the line between the colors, the top of the black being even-numbered values, and the bottom of the black being odd-numbered values. The tenths of feet and feet are numbered in black and red, respectively. The observer usually reads the rod directly while sighting through the telescope. This rod may be used with the level, transit, theodolite, and occasionally the hand level to measure difference in elevations.

Rod Targets. Conditions which hinder direct readings, such as poor visibility, long sights, and partially obstructed sights through brush or leaves, sometimes make it necessary to use rod targets (see Fig. 5-2). The target is also used to mark a rod reading when setting numerous points to the same elevation from one instrument setup.

Targets for the Philadelphia rod are usually oval, with the long axis at right angles to the rod and the quadrants of the target painted alternately red and white. The target is held in place on the rod by a thumb screw. It has a rectangular opening approximately the width of the rod and 0.15 ft (4.5 cm) high through which the face of the rod may be seen. A linear vernier scale for reading thousandths of a foot is mounted on the edge of the opening with the zero on the horizontal line of the target.

Hand Levels

The hand level, like all surveying levels, is an instrument which combines a level vial and a sighting device (see Fig. 5-3).

The locator hand level is so called because it is held by hand in front of the eye. For greater stability, it can be rested against a tree, rod, work tool, or any handy object. A horizontal line is provided in the sight tube as a reference line. The level vial is mounted on top of a slot in which a reflector is set at a 45° angle. This permits the observer to sight through the tube and see all at the same time the landscape or object, the position of the level bubble in the vial, and the index line. The distance over which a hand level is sighted is comparatively short, so no magnification is provided by the level.

The Abney topographic hand level is a more specialized type of hand level which has a graduated arc so that the vertical angle and the percentage of grade can be measured. This topographic level

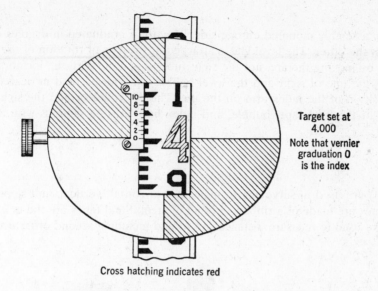

Target set at
4.000

Note that vernier
graduation 0
is the index

Cross hatching indicates red

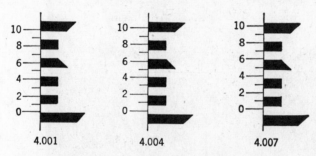

4.001   4.004   4.007

Fig. 5-2 Rod target. (*From P. Kissam, Surveying Practice, McGraw-Hill, New York, 1978. Courtesy of Keuffel and Esser Co.*)

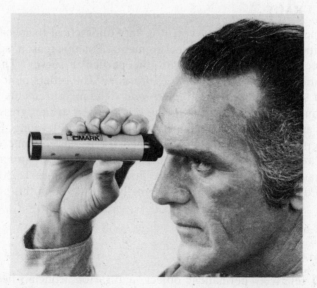

Fig. 5-3 Hand-held level. (*David White Instruments*)

has a reversible arc assembly mounted on one side. The arc is graduated in degrees on one side and percent of grade on the other. The level vial is attached at the axis of rotation of the index arm. The bubble is centered by moving the arc, not the sight tube as is done with the locator level. Thus, the difference between the line of sight and the level bubble axis can be read in degrees or percent of grade from the position of the index arm on the arc. The 45° reflector and the sighting mechanism (permitting a view of the landscape, bubble, and index line) are the same as with the locator hand level.

Tapes

Tapes (Fig. 5-4) are used in surveying to measure horizontal, vertical, and slope distances. The common survey tapes are made of a ribbon or band of steel. Steel tapes are the most accurate of all survey tapes and are used to measure distances up to and including second-order accuracy.

Fig. 5-4 Tape in field use. (*Midsouth Engineering Co.*)

5.4 ORDERS OF ACCURACY

When writing the specifications for a survey, it is very impractical to specify the exact degree of accuracy that is to be attained in each of the measurements. For this reason, specifications are based on the minimum degree of accuracy allowed for the particular survey. The range between the allowed degrees of accuracy is known as an *order of accuracy*. The orders of accuracy for surveys are called *first order*, *second order*, *third order*, and *lower order*. The measurements for first-order surveys are the most accurate, and the measurements for the other orders are progressively less accurate.

Orders of accuracy are specified for triangulation, traverse, and leveling. For the measurements made in mapping, orders of accuracy are also specified for the astronomic observations made to establish position and azimuth. As an example of the range between orders of accuracy, let us consider the allowed traverse position closure. The first order specifies 1:25 000 or better; second order, 1:10 000; third order, 1:5000. The surveys normally considered in this book require third- and lower-order accuracy.

5.5 LEVEL BENCH MARKS

A *bench mark* is a relatively permanent object, natural or artificial, bearing a marked point whose elevation is known. A bench mark may be further qualified as permanent, temporary, or supplementary. The purpose of a survey normally governs whether its stations will be permanently or temporarily marked. When it is known that the station may be reused over a period of several years, the station marker should be of a permanent type. A permanent bench mark is normally abbreviated BM, and a temporary or supplemental bench mark is abbreviated TBM.

5.6 DIFFERENTIAL LEVELING

In direct leveling, a horizontal line of sight is established using a sensitive level bubble in a level vial. The instrument is leveled and the line of sight is adjusted to be parallel to the level vial axis. When leveled, the line of sight of the instrument describes a horizontal plane if the instrument is rotated about its vertical axis. See Fig. 5-5. This procedure is referred to as differential or spirit leveling and is described in Example 5.1.

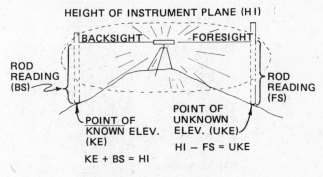

Fig. 5-5   Direct leveling.

EXAMPLE 5.1   Given: Bench mark 35 at elevation 154.375 m. Backsight (BS) on BM 36 gives a reading of 1.255 (65 m) to point A, then 65 m to point 2, where the foresight (FS) elevation reads 1.100. Then we have 75 m from point 2 to point B, BS reading 0.465; 90 m from point B to point 3, FS reading 2.095; 60 m from point 3 to point C, BS reading 0.130; 105 m from point C to point 4. BS reading from point C to point $3 = 0.130$, FS reading at point 4 is 0.245. Distance from point 4 to point D is 110 m. BS reading at point 4 is 3.765; distance from point D to point 5 (which is BM 36) is 50 m. FS reading at point 5 is 0.345. BM 36 has a known elevation of 156.205 m. Work out the survey.

Solution: First draw a sketch showing the stations and elevations (see Fig. 5-6).

Method: The leveling operation (see Fig. 5-6) consists of holding a rod vertically on a point of known elevation. A level reading known as a backsight (BS) is then made through the telescope on the rod; this gives the vertical distance from the ground elevation to the line of sight. By adding this backsight reading to the known elevation, the line-of-sight elevation, called the *height of instrument* (HI), is determined. Another rod is

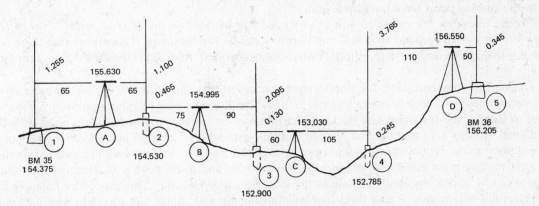

Fig. 5-6   Leveling operation.

placed on a point of unknown elevation, and a foresight (FS) reading is taken. By subtracting the foresight reading from the height of instrument, the elevation of the new point is established.

After the foresight is completed, the rod remains on that point and the instrument and the back rod are moved to forward positions. The instrument is set up approximately midway between the old and new rod positions. The new sighting on the back rod is now a backsight for the new height of instrument, and the sighting on the front rod is a foresight for a new elevation. The points on which the rods are held for the foresights and backsights are called *turning points* (TPs). This procedure is used as many times as necessary to transfer a point of known elevation to another distant point of unknown elevation. Other foresights to points not along the main line are known as *side shots*. Refer to Sec. 5.8 for the reasons for keeping backsight and foresight equal.

$$
\begin{array}{lr}
\text{BM 35 elevation} = & 154.375 \\
A + \text{BS} & +\ \ 1.255 \\
\hline
 & 155.630 \\
\text{HI} = & 155.630 \\
\text{TP at 2} - \text{FS} & -\ \ 1.100 \\
\hline
 & 154.530 \\
B + \text{BS} & +\ \ 0.465 \\
\hline
 & 154.995 \\
\text{TP at 3} - \text{FS} & -\ \ 2.095 \\
\hline
 & 152.900 \\
C + \text{BS} & +\ \ 0.130 \\
\hline
 & 153.030 \\
\text{TP at 4} - \text{FS} & -\ \ 0.245 \\
\hline
 & 152.785 \\
D + \text{BS} & +\ \ 3.765 \\
\hline
 & 156.550 \\
\text{BM 36 elevation} = & -\ \ 0.345 \\
\hline
 & 156.205
\end{array}
$$

BM 36 was given in the data as 156.205 m so the leveling problem is correct. The lower-order level notes that must be made in the field book are shown in Fig. 5-7.

5.7 SIGHT DISTANCES

Normally, for third and higher orders of leveling, sight distances are kept below 75 m (245 ft), except when it becomes necessary to pass an obstacle. For lower-order lines, the length of sight depends on the optical qualities of the instrument and atmospheric conditions, with the maximum being about 600 m (1968 ft) under ideal conditions.

Before leveling is begun, a reconnaissance of the terrain must be made. Probable locations of turning points and instrument setups can be noted. The slope of the terrain is a prime consideration in leveling. The normal instrument height at any setup is about 1.5 m (4.9 ft). On even downhill slopes, the ground where the instrument is set up must not be more than 1 to 1.5 m (3.3 to 4.9 ft) below the turning point for a level backsight.

On the foresight, the extended 4-m (13.1-ft) rod can be held on the ground about 2.5 m (8.2 ft) below the instrument ground level and still permit a reading to be taken. This means that there is a tendency to make foresight distances longer when going downhill. Backsights tend to be longer running uphill.

During the reconnaissance, the line of sight can be estimated by sighting through a hand level. This determines possible instrument and rod setups. The distances between these points are paced in order to balance the foresights and backsights. The procedure is to sight at the uphill point with the hand level, making sure that the line of sight is above ground level. The distance from the proposed turning point to the proposed instrument position is paced, and the same amount is paced to establish the next turning point. Once the distance between points and instrument is determined, this same amount can be paced repeatedly as long as the slope remains the same. This procedure balances the distances and makes sure that a level line will fall on the rod. Balancing foresight and backsight distances is very important in leveling, as is noted in Sec. 5.8.

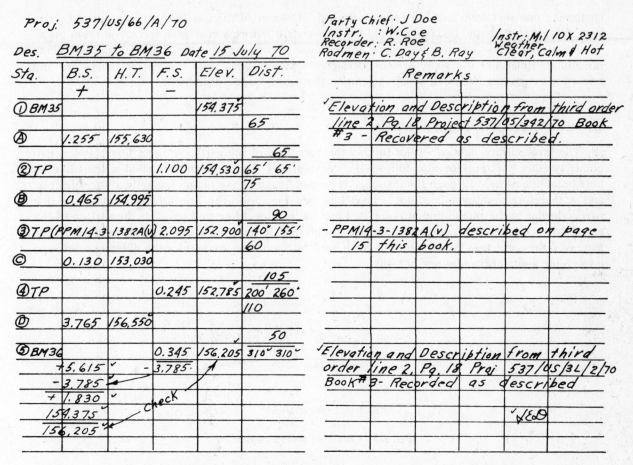

Proj: 537/US/66/A/70

Des. BM35 to BM36 Date 15 July 70

Party Chief: J Doe
Instr. : W. Coe
Recorder: R. Roe
Rodmen: C. Day & B. Ray

Instr: Mil 10X 2312
Weather
Clear, Calm & Hot

Sta.	B.S.	H.T.	F.S.	Elev.	Dist.	Remarks
	+		−			
① BM35				154.375		'Elevation and Description from third order line 2, Pg. 18, Project 537/05/342/70 Book #3 - Recovered as described.
					65	
Ⓐ	1.255	155.630				
					65	
② TP			1.100	154.530	65' 65'	
					75	
Ⓑ	0.465	154.995				
					90	
③ TP(PPM14-3-1382A(V)	2.095	152.900			140' 155'	- PPM14-3-1382A(v) described on page 15 this book.
					60	
Ⓒ	0.130	153.030				
					105	
④ TP			0.245	152.785	200' 260'	
					110	
Ⓓ	3.765	156.550				
					50	
⑤ BM36			0.345	156.205	310" 310"	'Elevation and Description from third order line 2, Pg. 18 Proj 537/05/3L/2/70 Book #3- Recorded as described
	+5.615		-3.785			
	-3.785					
	+1.830		check			✓EOD
	154.375					
	156.205					

Fig. 5-7 Lower-order level notes. (*Courtesy of U.S. Air Force, T O 00-25-103*)

5.8 BALANCING FORESIGHTS AND BACKSIGHTS

If the line of sight is not parallel to the axis of the bubble tube, rotating the instrument on its vertical axis will distort the horizontal plane into a conical plane above or below the horizontal. Unequal distances between backsight and foresight rod positions will cause an error which will increase in proportion to the difference in the distances (see Fig. 5-8). If the same sight (back or fore) is consistently longer, the error will accumulate. To eliminate this source of error, the level should be set up midway between the turning points. This is not always possible, and the next best method is to balance backsights and foresights at every opportunity. In practice, the distances from the back rod to the instrument and from the instrument to the front rod are measured at each setup station and then recorded. Separate running totals of backsights and foresights are kept; the two totals should be continually balanced, not left until the last few setups before closing the line, and not made up by

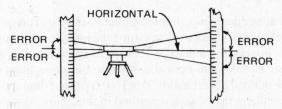

Fig. 5-8 Uneven distance error.

combining one very long and one very short sight. However, if the inequality is introduced by one long sight, then it should be compensated for in one sight before a change in refraction takes place.

Balancing backsights and foresights minimizes errors caused by the line of sight not being horizontal. Balancing the sums of the backsights and foresights does not correct curvature and refraction errors (see Sec. 5.9), which depend on the square of the distance.

5.9 CURVATURE AND REFRACTION

The level surface defined earlier follows the curvature of the earth. A direct line between two points on this level surface also follows the curvature of the earth and is called a *level line*. A horizontal line of sight through the telescope is perpendicular to the plumb line only at the telescope and is therefore a straight line, not a level line (see Fig. 5-9). The line *OH* is a horizontal line perpendicular to the plumb line at point *A*. Line *OL* is a level line that parallels the surface of the earth. At each point *OL* is perpendicular to a plumb line. In leveling, as the distance between points increases, the correction for curvature of earth must be applied to account for the difference between a level line and a horizontal line of sight.

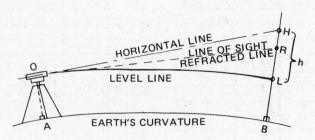

Fig. 5-9 Curvature and refraction.

Another correction which must be considered is the atmospheric refraction. The earth's atmosphere refracts, or bends, a ray of light because of differences in the density of the air between the instrument and the point being viewed. Figure 5-9 illustrates this. Due to the atmospheric density difference, the ray of light will follow the path *OR*. When viewed through the telescope, point *R* appears to be at point *H*.

In order to make corrections for curvature and refraction, the computations must first locate point *R* from point *H*, and then determine the correction to bring point *R* down to point *L* and establish a level line. In practice, the two corrections are combined. *RH* is about one-eighth of *LH*, and the value is given by the following formula:

$$h = 0.0000676M^2$$

where h = the correction in millimeters
          M = the distance between points in meters

5.10 ADJUSTING THE LEVEL

A check of the instrument's adjustments should be made before it is taken to the field. Adjustments should be checked every day before starting work, anytime the instrument is bumped or jolted, and at the end of each day's work. The instrument should be set up and approximately leveled over both pairs of screws. Since the check also includes the optical assembly, the cross hairs and objective should be focused sharply, using a well-defined object about 50 m (160 ft) away; then the parallax is removed. The check and adjustments are made in three steps described in the following subsections. They should be performed in the order listed. Do not perform these adjustments without your instructor's approval.

Level Vial

Adjustment of the level vial (see Fig. 5-10), makes the axis of the level bubble perpendicular to the axis of rotation (vertical axis).

Set the instrument over diametrically opposite leveling screws and center the bubble carefully (see Fig. 5-10, part 1).

Rotate the telescope 180° and note the movement of the bubble away from the center if the instrument is maladjusted (Fig. 5-10, part 2).

Bring the bubble half the distance back to the center of the vial by turning the capstan screws at the end of the vial (Fig. 5-10, part 3).

Relevel with the leveling screws (Fig. 5-10, part 4), and rotate the instrument 180°. Repeat the previous step if the bubble does not remain at the center of the vial.

Check the final adjustment by noting that the bubble remains in the center of the vial during the entire revolution about the vertical axis.

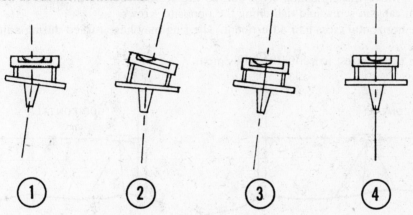

Fig. 5-10   Adjustments of the level vial.

Horizontal Cross Hairs

The horizontal cross hairs are adjusted (see Fig. 5-11) to make the horizontal hair lie in a plane perpendicular to the vertical axis.

Level the instrument carefully. Sight one end of the horizontal cross hair on a well-defined point about 50 m (160 ft) away. Turn the telescope slowly about its vertical axis, using the slow-motion screw. If the cross hair is in adjustment, the hair will stay on the point through its entire length. If it does not, loosen two adjacent reticle adjusting screws and turn the reticle by lightly tapping two opposite screws.

Sight on the point again, and if the horizontal hair does not follow the point through its entire length, turn the ring again.

Repeat this procedure as many times as necessary, until the cross hair stays on the point through its entire length. Then tighten the adjusting screws.

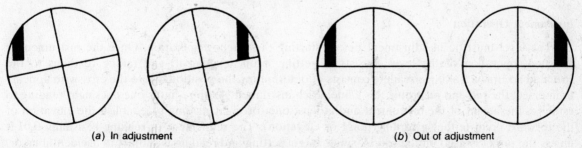

(*a*) In adjustment   (*b*) Out of adjustment

Fig. 5-11   Adjustment of horizontal cross hairs.

Line of Sight

The line-of-sight adjustment (see Fig. 5-12) makes the line of sight parallel to the axis of the level vial. This method is known as the *two-peg test*.

Set up the instrument (see Fig. 5-12, part 1); drive in a stake A about 50 m (160 ft) away; drive in another stake B at the same distance in the opposite direction.

Take a rod reading a on stake A and a rod reading b on stake B. With the instrument exactly halfway between the two stakes, $b - a$ is the true difference in elevation between the stakes.

Move the instrument close to stake A (see Fig. 5-12, part 2) so that the eyepiece swings within 10 mm of the rod.

Take a rod reading c on the stake A through the objective lens, and a rod reading d on the stake B in the normal manner. If the instrument is in adjustment, $d - c$ will equal $b - a$.

If the instrument is out of adjustment, calculate what the corrected reading e should be ($e = b + c - a$). Move the horizontal cross hair to the correct reading on stake B by loosening the correct vertical capstan screw and tightening the opposite screw.

Check the horizontal cross-hair adjustment. The ring may have turned during adjustment of the line of sight.

Rerun the two-peg test to verify the adjustment.

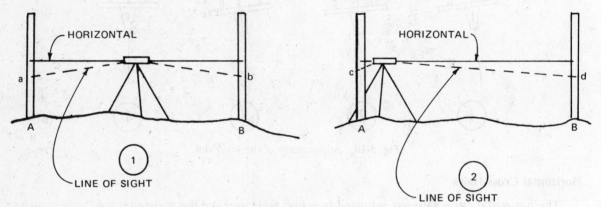

Fig. 5-12 Two-peg test.

5.11 FIELD PROCEDURE

The leveling operation requires the teamwork of both the instrumentperson and the rodperson at the moment of reading in order to achieve consistent results. The survey accuracy depends upon the refinement with which the line of sight can be made horizontal, the ability of the rodperson to hold the rod vertical, and the precision to which the rod is read. Accuracy with instruments using spirit level bubbles must also consider the adjustment of the level vial and the precision with which the bubble axis and the line of sight are made parallel. There should be no instrument settlement between the time of backsight and foresight at the instrument station.

Instrument Operation

The level must be in adjustment before starting the leveling operation. Once the instrument is adjusted, operation of the level consists of setting it up, leveling it, and taking readings to the specified accuracy. Taking a reading consists of determining the position where the crosswire appears to intersect the rod and recording this value. Each instrument setup requires one backsight reading to establish the height of the instrument and at least one foresight reading to establish the elevation of the forward point (either a turning point or elevation). The refinement of reading is usually 0.01 ft unless the target is used on the rod. A single target setting and reading is subject to accidental error. Additional foresights may be made to other points visible from the instrument setup if elevations of these points are also required. Depending on the type of survey and instruments used, either the center wire, all three crosswires, or the micrometer method may be used for taking readings.

One-Wire Method

In the one-wire method, only the middle crosswire is used. The instrumentperson, looking through the telescope, reads the value where the center wire appears to intersect the rod. If the survey requires more precise readings, a target and its vernier are used. In this case, the instrument-person sights through the instrument and signals the rodperson to move the target up or down until the crosswire bisects the horizontal line between the alternate red and white quadrants on the target.

When the bisection is achieved, the instrumentperson signals "OK," and the rodperson locks the target in position until the reading is complete. After the target has been locked in, a check reading is made to see that the target did not slip during the locking operation. The recorder should be near the level to record direct readings or near the rodperson to record the target setting operations. Immediately before taking any reading, the instrumentperson should check the level bubble and bring it back to center, if necessary.

The sequence for taking a level reading is as follows:

1. The level is set up and leveled.
2. The telescope is pointed so that its vertical crosswire is just off to one side of the rod, and the instrument is clamped.
3. The objective is focused and the parallax is removed.
4. The level bubble is checked for centering and adjusted if necessary.
5. The rod is read and the value recorded.
6. The bubble is rechecked for centering. If it is off center, it must be recentered and the reading repeated.
7. Once the instrumentperson is satisfied that the bubble has remained centered while the reading was taken, the intercept between the upper and lower wires is read to measure the distance from the level to the rod. This distance is used for balancing foresights and backsights and does not have to be read more closely than to the nearest centimeter.
8. The instrumentperson signals the rodperson to proceed to the next position.
9. The telescope is unclamped, revolved, pointed at the next rod position, and focused. Parallax is removed, the bubble centering is checked, the rod is read, and the bubble centering is rechecked.
10. These steps are repeated until the desired number of foresights are taken and a turning point established. The distance to the rod at the turning point is read and recorded. The rodperson then holds the position on the turning point.
11. The level is moved to the next setup position and the procedure repeated.

The target setting procedure requires the level operator to use hand or voice signals to the rodperson to move the target until the crosswire bisects the target. Figure 5-13 shows some of the more commonly used hand signals. If the rod must be extended, the target is set and the upper portion of the rod is moved up or down until the reading is set. Then the rod is clamped. After the instrumentperson has completed the operation and signals "OK," the rodperson reads the rod and vernier if necessary.

5.12   RECORDING

The field notebook is the permanent record of the survey; the notes must be clear and legible and must give only one possible interpretation. No survey recording is considered complete until the

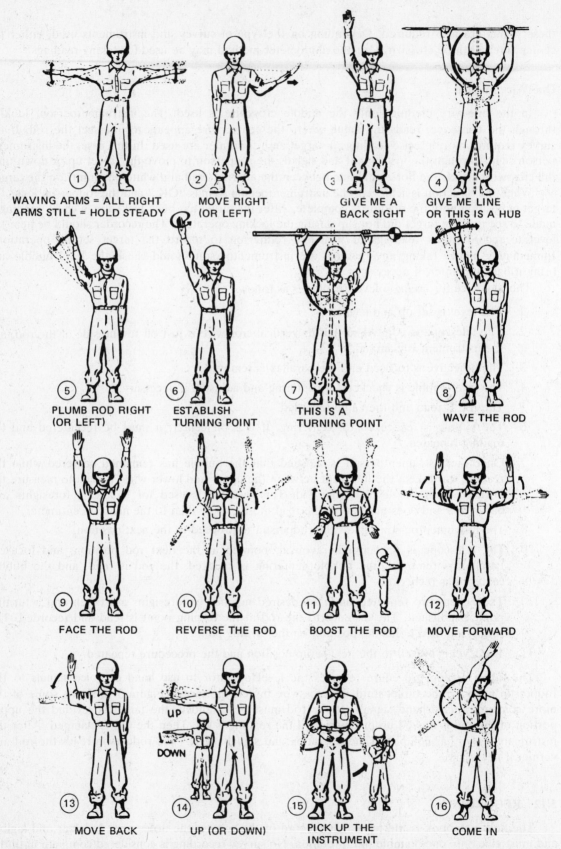

Fig. 5-13 Hand signals. (*From R. H. Wirshing and J. H. Wirshing, Civil Engineering Drafting, McGraw-Hill, New York, 1983, p. 317*)

notes and computations have been checked and initialed by the chief of party or the chief's designated representative.

There are differences in format for recording the three types of leveling readings (i.e., one-wire, three-wire, or micrometer method); only the one-wire method will be discussed here. The recording illustrated in Fig. 5-7 is based on the leveling operation shown in Fig. 5-6. Following is a description of Example 5.1 relative to the recordings required in the field book (refer to Fig. 5-7).

Level notes start with a known elevation or bench mark which is generally described from previous surveys. The identification of this point and its elevation are entered in the proper columns on the left-hand page. The right-hand page must show a reference to the source of the elevation and description. At times a level survey is run to establish grade, and the exact elevation above a datum plane is neither necessary nor readily available. In such a case, a more or less permanent point is selected as a starting point, and a fictitious value is assigned to it for use in the survey. This elevation and all elevations determined from it can be tied in to a known elevation at a later date. The notebook must describe this point and state that the elevation was assumed.

The first reading is a backsight (BS = 1.255), which is added to the elevation (154.375) to obtain the height of the instrument (HI = 155.630).

The next reading is a foresight (FS = 1.100), which is subtracted from the HI to get the elevation (154.530) of the next point (TP 2 in the sample notes).

The first instrument setup A was selected midway between bench mark 1 and turning point 2. In the distance column, 65 m appears on the left-hand side to show the distance from the bench mark to the instrument setup and on the right-hand side for the distance from the instrument to the turning point.

The instrument is moved to the next setup B while the rod remains at turning point 2. The backsight (0.465) and the distance of the rod (75 m) are read and recorded, and the HI is computed (154.995). The instrument is then pointed at turning point 3, and the foresight (2.095) and distance (90 m) are read and recorded. The elevation (152.900) is computed.

This method continues until the survey is tied to the next bench mark. The distances are balanced (the sum of the foresight distances should equal the sum of the backsight distances) as the survey progresses; balancing should be completed before the final bench mark is reached.

The recorder's computations can be verified by adding separately the backsight rod readings and the foresight rod readings. The difference between the two totals is the difference in elevation between the starting and final elevations. Adding (or subtracting) this total to or from the starting elevation should give the final elevation. Any disagreement is the result of an error in the calculations, which must be rechecked.

In the example, the curvature and refraction corrections would not be applied, since the correction for the longest distance (110 m) would be only 0.8 mm (0.03 in), which is too small to affect any of the readings.

EXAMPLE 5.2   Given: Figure 5-14 which shows (a) a plan and (b) a profile view of a bench mark leveling problem. BM 1 elevation = 30.476, BM 19 elevation = 22.181, BM 20 elevation = 18.557, and BM 21 elevation = 22.355. Back- and foresights are shown in the profile view of the level problem (Fig. 5-14b). Work out the levels in the note format of Fig. 5-15.

Solution: Starting at BM 1 enter the elevation of this BM in the elevation column. Then record the backsights in column 2, labeled plus (+). Add elevation and backsights to find height of instrument (HI) to be placed in the third column. The foresights in column 4, labeled minus (−), are subtracted from the HI column. This gives the elevations in column 6.

5.13   PROFILE LEVELING

The process of profile leveling obtains the elevations of a series of points along a continuous line. The results are plotted in the form of a continuous vertical cross section called a *profile*. The vertical scale is always made greater than the horizontal scale, usually at a ratio of 10:1. Example 5.3 gives the method for profile leveling.

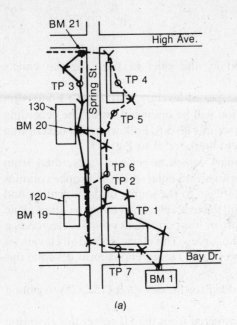

(a)

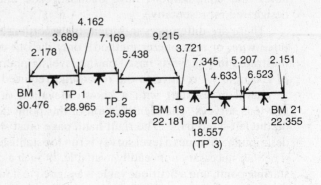

(b)

Fig. 5-14

Sta.	+	HI	−	Rod Elev.	
BM 1	2.178	32.654		30.476	B.M. Disk set in Curbing of Bay Dr.
TP 1	4.162	33.127	3.689	28.965	
TP 2	5.438	31.396	7.169	25.958	
BM19	3.721	25.902	9.215	22.181	"X" in step of Bldg. 120 SPRING ST.
BM 20	4.633	23.190	7.345	18.557	"X" in stone top step Bldg 130 SPRING ST.
TP 3	6.523	24.506	5.207	17.983	
BM21	4.528	26.883	2.151	22.355	☐ in Conc. Base Iron Fence
TP 4	5.812	26.517	6.178	20.705	S.W. Cor. Spring & High Ave.
TP 5	6.218	29.011	3.724	22.793	
BM 20	7.083	25.696	10.498	18.563	
TP 6	5.578	27.053	4.171	21.475	Arith. CK. 30.476 = BM 1
BM19	9.511	31.708	4.856	22.197	+ 73.620 = Σ BS 104.096
TP 7	8.235	33.622	6.321	25.387	− 73.613 = Σ F.S.
BM 1			3.139	30.483	30.483 Elev.
					Error + 0.007 R. Roe

Title row: B.M. Leveling – SPRING ST., Bay to High   π Roe   Rod Doe   Date 8/6/83   Fair 76°F

Fig. 5-15 Level notes.

EXAMPLE 5.3   Given: A series of level shots straight down the center of Robbins Street starting at Main Street from BM 7, which has an elevation of 40.476. BM 19, the final bench mark, has an adjusted elevation of 32.190. Figures 5-16 and 5-17 give the necessary data. Work out the survey.

Solution: Refer to Figs. 5-16 and 5-17. Marks are placed every 50 ft along the centerline desired. Each 100-ft point is a station and is numbered from zero. Points between stations are given a plus sign, i.e., the number of feet past the last station. The enumeration is shown in Fig. 5-16.

The level is set up near station $0 + 50$. The plus reading 2.587 is taken on BM 7. This is added to the elevation 40.476 to obtain the HI, 43.063. The rod is read on station $0 + 0$, 4.2. This is called a *rod reading*, or *rod shot*, and is placed in the rod column. It is subtracted from the HI to obtain the elevation of station $0 + 0$, 38.9. From this setup of the instrument, all rod shots are taken until the view is obstructed or a sight of over 150 ft is required. TP 1 is then established. A minus shot of 3.782 is taken on TP 1 and subtracted from the HI, 43.063, giving the elevation 39.281 for TP 1. The instrument is moved, the process is repeated between TP 1 and TP 2, and so on. The work must end on a bench mark of known elevation so that a check may be obtained.

The elevation of each station is computed by subtracting the rod shot from the "proper" HI. It is therefore essential that all the rod shots from one HI be recorded before the minus reading to the next TP. Also, the minus shot to the next TP should be taken after all the rod shots so that if the field check does not indicate a blunder, this is an immediate indication that the level was not disturbed at any HI.

These two considerations dictate the order of procedure; that is, all the rod shots should be taken at any HI before the minus sight to the next TP is taken.

Profile leveling is identical with bench mark leveling except that at many HIs a number of side, or rod, shots are taken. All rules for BM leveling apply.

5.14   TRIGONOMETRIC LEVELING

Trigonometric leveling applies the principles of trigonometry to determine differences in elevation (see Fig. 5-18). There are two applications of this method to surveying: on long lines of sight

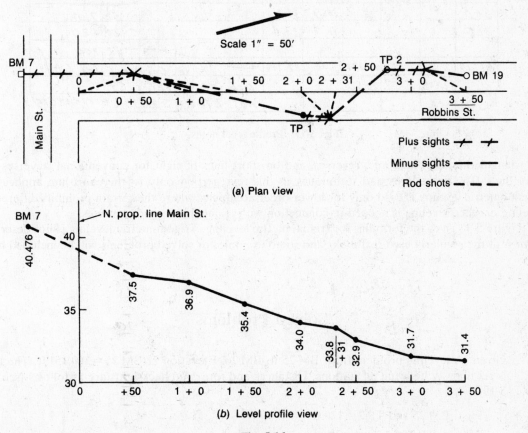

(a) Plan view

(b) Level profile view

Fig. 5-16

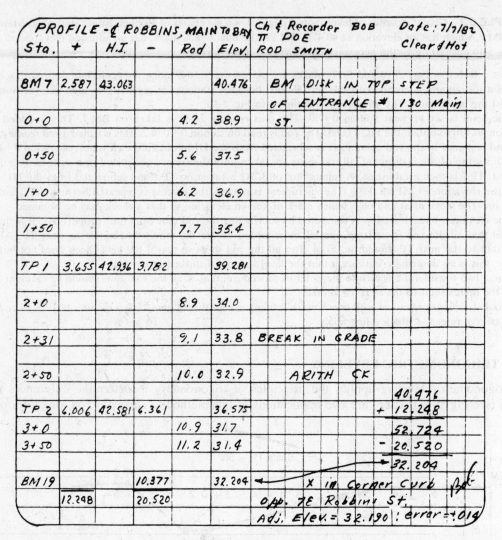

Fig. 5-17 Profile level notes.

for triangulation and electronic traverses, and on short lines of sight for conventional traverses and level lines. The procedures and techniques in this text pertain only to the short-line application. Trigonometric leveling is used only for lower-order accuracies where the terrain prohibits differential leveling or when leveling is needed in connection with triangulation and traverses.

Figure 5-18 gives the formulas for trigonometric leveling. Trigonometric leveling is not described in most of the regularly used textbooks and so no examples or solved problems will be included here.

Solved Problems

5.1   Given: A leveling problem from BM 25 to BM 26. Elevation at BM 25 = 160.151 ft. The level is set midway between all stations. Distances and fore- and backsights are as follows (see Fig. 5-19).

First: BM 25 and TP 2 = 120 ft

Second: TP 2 and TP 3 = 140 ft

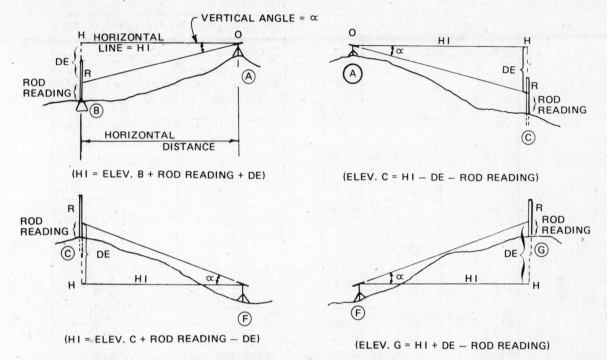

Fig. 5-18　Trigonometric leveling formulas.

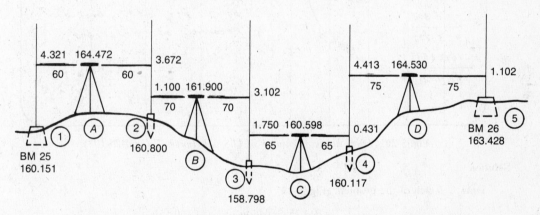

Fig. 5-19　Leveling operation. (*Courtesy of U.S. Air Force, T O 00-25-103*)

Third: TP 3 and TP 4 = 130 ft

Fourth: TP 4 and BM 26 = 150 ft

BS *A* to BM 25 = 4.321

FS *A* to TP 2 = 3.672

BS *B* to TP 2 = 1.100

FS *B* to TP 3 = 3.102

BS *C* to TP 3 = 1.750

FS *C* to TP 4 = 0.431

BS *D* to TP 4 = 4.413

FS *D* to BM 26 = 1.102

BM 26 has an elevation of 163.428 ft. Work out the level problem and show how it is entered in the field book (see Fig. 5-20).

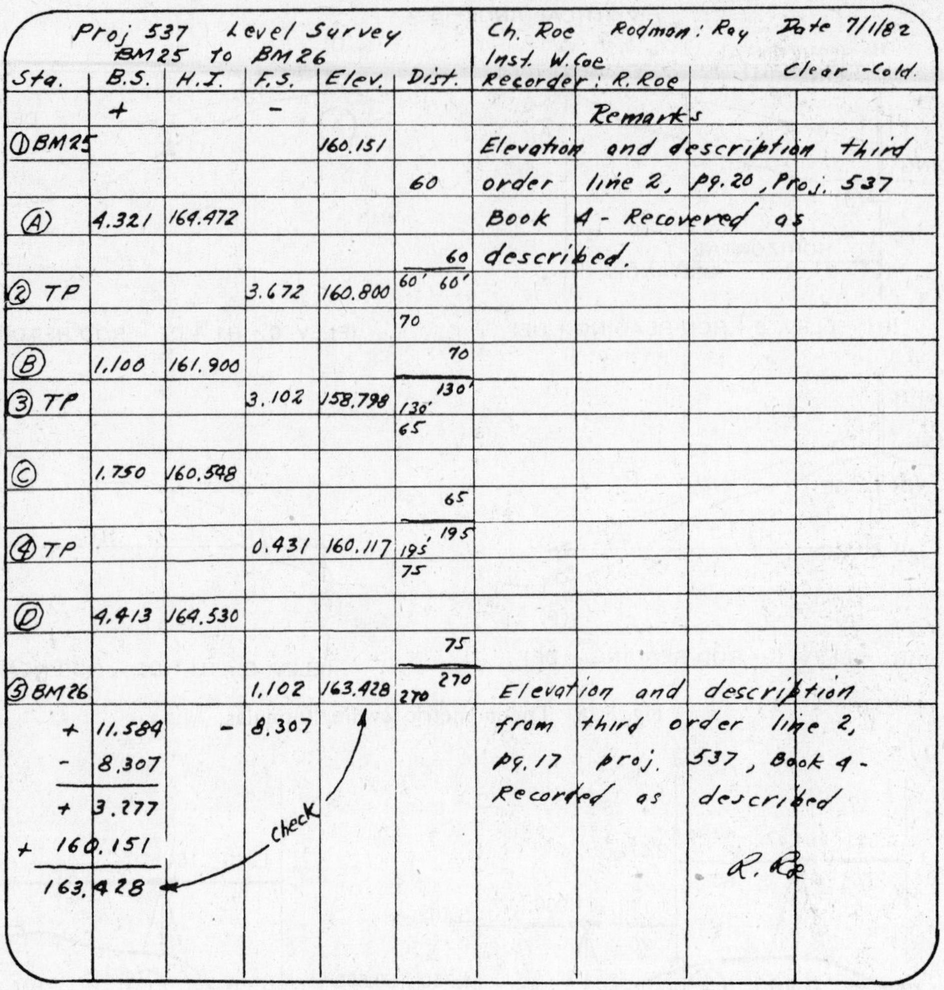

Sta.	B.S. +	H.I.	F.S. −	Elev.	Dist.	Proj 537 Level Survey BM25 to BM26 — Ch. Roe Rodman: Ray Date 7/1/82 Inst. W.Coe Recorder: R.Roe Clear–Cold / Remarks
① BM25				160.151		Elevation and description third
					60	order line 2, Pg.20, Proj. 537
Ⓐ	4.321	164.472				Book 4 - Recovered as
					60	described.
② TP			3.672	160.800	60' 60'	
					70	
Ⓑ	1.100	161.900			70	
③ TP			3.102	158.798	130 130' 65	
Ⓒ	1.750	160.548			65	
④ TP			0.431	160.117	195' 195 75	
Ⓓ	4.413	164.530			75	
⑤ BM26			1.102	163.428	270 270	Elevation and description from third order line 2, Pg.17 proj. 537, Book 4 - Recorded as described
	+ 11.589		− 8.307			R. Roe
	− 8.307					
	+ 3.277					
	+ 160.151					
	163.428					

check

Fig. 5-20   Level notes. (*Courtesy of U.S. Air Force, T O 00-25-103*)

Solution

Draw a sketch of the problem (Fig. 5-19).

BM 25 elevation =	160.151
+ BS	+ 4.321
HI =	164.472
− FS	− 3.672
Elevation at TP 2 =	160.800
+ BS	+ 1.100
HI =	161.900
− FS	− 3.102
Elevation at TP 3 =	158.798
+ BS	+ 1.750
HI =	160.548
− FS	− 0.431
Elevation at TP 4 =	160.117
+ BS	+ 4.413
HI =	164.530
− FS	− 1.102
BM 26 elevation =	163.428

Draw and label sketch (Fig. 5-19) and field book pages (Fig. 5-20).

5.2　　Given: Leveling problem from BM 10 at elevation of 145.250 to BM 11 at elevation of 148.325. Distances, backsights, and foresights follow. Work out the survey.

BM 10 to A = 50 ft	BS A to BM 10 = 4.250
A to TP 2 = 50 ft	FS A to TP 2 = 3.250
TP 2 to B = 45 ft	BS B to TP 2 = 1.250
B to TP 3 = 45 ft	FS B to TP 3 = 2.750
TP 3 to C = 51 ft	BS C to TP 3 = 1.750
C to TP 4 = 53 ft	FS C to TP 4 = 0.525
TP 4 to D = 73 ft	BS D to TP 4 = 4.100
D to BM 11 = 75 ft	FS D to BM 11 = 1.750

Solution

Draw a sketch of the problem. See Fig. 5-21. Work out the backsights and foresights.

BM 10 elevation =	145.250
+ BS	+ 4.250
HI =	149.500
− FS	− 3.250
Elevation at TP 2 =	146.250
+ BS	+ 1.250
HI =	147.500
− FS	− 2.750
Elevation at TP 3 =	144.750
+ BS	+ 1.750
HI =	146.500
− FS	− 0.525
Elevation at TP 4 =	145.975
+ BS	+ 4.100
HI =	150.075
− FS	− 1.750
BM 26 elevation =	148.325
BM 10 elevation =	145.250
+ sum of backsights	+ 11.350
HI =	156.600
− sum of foresights	− 8.275
Check of BMs:	148.325

Since BM elevation checks, the problem is correct as drawn.

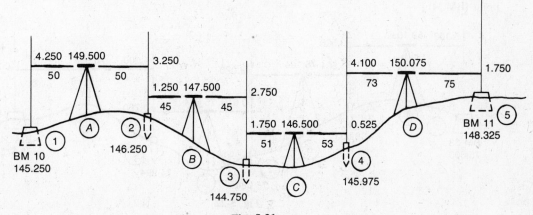

Fig. 5-21

5.3 Given: Leveling problem with four transit stations. BM 33 elevation is 75.00. Backsights and foresights are as follows.

Station	Backsight	Foresight
A	5.103	3.500
B	1.501	3.020
C	0.610	0.700
D	4.000	0.321

Find the elevation of BM 34.

Solution

Draw a sketch of the problem (see Fig. 5-22).

BM 33 elevation =	75.000
+ BS	+ 5.103
HI =	80.103
− FS	− 3.500
Elevation at TP 2 =	76.603
+ BS	+ 1.501
HI =	78.104
− FS	− 3.020
Elevation at TP 3 =	75.084
+ BS	+ 0.610
HI =	75.694
− FS	− 0.700
Elevation at TP 4 =	74.994
+ BS	+ 4.000
HI =	78.994
− FS	− 0.321
BM 34 elevation =	78.673

Sum of backsights =	11.214
Subtract sum of foresights =	− 7.541
Elevation difference =	3.673
Add BM 33 elevation =	+ 75.000
Arithmetic checks:	78.673

The BM elevation checks, by arithmetic computation. We do not know, however, if the BM elevation checks since this is not a closed loop. [There are two types of traverse (see Chap. 7)—*open* and *closed*. In a closed traverse the lines return to the starting point and form a closed polygon, so all angles and distances can be checked. Here there is no closed polygon, but the traverse is mathematically closed since it ends at a known point (BM 34).]

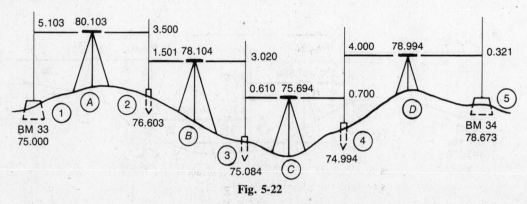

Fig. 5-22

5.4   Given: A line of levels to be run from BM 10 to BM 11. At BM 10 the elevation is 101.325. The level is set up at *A*, *B*, *C*, and *D*. Backsights and foresights are as follows.

Station	Backsight	Foresight
A	1.350	1.200
B	0.503	2.100
C	0.150	0.250
D	3.800	0.450

All the level setups are equally distant between back- and foresight points to reduce error to a minimum. Work out the levels to find the elevation of BM 10.

Solution

BM 10 elevation =	101.325
+ BS	+ 1.350
HI =	102.675
− FS	− 1.200
Elevation at TP 2 =	101.475
+ BS	+ 0.503
HI =	101.978
− FS	− 2.100
Elevation at TP 3 =	99.878
+ BS	+ 0.150
HI =	100.028
− FS	− 0.250
Elevation at TP 4 =	99.778
+ BS	+ 3.800
HI =	103.578
− FS	− 0.450
BM 11 elevation =	103.128

Sum of backsights =	5.803
Subtract sum of foresights =	− 4.000
Elevation difference =	1.803
Add BM 10 elevation =	+ 101.325
Arithmetic checks:	103.128

5.5   Given: BM 20 elevation = 51.275; level set at *A*, *B*, *C*, and *D*. Backsights and foresights are as follows.

Station	Backsight	Foresight
A	1.400	1.310
B	0.500	2.000
C	0.175	0.300
D	3.600	0.450

All the level setups are equally distant between back- and foresight points to reduce error. Work out the levels to find the elevation of BM 21.

Solution

BM 20 elevation =	51.275
+ BS	+ 1.400
HI =	52.675
− FS	− 1.310
Elevation at TP 2 =	51.365
+ BS	+ 0.500
HI =	51.865
− FS	− 2.000
Elevation at TP 3 =	49.865
+ BS	+ 0.175
HI =	50.040
− FS	− 0.300
Elevation at TP 4 =	49.740
+ BS	+ 3.600
HI =	53.340
− FS	− 0.450
BM 21 elevation =	52.890

Sum of backsights =	5.675
Subtract sum of foresights =	− 4.060
Elevation difference =	1.615
Add BM 20 elevation =	+ 51.275
Arithmetic checks:	52.890

5.6 Given: A line of levels to be run from BM 36 to BM 37. Elevation of BM 36 is 81.751. The level is set up at A, B, C, and D. Backsights and foresights are as follows.

Station	Backsight	Foresight
A	1.503	1.275
B	0.498	2.700
C	0.165	0.267
D	3.654	0.503

Work out the levels to find the elevation of BM 37.

Solution

BM 36 elevation =	81.751
+ BS	+ 1.503
HI =	82.254
− FS	− 1.275
Elevation at TP 2 =	81.979
+ BS	+ 0.498
HI =	82.477
− FS	− 2.700
Elevation at TP 3 =	79.777
+ BS	+ 0.165
HI =	79.942
− FS	− 0.267
Elevation at TP 4 =	79.675
+ BS	+ 3.654
HI =	83.329
− FS	− 0.503
BM 37 elevation =	82.826

Sum of backsights =	5.820
Subtract sum of foresights =	− 4.745
Elevation difference =	1.075
Add BM 36 elevation =	+ 81.751
Arithmetic checks:	82.826

5.7   Given: Levels to be run from BM 2 to BM 3. There are four level stations; BM 2 elevation = 89.123. Backsights and foresights are as follows.

Station	Backsight	Foresight
A	1.720	1.451
B	0.530	2.852
C	0.231	0.285
D	3.752	0.613

Work out the levels to find the elevation of BM 3.

Solution

BM 2 elevation =	89.123
+ BS	+ 1.720
HI =	90.843
− FS	− 1.451
Elevation at TP 2 =	89.392
+ BS	+ 0.530
HI =	89.922
− FS	− 2.852
Elevation at TP 3 =	87.070
+ BS	+ 0.231
HI =	87.301
− FS	− 0.285
Elevation at TP 4 =	87.016
+ BS	+ 3.752
HI =	90.768
− FS	− 0.613
BM 3 elevation =	90.155

Sum of backsights =	6.233
Subtract sum of foresights =	− 5.201
Elevation difference =	1.032
Add BM 2 elevation =	+ 89.123
Arithmetic checks:	90.155

5.8   A level run must be made from BM 31 to BM 32 to check the elevation at BM 32. BM 31 elevation is known to be 705.013. Backsights and foresights are as follows.

Station	Backsight	Foresight
A	2.001	1.666
B	0.798	3.001
C	0.210	0.479
D	3.854	0.806

Work out the levels to find the elevation of BM 32.

Solution

BM 31 elevation =	705.013
+ BS	+ 2.001
HI =	707.014
− FS	− 1.666
Elevation at TP 2 =	705.348
+ BS	+ 0.798
HI =	706.146
− FS	− 3.001
Elevation at TP 3 =	703.145
+ BS	+ 0.210
HI =	703.355
− FS	− 0.479
Elevation at TP 4 =	702.876
+ BS	+ 3.854
HI =	706.730
− FS	− 0.806
BM 32 elevation =	705.924

Sum of backsights =	6.863
Subtract sum of foresights =	− 5.952
Elevation difference =	0.911
Add BM 31 elevation =	+ 705.013
Arithmetic checks:	705.924

5.9   BM 104 elevation = 692.123. Backsights and foresights are as follows.

Station	Backsight	Foresight
A	2.052	1.750
B	0.810	3.210
C	0.300	0.510
D	3.579	0.910

Find the elevation of BM 105.

Solution

BM 104 elevation =	692.123
+ BS	+ 2.052
HI =	694.175
− FS	− 1.750
Elevation at TP 2 =	692.425
+ BS	+ 0.810
HI =	693.235
− FS	− 3.210
Elevation at TP 3 =	690.025
+ BS	+ 0.300
HI =	690.325
− FS	− 0.510
Elevation at TP 4 =	689.815
+ BS	+ 3.579
HI =	693.394
− FS	− 0.910
BM 105 elevation =	692.484

Sum of backsights =	6.741
Subtract sum of foresights =	− 6.380
Elevation difference =	0.361
Add BM 104 elevation =	+ 692.123
Arithmetic checks:	692.484

5.10 Given: Series of levels alternating back and forth across Front Avenue from BM 10 at elevation 15.610 through TP 1 at elevation 14.521, BM 11 at 11.841, BM 12 at 9.859 to a final point at house 99 on Front Avenue (see Fig. 5-23). Work out the problem in a field book.

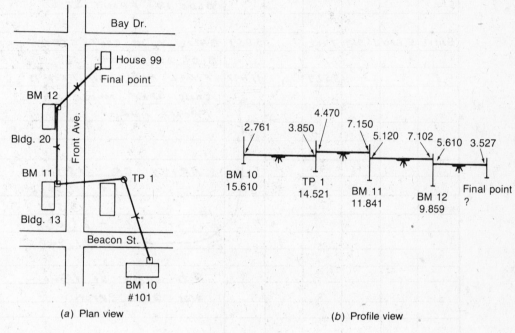

(a) Plan view   (b) Profile view

Fig. 5-23

Solution

See Fig. 5-24.

5.11 Given: The profile view of a series of levels on Beacon Street (see Fig. 5-25). From this series of levels find the elevation of the final bench mark (not given in the figure).

Solution

BM 23 elevation =	710.521
+ BS	+ 3.751
HI =	714.272
− FS	− 3.910
Elevation at TP 1 =	710.362
+ BS	+ 4.271
HI =	714.633
− FS	− 5.212
Elevation at TP 2 =	709.421
+ BS	+ 6.325
HI =	715.746
− FS	− 4.572
BM 24 elevation =	711.174

Sta.	+	HI	-	Rod	Elev.	BM, Leveling, Beacon to Bay	∧ Roe Date 6/7/82
							Rod Doe Fair 59°F
BM10	2.761	18.371			15.610	BM10 CHISELED X IN CONC.	
						FLOOR #101 BEACON ST.	
TP1	4.470	18.991	3.850		14.521		
BM11	5.120	16.961	7.150		11.841	BM 11 MARK IN CONC. FLOOR	
						BLDG. 13 , FRONT ST.	
BM12	5.610	15.469	7.102		9.859	BM12 X in CONC. STEP	
						BLDG.20 , FRONT ST.	
F. P.			3.527		11.992	FINAL POINT ON FRONT	
						CONC. STEP HOUSE 99	
						FRONT AVE,	

ARITH CK. 15.610
+ 17,961
33.571
− 21.629
11.992

Both agree so there
was zero error.

Fig. 5-24 Level notes.

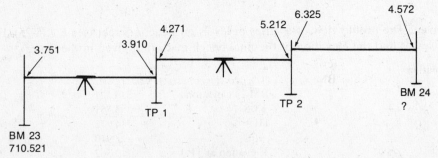

Fig. 5-25

Math check:	BM 23 elevation =	710.521
	Add sum of foresights	+ 14.347
		724.868
	Subtract sum of backsights	− 13.694
	Check:	711.174

5.12   Given: The profile view (see Fig. 5-26) of a series of levels on Dow Street. Find the elevation of the final point in the survey.

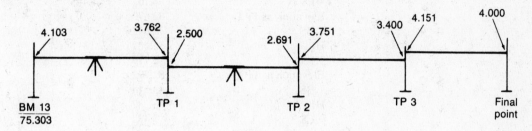

Fig. 5-26

Solution

BM 13 elevation =	75.303
+ BS	+ 4.103
HI =	79.406
− FS	− 3.762
Elevation at TP 1 =	75.644
+ BS	+ 2.500
HI =	78.144
− FS	− 2.691
Elevation at TP 2 =	75.453
+ BS	+ 3.751
HI =	79.204
− FS	− 3.400
Elevation at TP 3 =	75.804
+ BS	+ 4.151
HI =	79.955
− FS	− 4.000
Final elevation =	75.955

Math check:

BM 13 elevation =	75.303
Add sum of foresights	+ 14.505
	89.808
Subtract sum of backsights	− 13.853
Check:	75.955

5.13   Given: A profile view of the levels taken on Riverview Avenue (see Fig. 5-27). Find if the elevation given for BM 36 agrees with the elevation you get in working the survey. If not, how much is the error?

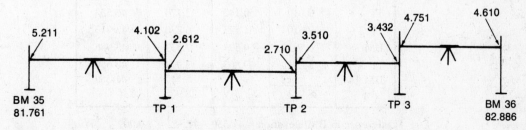

Fig. 5-27

Solution

BM 35 elevation =	81.761
+ BS	+ 5.211
HI =	86.972
− FS	− 4.102
Elevation at TP 1 =	82.870
+ BS	+ 2.612
HI =	85.482
− FS	− 2.710
Elevation at TP 2 =	82.772
+ BS	+ 3.510
HI =	86.282
− FS	− 3.432
Elevation at TP 3 =	82.850
+ BS	+ 4.751
HI =	87.601
− FS	− 4.610
BM 36 elevation =	82.991
BM 36 given elevation =	− 82.886
Error =	0.105

In Probs. 5.14 to 5.20 are given a series of levels as shown in the indicated figure. For each problem, give the complete form of the field notes, showing the difference in elevation and the check for each solution.

5.14   See Fig. 5-28a.

Solution

Station	+	HI	−	Elevation
BM 24	5.485	63.775		58.290
TP 1	4.982	66.116	2.641	61.134
TP 2	8.829	65.302	9.643	56.473
BM 38	8.636	66.296	7.642	57.660
TP 3	2.485	64.534	4.247	62.049
BM 24			6.240	58.294
	30.417		30.413	

Difference in BM elevation = 58.294 − 58.290 = 0.004   *Ans.*

Difference in BM elevation = sum of backsights − sum of foresights = 30.417 − 30.413 = 0.004   *Check*

5.15   See Fig. 5-28b.

Solution

Station	+	HI	−	Elevation
BM 12	7.901	49.253		41.352
TP 1	3.610	50.143	2.720	46.533
TP 2	7.610	51.523	6.230	43.913
BM 38	6.591	50.577	7.537	43.986
TP 3	3.211	49.867	3.921	46.656
BM 12			8.517	41.350
	28.923		28.925	

Difference in BM elevation = 41.350 − 41.352 = −0.002   *Ans.*

Difference in BM elevation = sum of backsights − sum of foresights = 28.923 − 28.925 = −0.002   *Check*

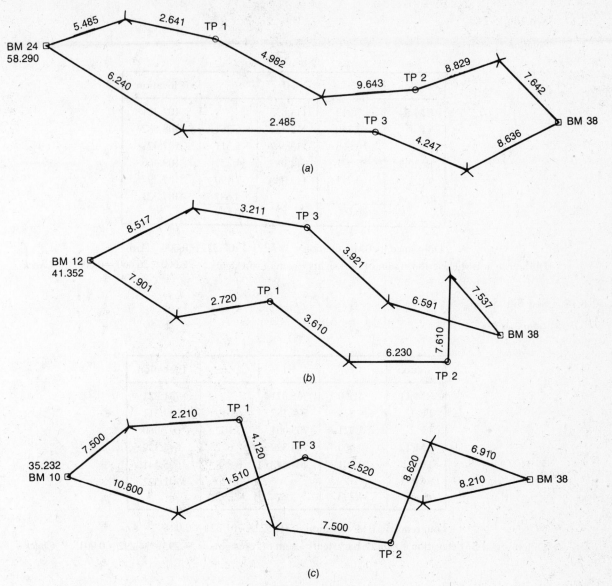

Fig. 5-28 BM leveling data.

5.16 See Fig. 5-28c.

Solution

Station	+	HI	−	Elevation
BM 10	7.500	42.732		35.232
TP 1	4.120	44.642	2.210	40.522
TP 2	8.620	45.762	7.500	37.142
BM 38	8.210	47.062	6.910	38.852
TP 3	1.500	46.052	2.520	44.542
BM 10			10.800	35.252
	29.960		29.940	

Difference in BM elevation = 35.252 − 35.232 = 0.020   *Ans.*

Difference in BM elevation = sum of backsights − sum of foresights = 29.960 − 29.940 = 0.020   *Check*

5.17   See Fig. 5-29*a*.

Solution

Station	+	HI	−	Elevation
BM 34	7.512	114.833		107.321
TP 1	3.750	112.332	6.251	108.582
TP 2	6.666	113.687	5.311	107.021
BM 37	5.521	114.987	4.221	109.466
TP 3	3.211	111.008	7.190	107.797
BM 34			3.687	107.321
	26.600		26.660	

Difference in BM elevation = 107.321 − 107.321 = 0.000   *Ans.*

Difference in BM elevation = sum of backsights − sum of foresights = 26.660 − 26.660 = 0.000   *Check*

5.18   See Fig. 5-29*b*.

Solution

Station	+	HI	−	Elevation
BM 11	4.500	705.821		701.321
TP 1	4.871	704.182	6.510	699.311
TP 2	5.921	703.001	7.102	697.080
BM 12	10.000	704.329	8.672	694.329
TP 3	9.001	704.442	8.888	695.441
BM 11			3.120	701.322
	34.293		34.292	

Difference in BM elevation = 701.322 − 701.321 = 0.001   *Ans.*

Difference in BM elevation = sum of backsights − sum of foresights = 34.293 − 34.292 = 0.001   *Check*

5.19   See Fig. 5-30*a*.

Solution

Station	+	HI	−	Elevation
BM 17	4.575	30.698		26.123
TP 1	5.250	30.848	5.100	25.598
TP 2	6.510	31.088	6.270	24.578
BM 18	4.881	31.169	4.800	26.288
TP 3	5.551	31.098	5.622	25.547
BM 17			4.972	26.126
	26.767		26.764	

Difference in BM elevation = 26.126 − 26.123 = 0.003   *Ans.*

Difference in BM elevation = sum of backsights − sum of foresights = 26.767 − 26.764 = 0.003   *Check*

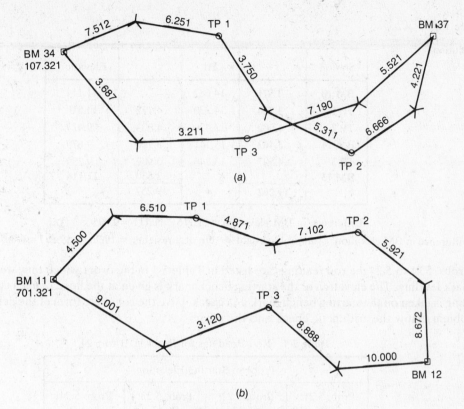

Fig. 5-29 BM leveling data.

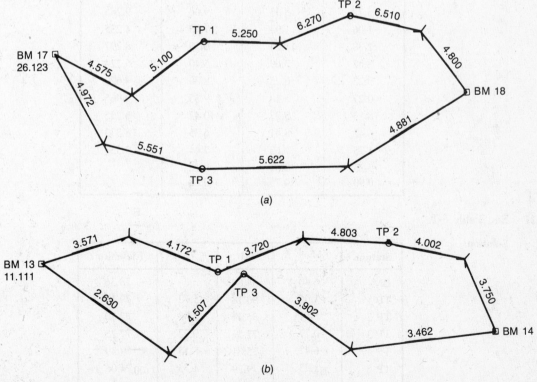

Fig. 5-30 BM leveling data.

5.20   See Fig. 5-30*b*.

Solution

Station	+	HI	−	Elevation
BM 13	3.571	14.682		11.111
TP 1	3.720	14.230	4.172	10.510
TP 2	4.002	13.429	4.803	9.427
BM 14	3.462	13.141	3.750	9.679
TP 3	4.507	13.746	3.902	9.239
BM 13			2.630	11.116
	19.262		19.257	

Difference in BM elevation = 11.116 − 11.111 = 0.005   *Ans.*

Difference in BM elevation = sum of backsights − sum of foresights = 19.262 − 19.257 = 0.005   *Check*

   In Probs. 5.21 to 5.24 the rod readings are listed in Table 5-1 in the order which they were taken in bench mark leveling. The elevation of the starting bench mark is given at the head of each column. The last reading is taken on the starting bench mark as a check. Give the complete form of the field notes for each problem. Show the arithmetic check and the errors.

Table 5-1   Rod Readings for Probs. 5.21 to 5.24

Problem Starting Elevation			
Prob. 5.21: 78.36	Prob. 5.22: 61.32	Prob. 5.23: 96.43	Prob. 5.24: 81.646
6.48	11.39	8.49	1.832
5.72	5.33	9.66	9.265
1.06	1.93	7.43	4.235
2.38	11.24	6.55	8.297
8.67	3.69	5.40	5.224
9.22	6.25	8.30	4.620
0.27	5.10	9.35	9.065
8.13	5.27	10.42	6.255
6.42	8.41	6.35	8.333
1.75	7.63	7.44	0.558
5.23	10.61	9.22	
0.90	4.75	3.88	

5.21   See Table 5-1.

Solution

Station	+	HI	−	Elevation
BM	6.48	84.84		78.36
TP 1	1.06	80.18	5.72	79.12
TP 2	8.67	86.47	2.38	77.80
TP 3	0.27	77.52	9.22	77.25
TP 4	6.42	75.81	8.13	69.39
TP 5	5.23	79.29	1.75	74.06
BM			0.90	78.39
	28.13		28.10	+0.03

5.22 See Table 5-1.

Solution

Station	+	HI	−	Elevation
BM	11.39	72.71		61.32
TP 1	1.93	69.31	5.33	67.38
TP 2	3.69	61.76	11.24	58.07
TP 3	5.10	60.61	6.25	55.51
TP 4	8.41	63.75	5.27	55.34
TP 5	10.61	66.73	7.63	56.12
BM			4.75	61.98
	41.13		40.47	+0.66

5.23 See Table 5-1.

Solution

Station	+	HI	−	Elevation
BM	8.49	104.92		96.43
TP 1	7.43	102.69	9.66	95.26
TP 2	5.40	101.54	6.55	96.14
TP 3	9.35	102.59	8.30	93.24
TP 4	6.35	98.52	10.42	92.17
	9.22	100.30	7.44	91.08
BM			3.88	96.42
	46.24		46.25	−0.01

5.24 See Table 5-1.

Solution

Station	+	HI	−	Elevation
BM	1.832	83.478		81.646
TP 1	4.235	78.448	9.265	74.213
TP 2	5.224	75.375	8.297	70.151
TP 3	9.065	79.82	4.620	70.755
TP 4	8.333	81.898	6.255	73.565
BM			0.558	81.340
	28.689		28.995	−0.306

In Probs. 5.25 and 5.26 are given sets of field data taken in the order given during profile leveling. Place each set in standard field book form, and draw the profile to the following scales: Horizontal, 1 in = 100 ft; vertical, 1 in = 10 ft.

5.25   Given: Data in Table 5-2.

Table 5-2   Data for Prob. 5.25

Station, Elevation	Point Sight	Rod	Point Sight	Rod
BM 23, 40.312	BM 23	4.516	5 + 0	2.4
	0 + 0	3.0	TP 2	2.201
	1 + 0	8.3	TP 2	4.016
	2 + 0	12.1	6 + 0	5.2
	3 + 0	11.4	7 + 0	9.3
	TP 1	7.872	8 + 0	10.9
	TP 1	5.280	9 + 0	10.7
BM 24, 33.050	4 + 0	4.90	BM 24	10.989

Solution

Table 5-3   Solution for Prob. 5.25

Station	+	HI	−	Rod	Elevation
BM 23	4.516	44.828			40.312
0 + 0				3.0	41.8
1 + 0				8.3	36.5
2 + 0				12.1	32.7
3 + 0				11.4	33.4
TP 1	5.280	42.236	7.872		36.956
4 + 0				4.90	37.3
5 + 0				2.4	39.8
TP 2	4.016	44.051	2.201		40.035
6 + 0				5.2	38.9
7 + 0				9.3	34.8
8 + 0				10.9	33.2
9 + 0				10.7	33.4
BM 29			10.989		33.062
	13.812		21.062		

Check:

BM 23 elevation =	40.312
Add sum of backsights	+ 13.812
	54.124
Subtract sum of foresights	− 21.062
	33.062
BM 24 elevation =	− 33.050
Error =	+ 0.012

See Fig. 5-31 for plotted elevations.

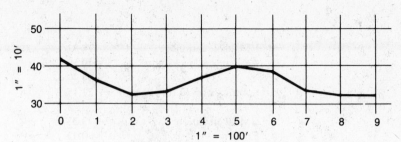

Fig. 5-31 Plotted elevations.

5.26 Given: Data in Table 5-4.

Table 5-4   Data for Prob. 5.26

Station, Elevation	Point Sighted	Rod
BM 28, 74.81	BM 27	11.39
	0 + 0	12.8
	1 + 0	8.3
	2 + 0	4.2
	3 + 0	3.2
	TP 1	7.50
	TP 1	3.29
	4 + 0	2.2
	5 + 0	8.0
	6 + 0	11.2
	TP 2	8.53
	TP 2	4.91
	7 + 0	7.3
	8 + 0	5.1
	9 + 0	3.7
BM 29, 73.23	BM 29	5.12

Solution

See Table 5-5.

Table 5-5   Solution for Prob. 5.26

Station	+	HI	−	Rod	Elevation
BM 28	11.39	86.20			74.81
0 + 0				12.8	73.4
1 + 0				8.3	77.9
2 + 0				4.2	82.0
3 + 0				3.2	83.0
TP 1	3.29	81.99	7.50		78.70
4 + 0				2.2	79.89
5 + 0				8.0	74.00
6 + 0				11.2	70.89
TP 2	4.91	78.37	8.53		73.46
7 + 0				7.3	71.17
8 + 0				5.1	73.37
9 + 0				3.7	74.77
BM 29			5.12		73.25
	19.59		21.15		

Check:

BM 28 elevation =	74.81
Add sum of backsights	+ 19.59
	94.40
Subtract sum of foresights	− 21.15
	73.25
BM 29 elevation	− 73.23
Error =	+ 0.02

See Fig. 5-32 for plotted elevations.

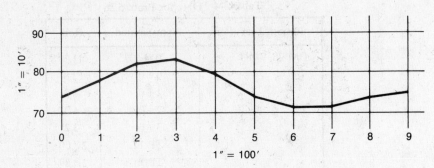

Fig. 5-32 Plotted elevations.

Chapter 6

Angle Measurement

6.1 GENERAL INFORMATION

Angles and directions are a fundamental part of surveying information. We must learn the various systems for measuring directions (horizontal angles, azimuths, bearings, and so forth) as well as the field procedures for making such measurements.

Surveying is the science of determining relative positions of points or objects on or near the earth's surface. From the study of geometry, it is evident that a point can be located by measuring only the distances from two known points. Surveyors find occasions when this two-distance method is very practical and even highly desirable. However, in many procedures which require locating a point, surveyors use a distance from one known point or sometimes only the directions (without any distances) from two known points. The discussion of angles and directions in this chapter is simplified in most instances to stay within the scope of plane surveying.

6.2 DIRECTION

In surveying and mapping, *direction* is the angular relationship of one line to another. The direction of a line is referred to some other definite line which acts as the zero value, and a dimension must be assigned to show the amount of change from the zero line.

A horizontal direction can be observed or measured in two ways—clockwise or counterclockwise. It is common practice in surveying to measure directions or angles clockwise (to the right) from the reference line, unless otherwise designated.

6.3 MEASURING ANGLES BY REPETITION

This chapter is concerned with measuring angles, and we may obtain more accuracy in our angle measurements by repeating the angle. Every surveyor must understand this procedure thoroughly. This method of repetition automatically eliminates, or substantially reduces, all residual errors in the geometry of the transit, with the exception of loose bearings. It also reduces the errors due to verniers that cannot be read precisely.

Procedure for Doubling an Angle

The summary steps for measuring a horizontal angle are repeated from Chap. 4 because of their importance here.

1. Free both motions and set the A vernier at zero with the upper motion (read the B vernier if necessary).
2. Point at the initial (left-hand) point, using the lower motion.
3. Point at the second point, using the upper motion.
4. Free the lower motion.
5. Read the A vernier (and the B vernier if necessary) clockwise.

Note that after you have turned the angle once and loosened the lower motion (step 4), the value of the angle is held at the A and B verniers. Nothing changes these values. The alidade may be turned but these readings remain. You may even turn the telescope upside down (reverse it); there is still no change.

It follows that, if the telescope is reversed and then aimed at the left-hand point *L*, the vernier will still give the value of the angle. So after step 5 has been completed, reverse the telescope and aim at *L* with the lower motion.

Now, aim the telescope at the second point with the upper motion. There will be added to the original reading a second value of the angle. This is shown schematically in Fig. 6-1. Reversing the telescope in this way eliminates many of the residual errors in the geometry of the transit.

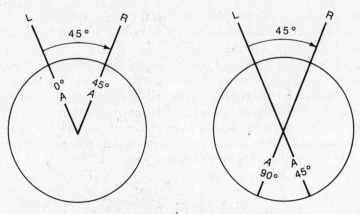

Fig. 6-1 Repeating an angle 1 DR.

Naturally you must now divide by 2 to obtain the desired angle. This cuts the reading errors in half.

The verniers are read after the first turn to give an approximate value of the angle. This is enough to give a rough check of the final angle computed.

The process of repeating an angle, with direction reversed (1 DR), is illustrated in Fig. 6-1. The *A* vernier starts at zero and is moved 45°. While the upper clamp is still tight, the telescope is reversed and pointed at *L* so that the *A* vernier still remains at 45° as in the figure on the right. When the angle is turned again, the *A* vernier moves to 90°.

EXAMPLE 6.1 See Fig. 6-2 and Table 6-1. Given: Initial pointing from *G* to *H* on *A* vernier is 0°00′00″; on *B* vernier, 180°00′40″. On pointing *G* to *I* the reading on the *A* vernier is 28°10′30″; on the *B* vernier it is 208°10′20″. What is the final angle?

Pointing		o	′	A″	B″	Ave″	Angles
				TRAVERSE ANGLES			
H	O.D.	0	00	00	40	20	
I	1.D.	28	10	30	20	25	28° 10′ 05″
I	1.D.R.	56	20	20	00	10	56° 09′ 50″
							28° 09′ 55″

D = Telescope Direct
R = Telescope Reversed

Obs. J. Doe   Date: 7/1/83
Record R. Roe   Clear
                                  75°F

Initial

÷ 2 = Final angle   28° 09′ 55″

Fig. 6-2 Field notes.

Table 6-1   Data for Example 6.1

Angle *HGI*: Telescope direct (D) and reversed (R)			
Pointing	A Vernier	B Vernier	Average
H (initial, or D) I (1 D)	0°00′00″ 28°10′30″	180°00′40″ 208°10′20″	0°00′20″ 28°10′25″
To compute angle, subtract initial average: 28°10′05″			
I (1 DR)	56°20′20″	236°20′00″	56°20′10″
To compute angle, subtract initial average: 56°19′50″ Divide by number of repetitions (2): 28°09′55″			

Solution: Write down the given data as in Table 6-1, noting the average for *H*: 00 + 40/2 = 20 seconds. Also the average for *I* is 30 + 20/2 = 25 seconds. Having an angle of 28°10′25″ you must, to complete the angle, subtract the initial average of 20 seconds, which gives 28°10′05″.

Now measure angle *I* with the direction reversed. The 20 seconds on the *A* vernier and 00 seconds on the *B* vernier average as follows: (20 + 00)/2 = 10 seconds, giving an angle of 56°20′10″. From this subtract the initial average of 20 seconds, which gives an angle of 56°19′50″. This angle now needs to be divided by 2 as we reversed the telescope and took two shots. So

$$\text{Final angle} = \frac{56°19′50″}{2} = 28°09′55″ \qquad Ans.$$

More Repetitions

More precision can be obtained by more repetitions. With the lower clamp released, repeat the procedure as many times as desired. Then divide the reading by the number of turns. The usual procedure is to turn the angle once, as in Example 6.1 and Probs. 6.1 to 6.4. The angle may, however, be turned once, twice, 6 times, or 12 times. The telescope is reversed after half the turns are completed. There is little advantage in repeating more than 12 times. Table 6-2 gives the names and symbols for these operations.

Table 6-2   Names and Symbols of Angle Operations

Number of Turns	Operation	Symbol
1	Once direct	1 D
2	Once direct and once reversed	1 DR
6	Three direct and three reversed	3 DR
12	Six direct and six reversed	6 DR

Steps in the Repeating Procedure

1. Free both motions.
2. Set the *A* vernier at zero with the upper motion. Read the *B* vernier.
3. Point at the left-hand point, using the lower motion.
4. Point at the second point, using the upper motion.
5. Free the lower motion and call out loud "one." If the number is called aloud, the number of turns will be remembered. The call should be made when the lower clamp is loosened.
6. Read the *A* and *B* verniers.

7. Reverse the telescope if the number called is half the number of turns to be used. In either case point at the left-hand point with the lower motion.

8. Repeat as desired, but do not read the vernier again until the total number of turns is completed.

EXAMPLE 6.2 Given: Angle BAC turned three times direct and three times reversed (3 DR). Readings of the transit are as follows:

AB:	A vernier = $0°00'00''$	B vernier = $180°00'15''$
AC:	A vernier = $126°32'45''$	B vernier = $306°32'55''$
C (3 DR):	A vernier = $39°15'50''$	B vernier = $218°15'50''$

Enter the data as they would appear in a field book. Determine the final angle.

Solution: See Fig. 6-3, where the data appear as required. In measuring an angle 3 DR, the circle may overrun several times. Therefore, a certain number of units of 360° should be added to the reading of the angle. Usually a certain number of units of 60° (which is $\frac{1}{6} \times 360°$) is added to one-sixth of the angle read. The number of units of 60° is chosen so as to make the final angle nearly equal to the value of the 1-D angle. In Fig. 6-3 for angle BAC:

$$\text{Angle read} = 39°15'42.5''$$

$$\begin{array}{lr} \text{Dividing by 6 turns} = & 6°32'37.1'' \\ \text{Add } 2(60°) & +120° \\ \hline & 126°32'37.1'' \quad \textit{Ans.} \end{array}$$

Check to see that the final angle is nearly equal to the 1-D angle of $126°32'42.5''$.

Note: Dividing the angle by 6 may be accomplished as follows:

$$\text{Angle to be divided} = 39°15'42.5''$$

1. Divide the degrees by 6, and use the remainder as the first digit of the minutes in the quotient. Thus: $39/6 = 6$ with a remainder of 3. The 6 is the number of degrees, and the remainder 3 is the first digit of the number of minutes. We now have $6°3?'$.

2. Divide the minutes by 6, using the quotient as the second digit of the minutes and the remainder as the first digit of the number of seconds. Thus: $15/6 = 2$ minutes with 3 as a remainder. We now have $6°32'3?''$.

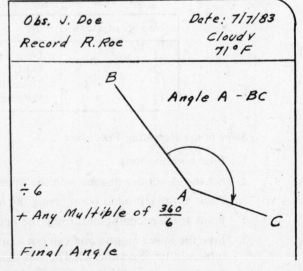

ANGLE at STATION A							Obs. J. Doe	Date: 7/7/83
Pointing	o	′	A″	B″	Ave″	Angle	Record R. Roe	Cloudy 71°F
B O.D.	0	00	00	15	7.5			
C 1.D	126	32	45	55	50	126° 32′ 42.5″		
C 3.D.R.	39	15	50	50	50	39° 15′ 42.5″	÷ 6	
						6° 32′ 37.1″	+ Any Multiple of $\frac{360}{6}$	
						+ 120		
						126° 32′ 37.1″	Final Angle	

Angle A – BC

Fig. 6-3 Field notes.

3. Divide the seconds by 6, using the result to fill out the seconds. Thus 42.5/6 = 7.1 seconds. Add to the amount already obtained. Thus the final angle = 6°32′37.1″.

When repeating an angle, always spot-check for blunders. The final angle should agree with the 1-D value within about 15 seconds, no matter how many times the angle is turned. The results of a 3-DR value can be relied upon to within 3 to 6 seconds of the true value.

6.4 PROLONGING A LINE PAST AN OBSTACLE BY ANGLES

Often a building, telephone pole, trees, and many other objects block survey lines. Two methods of extending lines past obstacles by the use of angles are (1) the equilateral-triangle method, and (2) the right-angle-offset method.

Equilateral-Triangle Method

See Fig. 6-4a. At point in the figure, a 120° angle is turned off from a backsight on *J*, and a distance *KL* of 70.00 ft (or any distance necessary, but not more than one tape length) is measured to locate point *L*. The transit is then moved to *L*; a backsight is taken on *K*; an angle of 60°00′ is put on the plates of the transit; and a distance *LM* = *KL* = 70.00 ft is laid off to mark point *M*. The transit is moved to *M*, backsighted on *L*, and an angle of 120° turned. The line of sight *MN* is now along *JK* prolonged if error-free work has been done. It might be well to measure the angles 1 DR.

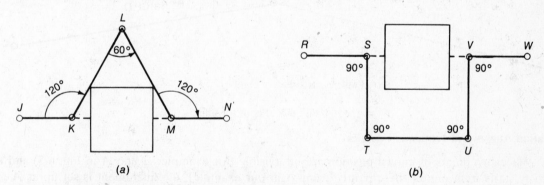

Fig. 6-4 Prolonging a line past an obstacle. (*From R. C. Brinker and P. R. Wolf, Elementary Surveying, IEP Publishing, Dun-Donnelley, New York, 1977*)

Right-Angle-Offset Method

Set the transit up at points *S, T, U, V* (Fig. 6-4b; 90° angles should be turned off at each hub. Distances *TU* and *ST*, which is equal to *UV*, need only clear the obstacle, but if possible try to provide longer lengths which increase the accuracy of sighting.

EXAMPLE 6.3 See Fig. 6-4. Given: Line *JK* in Fig. 6-4a and line *RS* in Fig. 6-4b, which are obstructed by a 40-ft square building. Prolong the lines *JK* and *RS* by the equilateral-triangle method and the right-angle-offset method.

Solution: As shown in Fig. 6-4a line *JK* is stopped short of the obstacle at *K*, and a 120° angle is turned and a 60.00-ft distance is laid out to establish point *L*. From here a 60° interior angle is set on the transit and another 60.00 ft is measured to establish point *M*. Here turn 120° and line *MN* is line *JK* prolonged.

Figure 6-4b adequately depicts the right-angle-offset method of passing an obstacle.

6.5 KINDS OF HORIZONTAL ANGLES

An angle can be defined as the difference in the values of two directions observed from the same initial direction. At times, the directions to several survey features must be observed from one instrument setup, or the changing directions of a series of connected lines must be determined. The series of lines might represent a boundary which starts from one point, extends around a specific area, and returns (ties in) to the starting point. For example, the lines might represent a road centerline which starts at one point and extends (traverses) to another location or point some distance away. Direction and angle equipment measure the changes of direction (angles) between the lines in the series. In some types of surveys, only selected angles are observed.

To distinguish the different ways the angles are observed in traverse and construction surveys, they are given special names, such as station angles, explement angles, and deflection angles (see Fig. 6-5). In triangulation, only the differences in directions are observed and used as angles in the computations.

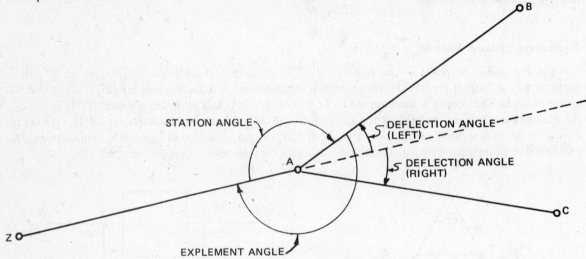

Fig. 6-5 Setup for angles.

Station Angles and Explement Angles

When two points of known position are intervisible (for example, Z and A in Fig. 6-5) and the survey starts from one of these points (from A in our example), the instrument is set up at A and backsighted on Z as the zero direction. The angle is turned to point B, the first point of the new line. As discussed, the angle is normally turned and read clockwise (to the right). This angle is referred to as the *station angle*. Generally, surveys require at least one direct and one reverse pointing on each station to observe the angle. A direct pointing is made with the telescope in the direct position. A reverse pointing is made with the telescope in the reverse position (plunged). The direct/reverse (D/R) measurement is made to eliminate the collimation error in the instrument.

For a complete direct/reverse reading, the station angle is read; then the telescope is reversed and pointed on the forward point (B in Fig. 6-5) and read. The angle to the starting point (Z) is turned to the right and read; this angle is called the *explement angle* and should equal 360° minus the station angle.

EXAMPLE 6.4 Given: A station angle of 150°58′20″ in Fig. 6-5. Find the explement angle for this station angle.

Solution: The theoretical sum of the station angle and explement angle always equals 360°. So subtract the station angle from 360° to obtain the explement angle.

$$
\begin{array}{l}
359°59′60″ \quad (360°)\\
\underline{-\ 150°58′20″}\\
209°01′40″ = \text{explement angle} \qquad Ans.
\end{array}
$$

Deflection Angles

Some surveys use deflection angles in their computations. These can be computed from the station or explement angles, or they can be turned directly by using a transit or 1-minute theodolite. The instrument is set up at A (Fig. 6-5), and a backsight is made on Z as is done for station angles. The telescope is plunged first instead of being rotated horizontally. The angle to B is turned to the left. This is called a *left deflection angle*; it is recorded with an L or minus sign in front of the angular value. Figure 6-5 shows a left deflection angle to B and a *right deflection angle* to C, if C were the required point.

Right deflection angles are recorded with an R or plus sign in front of the angular value. When turning and reading deflection angles, special care must be taken to distinguish left and right values and to record them properly.

Interior and Exterior Angles

Some surveys follow the border of a figure or area and close (tie in) to the starting point. The angles which are inside the figure are referred to as the *interior angles*, while their explements are called *exterior angles*. Depending on the direction in which the survey is run, either the interior or exterior angles may be read as the direct or station angles. The other angles (exterior or interior) are then explement angles. Interior and exterior angles are not recorded in surveys; angles are read and recorded as direct and reverse directions.

6.6 CLOSING THE HORIZON

If you wish to check the sum of the angles around a point, you may do so by a procedure known as *closing the horizon*. This simply means that all the angles around a point should add up to 360°00′00″.

EXAMPLE 6.5 Given: Angles as shown in Fig. 6-6. At B the plate reads 7°26′ and the magnetic bearing is N33°38′E. At C the plate reads 49°38′ and the magnetic bearing is N75°53′E. At D the plate reads 99°04′ and the magnetic bearing is S54°41′E. Close the horizon and state the closure error.

Solution: The problem is solved in Table 6-3; the closure error is 0°00′.

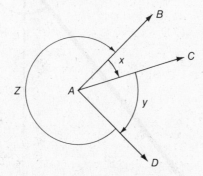

Fig. 6-6   Closing the horizon. (*From R. C. Brinker and P. R. Wolf, Elementary Surveying, IEP Publishing, Dun-Donnelly, New York, 1977*)

6.7 AZIMUTH

One way to describe directions in surveying is by using azimuths. The *azimuth* of a line is the horizontal direction measured clockwise from a zero direction which points north from the station occupied. Every line has two azimuths, depending on the observer's position. For example, in Fig. 6-7 a survey is progressing from A toward B. Angle a is the *forward azimuth* for this line. To designate

Table 6-3   Field Notes: Closing the Horizon (Example 6.5)

Point Sighted	Plate Reading	Angle	Magnetic Bearing	Angle Computed from Bearing
Transit at point *A*				
B	7°26′		N33°38′E	
C	49°38′	42°12′		42°15′
C	49°38′		N75°53′E	
D	99°04′	49°26′		49°26′
D	99°04′		S54°41′E	
B	7°26′	268°22′		268°19′
		360°00′		

Closure = 0°00′

the direction from *B* to *A*, angle *b* is used; angle *b* is known as the *back azimuth* of the line. In plane surveying, the forward and back azimuths of a line always differ by exactly 180°. The zero azimuth line can be based on *true*, *grid*, or *magnetic* north. Forward azimuths are converted to back azimuths, and vice versa, by adding or subtracting 180°.

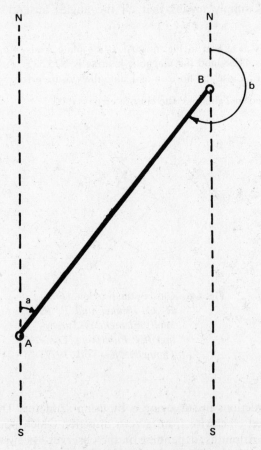

Fig. 6-7   Azimuth survey angles.

For example, if the azimuth in Fig. 6-8 which is labeled 45° were to be taken backward, it would be 45° + 180°, or 225°. Also if the azimuth for 217°43′ were reversed, it would be 217°43′ − 180° = 37°43′.

Azimuths are measured from north or south, and the angles vary from 0 to 360° (see the inner circle of Fig. 6-8).

Informational note: For convenience in using tables, hand-held calculators can accept values from 0 to 360°.

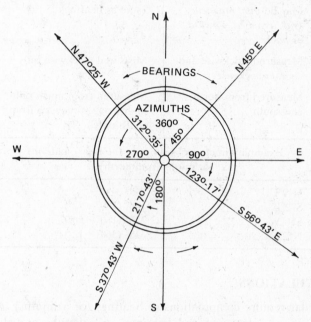

Fig. 6-8　Bearings and azimuths.

6.8 BEARINGS

The use of bearings in measurements automatically keeps the angles below 90° (see the outer circle of Fig. 6-8).

The *bearing* of a line is its direction (within a quadrant) with reference to a meridian (i.e., north or south line). Bearings are measured clockwise or counterclockwise, depending on the quadrant, from either the north or the south line. A bearing is identified by first naming the end of the meridian from which it is reckoned (north or south), then the angle value, and finally the direction (east or west) from the meridian. For example, a line in the southwest quadrant making an angle of 37°43′ with the south reference meridian has a bearing of S37°43′W. An angle in the northwest quadrant 47°25′ from the north meridian has a bearing of N47°25′W. Bearings never exceed 90°.

Bearings, like azimuths, can be *true*, *grid*, or *magnetic*, depending on the reference meridian. If the meridian is a true north-south line, the bearing is *true*. Grid north-south lines as reference result in grid bearings, and using magnetic north or south gives magnetic bearings.

The compass included with the transit and the 1-minute theodolite can be used to read magnetic bearings.

6.9 COMPARISON OF BEARINGS AND AZIMUTHS

Bearings and azimuths are encountered in most surveying operations. Table 6-4 shows a comparative summary of their properties. Bearings can easily be computed from azimuths by noting the quadrant in which the azimuth falls. You can then convert the bearing to an azimuth as shown in Table 6-4.

Table 6-4 Explanation of Bearings and Azimuths

Bearings	Azimuths
Vary from 0 to 90°	Vary from 0 to 360°
Require two letters and a numerical value	Require only a numerical value
May be *true*, *magnetic*, *grid*, *assumed*, *forward*, or *back*	Same as bearings
Measured clockwise and counterclockwise	Measured clockwise only
Measured from north and south	Measured from north only in any one survey, or from south only
Example directions for lines in the four quadrants (azimuths from north):	
N49°E	49°
S72°E	108° (180° − 72°)
S56°W	236° (180° + 56°)
N12°W	348° (360° − 12°)

6.10 BEARING CALCULATIONS

Traverses in particular require computations of bearings (or azimuths). A traverse consists of a series of angles and distances, distances and bearings, or azimuths and distances which connect successive instrument points. The boundary lines of property form a closed polygon, or a *closed traverse*. Highway surveys from city to city are normally *open traverses*.

All computations on bearing lines will be simplified if a sketch is made for each line showing all the data. Figure 6-9 is a closed polygon with counterclockwise interior angles turned to the left. Interior angles are usually turned clockwise and are called *angles to the right*. It is recommended that surveys be made with angles to the right to prevent the confusion that might result if angles were turned both right and left.

EXAMPLE 6.6 See Figs. 6-9 and 6-10. Given: Line *AB* bearing N45°35′E. Angle at *B* turned left (counterclockwise) of 127°11′. Find the bearing of line *BC*.

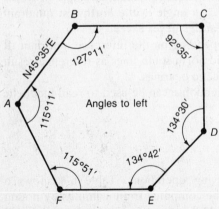

Fig. 6-9 Counterclockwise interior angles on a closed polygon.

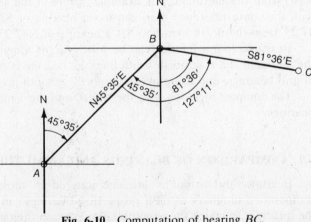

Fig. 6-10 Computation of bearing *BC*.

Solution: Draw a sketch including all known and calculated data (Fig. 6-10). The best way to draw the sketch is to use a protractor and put all angles in to scale. This will catch any mistakes you may make, as the scaled sketch should put you into the proper quadrant and allow you to assign the proper directions to the bearings.

$$45°35' \text{ at point } A = \text{angle at } B$$

So: Angle $ABC - 45°35' = 81°36'$

This is the angle turned from a north-south line so by adding the directions, we find:

$$\text{Bearing of line } BC = S81°36'E$$

6.11 COMPUTING AZIMUTHS

Azimuths are easier to work with than bearings. Because sines and cosines of azimuth angles automatically provide correct algebraic signs for latitude and departures (see Chap. 7), many surveyors prefer their use in computing traverses.

As with bearings a sketch should be made for azimuth calculations.

EXAMPLE 6.7   See Figs. 6-9 and 6-11. Find azimuths from north of AB and BA, BC.

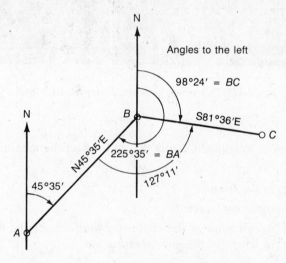

Fig. 6-11   Computation of azimuth of BC.

Solution: Draw a sketch (Fig. 6-11). Azimuth BA is obtained by adding 180° to azimuth AB. Then counterclockwise angle B, 127°11′, is subtracted from azimuth BA to get azimuth BC.

$$
\begin{aligned}
&+\ \ 45°35' = AB \\
&+180°00' \\
\hline
&\ \ 225°35' = \text{azimuth of } BA \qquad Ans.
\end{aligned}
$$

$$
\begin{aligned}
&\ \ 225°35' = \text{azimuth of } BA \\
&-127°11' \\
\hline
&\ \ \ \ 98°24' = \text{azimuth of } BC \qquad Ans.
\end{aligned}
$$

Figure 6-12 will give the student a good picture of the difference between a level and a transit. We are using a transit for all the work in this chapter.

6.12 FACTORS AFFECTING ANGLE OBSERVATIONS

There are three types of errors that affect angle observations. These errors may be due to the instrument, the observer, or the environment. The surveyor should try to reduce these errors or avoid them completely.

Fig. 6-12 Difference between a transit and a level.

Instrument Errors

There are several types of instrument errors that may affect the measured value for a horizontal angle:

1. Adjustment errors in the transit
2. Eccentricity of the horizontal circle
3. Small errors in the graduation of the circle, the verniers, or the micrometer scale; and an error in the apparent length of the micrometer scale

Except for the last error, the effect of all instrument errors can be eliminated or minimized by proper adjustment of the instrument and by using systematic observing procedures.

Personnel Errors

There are several personnel errors that affect horizontal angle measurements:

1. Errors in centering the instrument and targets over their stations
2. Errors in leveling the instrument
3. Errors in pointing the instrument
4. Errors in reading the circle and verniers
5. Errors in judging coincidence for readings

These errors can all be minimized by using correct procedures and by proper training.

Environmental Errors

Environmental errors affecting horizontal angle measurements are due to differential temperatures within the instrument, horizontal refraction of the line of sight, and phase.

6.13 COMMON MISTAKES

Common mistakes in angle measurements are as follows:

1. Mixing bearings and azimuths
2. Adopting an assumed reference line which is difficult to reproduce
3. Orienting an instrument by resighting on magnetic north
4. Confusing magnetic and true bearings
5. Not including the last angle to recompute the starting bearing as a check

Solved Problems

6.1 Given: Angle *BAC*; see Fig. 6-13 and Table 6-2 for reference. With shot *AB* the *A* vernier is 0°00′00″, and the *B* vernier is 180°00′00″. The telescope is used once direct (1 D) and once reversed (1 DR). In shot *AC* the *A* vernier is 30°10′30″ and the *B* vernier is 210°10′20″. For the 1-DR reading the *A* vernier is 60°20′25″ and the *B* vernier is 240°20′10″. Compute the final angle.

Solution

Put the information into table form, computing the average for each reading.

Point	*A* Vernier	*B* Vernier	Average	Angle	Explanation
B (D)	0°00′00″	180°00′30″	0°00′15″		
C (1 D)	30°10′30″	210°10′20″	30°10′25″	30°10′10″	Subtract initial average
C (1 DR)	60°20′25″	240°20′10″	60°20′17.5″	60°20′02.5″	Subtract initial average

Divide the angle by the number of repetitions to obtain the final angle:

$$\text{Final angle} = \frac{\text{1-DR angle}}{\text{number of repetitions}} = \frac{60°20′02.5″}{2} = 30°10′01″ \qquad Ans.$$

Compare the answer with the 1-D angle of 30°10′10″.

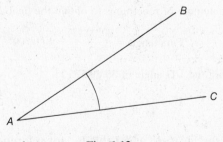

Fig. 6-13

6.2 Given: Angle *BAC*; see Fig. 6-13 and Table 6-2. With shot *AB*, the *A* vernier reads 0°00′00″ and the *B* vernier reads 180°00′40″. The telescope is used once direct and once reversed (1 DR). With shot *AC* the *A* vernier reads 36°20′30″ and the *B* vernier reads 216°20′20″. For the 1-DR reading the *A* vernier is 72°40′58″ and the *B* vernier is 252°40′10″. Compute the final angle.

Solution

Put the information into table form, computing the average for each reading.

Point	A Vernier	B Vernier	Average	Angle	Explanation
B (D)	0°00′00″	180°00′40″	0°00′20″		
C (1 D)	36°20′30″	216°20′20″	36°20′25″	36°20′05″	Subtract initial average
C (1 DR)	72°40′58″	252°40′10″	72°40′34″	72°40′14″	Subtract initial average

Divide the angle by the number of repetitions to obtain the final angle:

$$\text{Final angle} = \frac{72°40′14″}{2} = 36°20′07″ \qquad Ans.$$

Compare the answer with the 1-D angle of 36°20′05″.

6.3   Given: Angle BAC; see Fig. 6-13 and Table 6-2 for reference. Shot AB: A vernier = 0°00′00″; B vernier = 180°00′36″. Shot AC: A vernier = 40°00′20″; B vernier = 220°00′10″. The telescope is used once direct and once reversed. The 1-DR reading is as follows: A vernier = 80°00′40″; B vernier = 260°00′30″. Compute the final angle.

Solution

Put the information into table form, computing the average for each reading.

Point	A Vernier	B Vernier	Average	Angle	Explanation
B (D)	0°00′00″	180°00′36″	0°10′18″		
C (1 D)	40°00′20″	220°00′10″	40°00′15″	39°59′57″	Subtract initial average
C (1 DR)	80°00′40″	260°00′30″	80°00′35″	80°00′17″	Subtract initial average

Divide the angle by the number of repetitions to obtain the final angle:

$$\text{Final angle} = \frac{80°00′17″}{2} = 40°00′09″ \qquad Ans.$$

Compare the answer with the 1-D angle of 39°59′57″.

6.4   Given: Angle BAC; see Fig. 6-13 and Table 6-2 for reference. Shot AB: A vernier = 0°00′00″; B vernier = 180°00′18″. Shot AC: A vernier = 60°00′30″; B vernier = 240°00′10″. The telescope is used once direct and once reversed. The 1-DR reading is as follows: A vernier = 120°00′58″; B vernier = 120°00′54″. Compute the final angle.

Solution

Put the information into table form, computing the average for each reading.

Point	A Vernier	B Vernier	Average	Angle	Explanation
B (D)	$0°00'00''$	$180°00'18''$	$0°00'09''$		
C (1 D)	$60°00'30''$	$240°00'10''$	$60°00'20''$	$60°00'11''$	Subtract initial average
C (1 DR)	$120°00'58''$	$300°00'50''$	$120°00'54''$	$120°00'45''$	Subtract initial average

$$\text{Final angle} = \frac{120°00'45''}{2} = 60°00'22.5'' \qquad Ans.$$

Compare the answer with the 1-D angle of $60°00'11''$.

6.5   Given: Angle $BAC = 144°01'58''$ for the calculated angle for the 1-D reading [this shows in line C (1 D) in the field notes]. Line C (3 DR) gives a reading of $144°11'44''$. Work out angle BAC for the transit reading (turned six times).

Solution

Divide $144°11'44''$ (the 3-DR reading) by 6.

1. $144/6 = 24°$ with 0 to carry to minutes.
2. $11/6 = 1$ minute with 5 remainder to carry to seconds.
3. $44/6 = 7.3$ seconds. Precede by the remainder of 5 from step 2 to get 57.3 seconds.

So: Angle $= 24°01'57.3''$

Finally, add $2(60) = 120°$ to the above angle.

Angle obtained by six turns $= 120° + 24°01'57.3'' = 144°01'57.3'' \qquad Ans.$

This compares closely with the 1-D reading of $144°01'58''$.

6.6   Given: Angle $BAC = 156°01'25''$ for the calculated angle for the 1-D reading [this shows in line C (1 D) in the field notes]. Line C (3 DR) gives a reading of $216°07'58''$. Work out angle BAC for the transit reading (turned six times).

Solution

Divide $216°07'58''$ (the 3-DR reading) by 6.

1. $216/6 = 36°$ with 0 remainder for the minutes reading.
2. $07/6 = 1$ minute with 1 remainder for the seconds reading.
3. $58/6 = 9.7$ for the continuation of the seconds. Precede by the remainder of 1 from step 2 to get 19.7 seconds.

So: Angle $= 36°01'19.7''$

Add $2(60) = 120°$ to get: $120° + 36°01'19.7'' = 156°01'19.7''$

Angle obtained by six turns $= 156°01'19.7'' \qquad Ans.$

This compares favorably with the 1-D reading of $156°01'25''$.

6.7   Given: A 1-D calculated angle BAC of $164°01'28''$. Line C (3 DR) in the field book shows a calculated angle of $264°10'10''$. Work out angle BAC for the 3-DR transit turning.

Solution

Divide 264°10'10" by 6.

1. 264/6 = 44° with 0 to carry to minutes.
2. 10/6 = 1 minute with 4 to carry to seconds.
3. 10/6 = 1.7 seconds to add to the 4 carried from step 2.

So: Angle = 44°01'41.7"

$$44°01'41.7'' + 2(60) = 44°01'41.7'' + 120° = 164°01'41.7''   \quad Ans.$$

This compares favorably with the 1-D reading of 164°01'28".

6.8   Given: A 1-D calculated angle *BAC* of 190°06'46". Line *C* (3 DR) in the field book shows a calculated angle of 60°40'20". Work out angle *BAC* for the 3-DR transit turning.

Solution

Divide 60°40'20" by 6.

1. 60/6 = 10° with 0 carried to minutes.
2. 40/6 = 6 minutes with 4 carried to seconds.
3. 20/6 = 3.3 seconds to add to 4 carried from step 2.

So: Angle = 10°06'43.3"

$$10°06'43.3'' + 3(60) = 10°06'43.3'' + 180° = 190°06'43.3''   \quad Ans.$$

This compares favorably with the 1-D reading of 190°06'46".

6.9   Given: A 1-D calculated angle *BAC* of 130°11'01". Line *C* (3 DR) in the field book shows a calculated angle of 61°06'08". Work out angle *BAC* for the 3-DR transit turning.

Solution

Divide 61°06'08" by 6.

1. 61/6 = 10° with remainder 1 to carry to the minutes column.
2. 06/6 = 1 minute to add to 1 carried from step 1, and 0 to carry to the seconds column. (So minutes = 11.)
3. 08/6 = 1.3 seconds to finish out the seconds column.

So: Angle = 10°11'01.3"

$$10°11'01.3'' + 2(60) = 10°11'01.3'' + 120° = 130°11'01.3''   \quad Ans.$$

This compares favorably with the 1-D reading of 130°11'01".

6.10   Given: A station angle of 39°37'25" (see Fig. 6-5). Find the explement angle for this station.

Solution

$$\text{Station angle} + \text{explement angle} = 360°$$

$$
\begin{array}{r}
359°59'60'' \\
- \quad 39°37'25'' \\
\hline
320°22'35'' = \text{explement angle} \qquad Ans.
\end{array}
$$

6.11   Given: Station angle of 127°56′59″. Find the explement angle.

> **Solution**   Station angle + explement angle = 360°
>
> $$\begin{array}{r} 359°59′60″ \\ -\;127°56′59″ \\ \hline 232°03′01″ = \text{explement angle} \qquad \textit{Ans.} \end{array}$$

6.12   Given: A transit stationed at A; the sighting of the transit pointing to B is 10°20′ (plate reading). At this point we have a magnetic bearing of N20°30′E. We turn an angle of 40°40′, giving a plate reading pointing toward C of 51°00′. Here our magnetic bearing reads N61°23′E. Close the horizon for this station.

> **Solution**
>
> See the accompanying table.

Point Sighted	Plate Reading	Angle	Computed Magnetic Bearing	Angle Computed from Bearing
Transit at point A				
B	10°20′		N20°30′E	
C	51°00′	40°40′	N61°23′E	40°53′
B	10°22′	319°22′	N20°30′E	
		359°62′		

Since the final angle = 359°62′ = 360°02′,

$$\text{Closure} = 360°02′ - 360° = 0°02′$$

Thus the horizon closure angle was 319°22′ and the error of closure was 0°02′.

6.13   Given: Transit at A. Plate reading as transit points to B is 0°25′, magnetic bearing is N0°59′E. The transit is turned to point to C. The plate reads 75°25′, and the magnetic bearing at this pointing is N76°20′E. Close the horizon for this station.

> **Solution**
>
> See the accompanying table.

Point Sighted	Plate Reading	Angle	Computed Magnetic Bearing	Angle Computed from Bearing
Transit at point A				
B	0°25′		N0°59′E	
C	75°25′	75°00′	N76°20′E	75°21′
B	0°25′	285°00′	N0°59′E	
		360°00′		

Since the final angle is 360°,

$$\text{Closure} = 00°00′$$

6.14   Given: Line *BC* bearing S81°36′E. Angle turned left at *C* (counterclockwise) = 92°35′. Find the bearing of line *CD*.

Solution

First draw a sketch (Fig. 6-14; refer to Fig. 6-9), and include all given and calculated angles. It can be seen from the sketch that subtracting the sum of 81°36′ + 92°35′ from the 180° of the north-south line at *C* will give the bearing angle off the north-south line. So:

$$180°00′ - (81°36′ + 92°35′) = 05°49′$$

Add the south and west directions to the angle for a bearing on line *CD* of S05°49′W.   *Ans.*

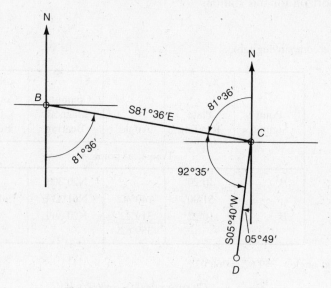

Fig. 6-14 Computation of bearing *CD*.

6.15   Given: Line *CD* bearing S05°49′W and an angle turned left at *D* of 134°30′. Find the bearing of line *DE*.

Solution

Draw a sketch (Fig. 6-15; refer to Fig. 6-9). It can be seen that 134°30′ − 05°49′ = 128°41′. This is the bearing of the north-south line but it needs to be substracted from the 180° angle of a straight line with *D* as its center. So:

$$180° = \quad\quad 179°60′$$
$$\underline{- \ 128°41′}$$
$$51°19′$$

Thus since line *DE* is in the southwest quadrant,

Bearing of line *DE* = S51°19′W   *Ans.*

6.16   Given: Line *DE* bearing S51°19′W; angle *DEF* = 134°42′ turned left at *E*. Find the bearing of line *EF*.

Solution

Draw a sketch (Fig. 6-16; refer to Fig. 6-9). Computations for the angles:

$$134°42′ \text{ turned left}$$
$$\underline{- \ 51°19′ \text{ off N-S line}}$$
$$83°23′$$

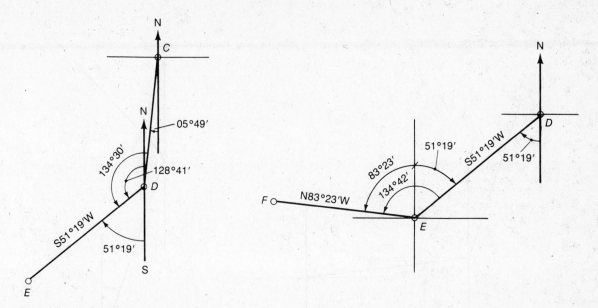

Fig. 6-15 Computation of bearing *DE*.   **Fig. 6-16** Computation of bearing *EF*.

Thus since line *EF* is in the northwest quadrant,

Bearing of line *EF* = N83°23′W   *Ans.*

6.17 Given: Line *EF* bearing N83°23′W and an angle of 115°51′ turned left at *F*. Find the bearing of line *FA*.

Solution

Draw a sketch (Fig. 6-17; refer to Fig. 6-9).

$$
\begin{array}{r}
89°60′ \\
-\ 83°23′ \\
\hline
06°37′
\end{array}
$$

It can be seen from the sketch that a 90° quadrant angle should be added to 06°37′. So:

$$
\begin{array}{r}
90° \\
+\ 06°37′ \\
\hline
96°37′
\end{array}
$$

This angle should be subtracted from the 115°51′ turned left at *F*:

$$
\begin{array}{r}
115°51′ \\
-\ 96°37′ \\
\hline
19°14′
\end{array}
$$

Since line *FA* is in the northwest quadrant,

Bearing of line *FA* = 19°14′W   *Ans.*

6.18 Given: Line *FA* bearing N19°14′W and angle turned left at *A* = 115°11′. Find the bearing of line *AB*.

Solution

Draw a sketch (Fig. 6-18; refer to Fig. 6-9).

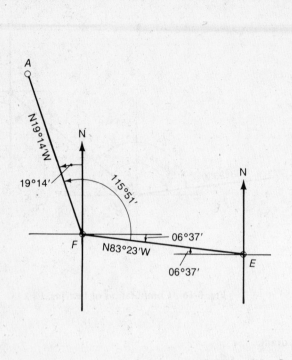

Fig. 6-17 Computation of bearing *FA*.   **Fig. 6-18** Computation of bearing *AB*.

To find the bearing off the N-S line, first add:

$$\begin{array}{r} 19°14' \\ + 115°11' \\ \hline 134°25' \end{array}$$

This angle is subtracted from the 180° of the straight line with center at *A*:

$$\begin{array}{r} 179°60' \\ - 134°25' \text{ line } AB \\ \hline 45°35' \end{array}$$

Since line *AB* is in the northeast quadrant,

          Bearing of line *AB* = N45°35′E   *Ans.*

Note: This angle is the same as the first traverse of line *AB* in Fig. 6-9. This proves that all the bearings found in Probs. 6.14 to 6.18 are correct.

6.19   Given: Azimuth *BC* = 98°24′ to the left, and angle *C* = 92°35′ to the left. (Use Fig. 6-9 as a reference.) Find azimuths from north of *CB* and *CD*, *DC*.

Solution

      Draw a sketch (not shown in text).

BC =	98°24′
Add 180°	+ 180°00′
Azimuth of *CB* =	278°24′
Subtract angle *C*	− 92°35′
Azimuth of *CD* =	185°49′

6.20 Given: CD with azimuth of 185°49′; angle at D is 134°30′ left. Find azimuths from north of lines DC and DE.

Solution

See Fig. 6-9.

Azimuth of CD =	185°49′
	+ 180°00′
Azimuth of DC =	365°49′
Subtract angle D	− 134°30′
Azimuth of DE =	231°19′

6.21 Given: DE with azimuth of 231°19′. Angle at E is 134°42′ left. Find azimuths from north of lines ED and EF.

Solution

See Fig. 6-9.

Azimuth of DE =	231°19′	
	+ 180°00′	
Azimuth of ED =	411°19′	
Subtract angle E	− 134°42′	
Azimuth of EF =	276°37′	(Checks with bearing of EF)

6.22 Given: EF with azimuth of 276°37′ and angle at F = 115°51′ left. Find azimuths from north of lines FE and FA.

Solution

See Fig. 6-9.

Azimuth of EF =	276°37′	
	+ 180°00′	
Azimuth of FE =	456°37′	
Subtract angle F	− 115°51′	
Azimuth of FA =	360°46′	(Checks with bearing of FA)

6.23 Given: Azimuth of line FA = 340°46′ and angle at A = 115°11′. Find azimuths from north of lines AF and AB.

Solution

See Fig. 6-9.

Azimuth of FA =	340°46′
	− 180°00′
Azimuth of AF =	160°46′
Subtract angle A	− 115°11′
Azimuth of AB =	45°35′

This checks with the starting azimuth.

Supplementary Problems

6.24   Given: A 1-D calculated angle *BAC* of 143°27'22". Line *C* (3 DR) in the field book shows a calculated angle of 140°44'44". Work out angle *BAC* for the 3-DR transit turning.   *Ans.*   143°27'27.3"

6.25   Given: Explement angle of 212°12'35". Find the station angle.   *Ans.*   147°47'25"

6.26   Given: Station angle of 180°07'28". Find the explement angle.   *Ans.*   179°52'32"

6.27   Given: Station angle of 193°56'57". Find the explement angle.   *Ans.*   166°03'03"

Chapter 7

Traverses

7.1 DEFINITIONS

A *traverse* is a continuous series of lines called *courses* whose lengths have been determined by field measurements. The courses run between a series of points called *traverse stations*. Traverses can be open or closed. Open traverses (Fig. 7-1) end without closure. Open traverses are used on route surveys, but are to be avoided, if possible, because they cannot be properly checked. With open traverses, measurements should be repeated to guard against mistakes.

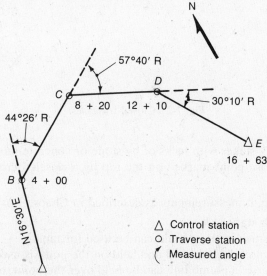

Fig. 7-1 Open traverse.

Closed traverses are of two kinds:

1. *Loop traverses*, which close on themselves (see Fig. 7-2a).
2. *Connecting traverses*, which begin at a known direction and position and end at another known position and direction (see Fig. 7-2b).

Both the angles and the measured lengths in a closed traverse may be checked.

7.2 USE OF TRAVERSES

Traverses are used to find the accurate positions of a small number of marked stations. From these stations, many less precise measurements can be made to features to be located, without accumulating accidental errors. Thus traverses usually serve as control surveys. When drawing construction plans, the stations can be used as beginning points from which to lay out work. When new construction of any kind is to be made, a system of traverse stations in the area must be established and surveyed. Efforts to avoid the setup of a traverse are costly and often necessitate serious revisions of the plans.

7.3 TRAVERSE FIELD WORK

Traverse field work may be described by a series of steps as follows:

1. Choose positions for stations as close as possible to the objects to be located.

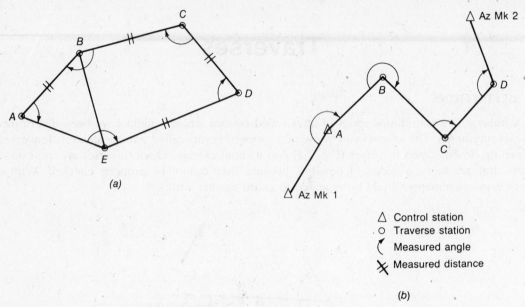

Fig. 7-2 Closed traverses.

2. Mark the stations by stakes with tacks or by stone or concrete monuments set flush with the ground, with a precise point marked on the top by a chiseled cross, drilled hole, or bronze tablet.

3. Make angle and length measurements as described in Chaps. 3 to 6.

4. Place signals at each station.
   (*a*) A range pole stuck in the ground can be used for taping.
   (*b*) A range pole carefully balanced and held on the point is used for measuring angles.
   (*c*) For a short course, a plumb bob can be held over the point, or a pencil can be balanced on the point for angle measurements.

Forward Direction

To be consistent, it is necessary to assume which are the forward and backward directions for any traverse. The order in which measurements are made usually is called the forward direction. Loop traverses should be measured counterclockwise around the loop.

Direction of Angle Measurement

The angles of a traverse should be measured clockwise from the backward direction to the forward direction. In highway surveys and in other connecting or open traverses, measuring deflection angles is the method of measurement. A *deflection angle* is the angle between the forward prolongation of the back course and the forward course (see Fig. 7-3). A deflection angle can also be described as the change of direction of the traverse at a station. You must properly record the direction of the deflection angles as left or right; otherwise a mistake will result.

Measuring Deflection Angles

Measure the total angle from the back line to the forward line and subtract 180° from the result. If the difference is positive, the angle is a right deflection angle. If the difference is negative, the angle is a left deflection angle. Some recommendations for measuring deflection angles are as follows:

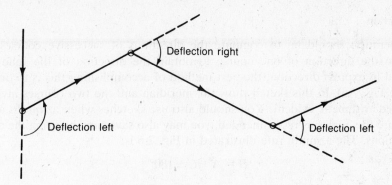

Fig. 7-3 Direction of angle measurement.

1. Set the vernier to zero.

2. Take the backsight with the telescope reversed, using the lower motion.

3. Take the foresight with the telescope direct, using the upper motion.

4. Record the clockwise or counterclockwise angle, whichever is less than 180°. If clockwise, it is right deflection; if counterclockwise, left.

5. To achieve a 1-DR reading leave the vernier as it is after step 4 and take the backsight with the telescope direct, using the lower motion.

6. Take the foresight with the telescope reversed, using the upper motion.

7.4 TRAVERSE COMPUTATIONS

It is generally necessary to reduce the field data of the traverse to the form of rectangular coordinates of the stations. The examples and problems in this chapter show how this is accomplished. A sketch of the traverse must be drawn to scale. Show the names of each of the traverse stations (see Fig. 7-4). This sketch will help you check for any blunders in your computations.

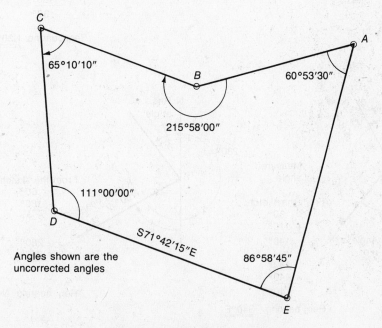

Fig. 7-4 Traverse computation.

Computing Direction

The first operation should be to compute the directions of successive courses by applying a traverse angle to the direction of one course to obtain the direction of the following course. If bearings are used to express direction, the best method of accomplishing this is to draw a sketch for each station (see Fig. 7-5). In this sketch show the meridian and the two courses involved. This will make the required arithmetic evident. You should also use sketches when azimuths are used.

If you measure the angles as recommended, you may also save time by using the azimuth rule for computing directions. The azimuth rule illustrated in Fig. 7-6 is:

$$B = A + C \pm 180°$$

where B = azimuth of next course
A = azimuth of previous course
C = traverse angle

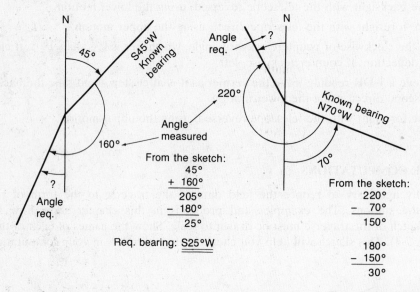

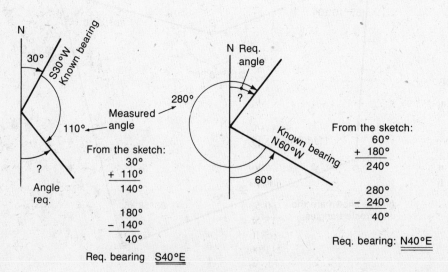

Fig. 7-5 Computing direction.

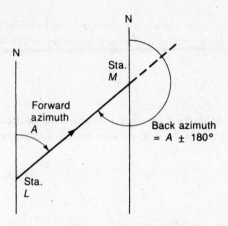
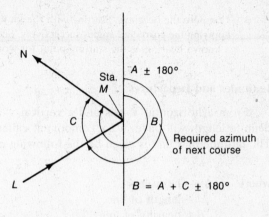

Fig. 7-6 Azimuth rule for direction: Add the traverse angle to the previous azimuth and add or subtract 180°. (*From P. Kissam, Surveying Practice, McGraw-Hill, New York, 1978*)

Looking at Fig. 7-6, notice that $A \pm 180°$ is the back azimuth of the previous course. C can then be added, for it is measured in a way that it expresses the increase in azimuth from the back azimuth of the previous course to the forward azimuth of the next course. Often it is required to add or subtract 360° to avoid negative angles or angles over 360°.

EXAMPLE 7.1 Given: A loop traverse *ABCDE* with a bearing on course *DE* of S71°42′15″E; and the following angles: $A = 60°53′30″$, $B = 215°58′00″$, $C = 65°10′10″$, $D = 111°00′00″$, $E = 86°58′45″$. (1) Compute angular error, (2) adjust the angles, and (3) compute the bearings.

Solution: First draw a sketch to scale of the problem (see Fig. 7-4).

1. Compute the angular error as follows. By definition,

$$\text{Sum of the angles of a closed polygon} = (n - 2)(180°)$$

where n = number of angles. So the number of angles in the traverse should equal

$$(5 - 2)(180°) = 540°$$

Find the sum of the angles as measured.

$$
\begin{aligned}
A &= 60°53′30″ \\
B &= 215°58′00″ \\
C &= 65°10′10″ \\
D &= 111°00′00″ \\
E &= 86°58′45″ \\
\hline
\text{Sum} &= 540°00′25″
\end{aligned}
$$

Thus $540°00′25″ - 540° = 25$ seconds error

Since there are five angles, the error should be proportioned equally.

$$\frac{25 \text{ seconds}}{5 \text{ angles}} = 05 \text{ seconds error per angle}$$

Normally up to 1 minute of error is allowable per angle, so this error is quite acceptable.

2. Adjust the angles. Subtract 5 seconds from each of the angles in the traverse.

$$
\begin{aligned}
A &= 60°53′30″ - 05″ = 60°53′25″ \\
B &= 215°58′00″ - 05″ = 215°57′55″ \\
C &= 65°10′10″ - 05″ = 65°10′05″ \\
D &= 111°00′00″ - 05″ = 110°59′55″ \\
E &= 86°58′45″ - 05″ = 86°58′40″ \\
\hline
\text{Sum} &= 540°00′00″ \quad \text{Check}
\end{aligned}
$$

3. Compute the bearings. Starting with the known bearing of $DE = S71°42'15''E$, compute the bearings by applying the corrected angles successively. See Table 7-1. Notice in Table 7-1 the traverse leg, which has a known bearing, is the starting point for working out the traverse.

Latitudes and Departures

From a horizontal x axis and a vertical y axis a line of certain bearing (or azimuth) and of a definite length will have: a Δy component called a *latitude* and a Δx component called a *departure*. The latitude of a line is given by the following equation:

$$\Delta y = l \cos \beta \qquad (7-1)$$

where $\Delta y =$ latitude
           $l =$ length of line
           $\beta =$ bearing of line

The departure of a line is given by the following equation:

$$\Delta x = l \sin \beta \qquad (7-2)$$

where $\Delta x =$ departure
           $l =$ length of line
           $\beta =$ bearing of line

Note: For latitudes, the north (N) direction is positive; the south (S) direction is negative. For departures, the east (E) direction is positive; the west (W) direction is negative.

In the examples and problems which follow, use a calculator to compute the trigonometric functions. Logarithmic computations will not be used here as they are more cumbersome; computers and calculators are so universally available to all survey parties that logarithms are rarely used.

Adjusting Latitudes and Departures

There are two methods for adjusting latitudes and departures, that is, ensuring that the sums of the latitudes and departures equal zero: the compass rule and the transit rule. Both methods are described below.

The compass rule applies corrections in proportion to the lengths of the courses. The equation is as follows (*correction* indicates the correction to a latitude or departure):

$$\text{Correction} = \frac{C}{L} S \qquad (7-3)$$

where $C =$ total error in sum of latitudes or departures *with sign changed*
           $L =$ total length of survey
           $S =$ length of the particular course

The compass rule is more mathematically correct than the transit rule; however, it changes the latitudes and departures in such a way that both the bearings and lengths of the courses are changed.

The transit rule applies corrections in proportion to the lengths of the latitudes and departures. The equation is as follows:

$$\text{Correction} = \frac{C}{l} s \qquad (7-4)$$

where $C =$ total error in sum of latitudes or departures *with sign changed*
           $l =$ sum of the latitude or departure *without regard to sign*
           $s =$ length of particular latitude or departure

The transit rule changes latitudes and departures in such a way that course lengths are slightly changed but bearings remain nearly the same.

Table 7-1   Computation of Bearings for Example 7.1

	$DE = S71°42'15''E$ \qquad $86°58'40''$ $-\ 71°42'15''$ $15°16'25''$ $EA = N15°16'25''E$
	$60°53'25''$ $+\ 15°16'25''$ $76°09'50''$ $AB = S76°09'50''W$
	$76°09'50''$ $+215°57'55''$ $292°07'45''$ $359°59'60''$ $-292°07'45''$ $67°52'15''$ $BC = N67°52'15''W$
	$67°52'15''$ $-\ 65°10'05''$ $02°42'10''$ $CD = S02°42'10''E$
	$110°59'55''$ $-\ 02°42'10''$ $108°17'45''$ $179°59'60''$ $-108°17'45''$ $71°42'15''$ $DE = S71°42'15''E$ \quad Bearings check

With both the compass and transit rules, note the following:

1. The sum of the corrections for latitude must be equal to the error in latitude *with its sign changed*.

2. The sum of the corrections for departures must be equal to the error in departure *with its sign changed*.

Sometimes because of the rounding off of the computed corrections, it may be necessary to slightly change one of the two corrections to create this relationship. Usually the changes will be applied to the largest values.

EXAMPLE 7.2   Given: A loop traverse *ABCDE* with the following distances, bearings, and coordinates: Point *E* is located at coordinates N100.00, E1400.00. Interior angles are:

$$A = 68°50'20'' \qquad D = 107°19'43''$$
$$B = 212°28'15'' \qquad E = 91°57'30''$$
$$C = 59°24'12''$$

DE has bearing S84°50′00″E. (1) Compute angular error; (2) adjust the angles; (3) compute bearings; (4) compute the latitudes and departures; (5) compute the error; (6) compute the accuracy; (7) adjust the latitudes and departures; (8) compute the adjusted latitudes and departures; (9) compute the coordinates.

Solution: First a sketch must be drawn to scale. The larger you make the sketch the more accuracy you will attain (see Fig. 7-7). If you have it, 12 by 18 in paper gives a size which will give very good accuracy. The tangent method of plotting traverse angles (see Chap. 14) is easy and accurate; alternatively a protractor may be used. Assume we are using a sheet of paper 12 by 18 in; $1\frac{1}{2}$ in from the left edge draw a vertical grid line. Then draw vertical grid lines at 500 (5 in to the right of the first line), 1000 (10 in to the right of the first line), and 1500 (15 in to the right of the first line). Now put in the horizontal grid lines. At the bottom of the sheet draw a horizontal grid line up 1 in from the bottom and label it 0. The 500 grid line will be vertically up 5 in above the zero line. One more horizontal line should be drawn at 1000 (10 in above the zero grid line). You should now have a grid that is labeled 0 horizontally and 0 vertically at the lower left-hand corner (see Fig. 7-7) and that is large enough to contain the entire traverse.

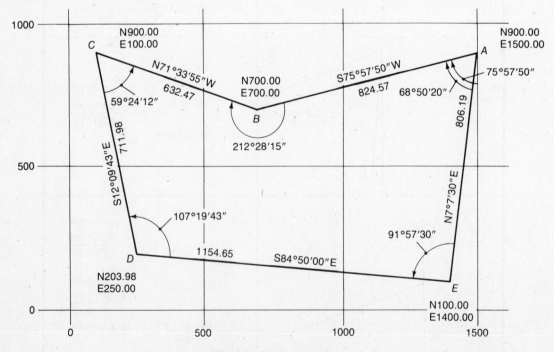

Fig. 7-7 Loop traverse.

Station E is given in the problem data. Plot it on the grid (Fig. 7-7) at N100.00 and E1400.00. Next plot the angles, using either the tangent method, which is quite accurate, or a protractor, which is less accurate but faster. The bearing of line DE is known to be S84°50′00″E, so a back line from E can be plotted. Draw a sketch of point DEA (see Table 7-2) to get the required angles for the grid sketch.

First we must see what the angular error is in the traverse.

1. Compute the angular error. The sum of the angles should be:

$$(n-2)(180°) = (5-2)(180°) = 540°$$
$$A = 68°50′22″$$
$$B = 212°28′17″$$
$$C = 59°24′14″$$
$$D = 107°19′45″$$
$$E = 91°57′32″$$
$$\text{Sum} = 540°00′10″$$

So:
$$\text{Error per angle} = \frac{10 \text{ seconds}}{5} = 02 \text{ seconds}$$

This is a very small error and is thus acceptable.

2. Adjust the angles. Correct each angle by the same amount, as the chances for error are the same.

$$A = 68°50′22″ - 02″ = 68°50′20″$$
$$B = 212°28′17″ - 02″ = 212°28′15″$$
$$C = 59°24′14″ - 02″ = 59°24′12″$$
$$D = 107°19′45″ - 02″ = 107°19′43″$$
$$E = 91°57′32″ - 02″ = 91°57′30″$$
$$\text{Sum} = 540°00′00″ \qquad \text{Check}$$

3. Compute bearings. Starting with the known bearing $DE = $ S84°50′00″E, compute the bearings by applying the corrected angles successively (see Table 7-2).

4. Compute the latitudes and departures. See Table 7-3. Use Eqs. (7-1) and (7-2) and a calculator to find the trigonometric functions.

5. Compute the error. The traverse begins and ends at the same point, so the sum of the latitudes and the sum of the departures should both be zero. By adding the columns, the error can be found. The error in latitude is -0.19; the error in departure is -0.03. The total error is the square root of the sum of the squares of these values.

$$\text{Total error} = \sqrt{(0.19)^2 + (0.03)^2} = \sqrt{0.0370} = 0.192$$

6. Compute the measure of accuracy, which is the ratio of the total error to the total length of the survey. The sum of the lengths of the course is 4129.86; hence

$$\text{Accuracy} = 0.192 : 4129.86 = 1 : 21\,510$$

The accuracy of the usual traverse with a 1-minute transit is about 1 : 3000. If the ratio is smaller a blunder may exist. This survey is acceptable.

7. Adjust the latitudes and departures. There are two methods for this procedure as explained in Sec. 7.4.

   (a) The compass rule. Apply corrections in proportion to the lengths of the course (see Table 7-4). Use Eq. (7-3). Round off the total length of the survey of 4130 for your computations. From Table 7-4 add the latitude and departure corrections to Table 7-3.

   (b) The transit rule. Corrections are applied to latitudes and departures in proportion to the lengths of the latitudes and departures. See Table 7-5. Use Eq. (7-4).

8. Compute the adjusted latitudes and departures. Add the corrections algebraically to the unadjusted latitudes and departures; see Table 7-3.

9. Compute the coordinates. Choose your coordinates so that all the coordinates will be plus. Point E, the most southerly point, is given a y, or north, coordinate of 100.00; and point C, the most westerly point, is given an x, or east, coordinate of 100.00. The coordinates are computed by successive algebraic addition of the adjusted latitudes and departures. Check can be made by carrying computations around

Table 7-2   Computation of Bearings for Example 7.2

 (a)	$DE = $ S84°50′00″E $91°57′30″$ $\underline{-\ \ 84°50′00″}$ $7°07′30″$ $EA = $ N7°07′30″E
 (b)	$68°50′20″$ $\underline{+\ \ 7°07′30″}$ $75°57′50″$ $AB = $ S75°57′50″W
 (c)	$75°57′50″$ $\underline{+\ 212°28′15″}$ $288°26′05″$ $359°59′60″$ $\underline{-\ 288°26′05″}$ $71°33′55″$ $BC = $ N71°33′55″W
 (d)	$71°33′55″$ $\underline{-\ 59°24′12″}$ $12°09′43″$ $CD = $ S12°09′43″E
 (e)	$107°19′43″$ $\underline{-\ 12°09′43″}$ $95°10′00″$ $179°60′00″$ $\underline{-\ 95°10′00″}$ $84°50′00″$ $DE = $ S84°50′00″E Bearings check

Table 7-3 Computations for Example 7.2

Station	Corrected Bearing and Length	Cos, Sin	Unadjusted N-S Lat.	Unadjusted E-W Dep.	Correction Lat.	Correction Dep.	Adjusted Coordinates Lat.	Adjusted Coordinates Dep.
A							900.00	1500.00
	S75°57'50"W	0.24253	− 199.98	− 799.95	−0.04	−0.01	− 199.94	− 799.94
	824.57	0.97014						
B							700.06	700.06
B	N71°33'55"W	0.31633						
	632.47	0.94868	+ 200.07	− 600.01	−0.03	0.00	+ 200.04	− 600.01
C							900.10	100.05
C	S12°09'43"E	0.21068						
	711.98	0.97756	− 696.00	+ 150.00	−0.03	0.00	− 695.97	+ 150.00
D							204.13	250.05
D	S84°50'00"E	0.09028						
	1154.65	0.99592	− 104.25	+ 1149.93	−0.05	−0.01	− 104.20	+ 1149.92
E							99.93	1399.97
E	N7°07'30"E	0.12403						
	806.19	0.99228	+ 799.97	+ 100.00	−0.04	−0.01	+ 799.93	+ 99.99
Plus sums			1000.09	1399.93			899.86	1499.96
Minus sums			− 1000.23	− 1399.96				
Sums	4129.86		− 0.19	− 0.03				

Error in dep. = −0.03

Total error = 0.192   Error in lat. = −0.19

Total error = $\sqrt{(-0.19)^2 + (-0.03)^2} = 0.192$

Table 7-4   Adjusting Latitudes and Departures by the Compass Rule (Example 7.2)

Course	Correction to Latitude	Correction to Departure
AB	$\dfrac{-0.19}{4130}(825) = -0.04$	$\dfrac{-0.03}{4130}(825) = -0.01$
BC	$\dfrac{-0.19}{4130}(632) = -0.03$	$\dfrac{-0.03}{4130}(632) = 0.00$
CD	$\dfrac{-0.19}{4130}(712) = -0.03$	$\dfrac{-0.03}{4130}(712) = 0.00$
DE	$\dfrac{-0.19}{4130}(1155) = -0.05$	$\dfrac{-0.03}{4130}(1155) = -0.01$
EA	$\dfrac{-0.19}{4130}(806) = -0.04$	$\dfrac{-0.03}{4130}(806) = -0.01$
	−0.19	−0.03

Table 7-5   Adjusting Latitudes and Departures by the
Transit Rule (Example 7.2)

Course	Correction to Latitude	Correction to Departure
AB	$\dfrac{-0.19}{2000}(200) = -0.02$	$\dfrac{-0.03}{2800}(800) = -0.01$
BC	$\dfrac{-0.19}{2000}(200) = -0.02$	$\dfrac{-0.03}{2800}(600) = -0.01$
CD	$\dfrac{-0.19}{2000}(696) = -0.06$	$\dfrac{-0.03}{2800}(150) = 0.00$
DE	$\dfrac{-0.19}{2000}(104) = -0.01$	$\dfrac{-0.03}{2800}(1500) = -0.01$
EA	$\dfrac{-0.19}{2000}(800) = -0.08$	$\dfrac{-0.03}{2800}(100) = 0.00$
	-0.19	-0.03

Table 7-6   Example 7-2 Survey Adjusted by Transit Rule

Station	Unadjusted		Correction		Adjusted Coordinates	
	Lat.	Dep.	Lat.	Dep.	Lat.	Dep.
A					900.00	1500.00
	− 199.98	− 799.95	− 0.02	− 0.01	− 199.96	− 799.94
B					700.04	700.06
B	+ 200.07	− 600.01	− 0.02	− 0.01	+ 200.05	− 600.00
C					900.09	100.06
C	− 696.00	+ 150.00	− 0.06	0.00	− 695.94	+ 150.00
D					204.15	250.06
D	− 104.25	+ 1149.93	− 0.01	− 0.01	− 104.24	+ 1149.92
E					99.91	1399.98
E	+ 799.97	+ 100.00	− 0.08	0.00	+ 799.89	+ 100.00
					899.80	1499.98
Absolute sums	2000.27	2799.89	− 0.19	− 0.03		
Plus sums	1000.04	1399.93				
Minus sums	− 1000.23	− 1399.96				
	− 0.19	− 0.03				

to the starting point, which should have the same coordinates as before. Table 7-6 shows the adjustment of the survey by the transit rule; see also Table 7-5.

7.5 WORKING A CONNECTING TRAVERSE

Figure 7-8 is a sketch required to work out a connecting traverse beginning at the known positions of triangulation stations Mouse and Fox. The closing station is Rat, with the direction of Rat-Duck. The data on the figure are triangulation data and field data. The coordinates of the triangulation stations must be held fixed and the traverse adjusted to them. Example 7.3 describes how to work out a traverse such as this.

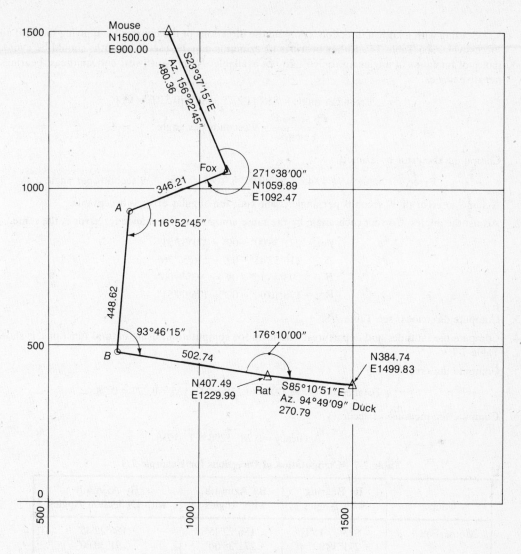

Fig. 7-8 Traverse.

EXAMPLE 7.3 See Fig. 7-8. Given: Coordinates of Mouse = N1500.00, E900.00. Distance Mouse to Fox = 480.36, with a bearing of S23°37′15″, azimuth 156°22′45″. Angle at Fox = 271°38′00″; distance Fox to A = 346.21; angle at A = 116°52′45″, distance A to B = 448.62; angle at B = 93°46′15″, distance B to Rat = 502.74; distance Rat to Duck = 270.86, bearing S85°11′27″, azimuth = 94°48′33″. Coordinates of Rat = N407.49, E1229.99. Work out the traverse.

Solution: See Fig. 7-8.

1. Compute the angular error. Compute the directions of the fixed lines upon which the traverse begins and closes.

$$\tan(\text{Mouse-Fox direction}) = \frac{900.00 - 1092.47}{1500.00 - 1059.89} = \frac{-192.47}{440.11} = -0.437322$$

Mouse-Fox bearing = S23°37′15″E   Azimuth = 156°22′45″

$$\tan(\text{Rat-Duck direction}) = \frac{1229.99 - 1499.83}{407.49 - 384.74} = \frac{-269.84}{22.75} = -11.86110$$

Rat-Duck bearing = S85°10′51″E   Azimuth = 94°49′09″

Starting with a known direction, compute the directions of the courses by applying the field angles successively; see Table 7-7. Either bearings or azimuths may be used. Azimuths should be avoided if tables of functions of angles up to 360° are not available. *Note*: Northwest and southeast bearings are negative angles.

$$\text{Error per angle} = S85°11'27''E - S85°10'51''E = 0'36''$$

$$\frac{36 \text{ seconds}}{4 \text{ angles}} = 9 \text{ seconds per angle}$$

Computing the error by azimuths:

$$\text{Error per angle} = 94°49'45'' - 94°49'09'' = 0'36'' \quad \text{or} \quad 9 \text{ seconds per angle}$$

Assume an error of 30 seconds per angle is allowed; the angular error is acceptable.

2. **Adjust the angles.** Correct each angle by the same amount, as the chance for error is the same.

$$\text{Fox} = 271°38'00'' - 09'' = 271°37'51''$$
$$A = 116°52'45'' - 09'' = 116°52'36''$$
$$B = 93°46'15'' - 09'' = 93°46'06''$$
$$\text{Rat} = 176°10'00'' - 09'' = 176°09'51''$$

3. **Compute directions;** see Table 7-8.

4. **Compute the latitudes and departures.** The form for computation, using natural functions, is shown in Table 7-9.

5. **Compute the error.**

$$\text{Total error} = \sqrt{(0.12)^2 + (0.25)^2} = \sqrt{0.125} = 0.277 = 0.28$$

6. **Compute the measure of accuracy.**

$$\text{Accuracy} = 0.28 : 1298 = 1 : 4636$$

Table 7-7 Computation of Directions for Example 7.3

Course	By Bearing with Angles	By Azimuth with Angles	By Azimuth with Deflection Angles
Mouse-Fox	S23°37'15"E + 271°38'00"	156°22'45" + 271°38'00"	156°22'45" 91°38'00"
	248°00'45" − 180°	428°00'45" − 180°	
Fox-*A*	S68°00'45"W + 116°52'45"	248°00'45" + 116°52'45"	248°00'45" − 63°07'15"
	184°53'30" − 180°	364°53'30" − 180	
AB	S04°53'30"W + 93°46'15"	184°53'30" + 93°46'15"	184°53'30" − 86°13'45"
	98°39'45" − 179°59'60"	278°39'45" − 180°	
B-Rat	S81°20'15"E + 176°10'00"	98°39'45" + 176°10'00"	98°39'45" − 3°50'00"
	94°49'45" − 179°59'60"	274°49'45" − 180°	
Rat-Duck Rat-Duck fixed	S85°10'15"E − 85°10'51"	94°49'45" − 94°49'09"	94°49'45" − 94°49'09"
Error	0°00'36"	0°00'36"	0°00'36"

**Table 7-8 Computation of Directions
for Adjusted Angles (Example 7.3)**

Course	By Bearing	By Azimuths
Mouse-Fox	S23°37′15″E + 271°37′51″	156°22′45″ + 271°37′51″
	248°00′36″ − 180°	428°00′36″ − 180°
Fox-A	S68°00′36″W + 116°52′36″	248°00′36″ + 116°52′36″
	184°53′12″ − 180°	364°53′12″ − 180°
AB	S04°53′12″W + 93°46′06″	184°53′12″ + 93°46′06″
	98°39′18″ − 179°59′60″	278°39′18″ − 180°
B-Rat	S81°20′42″E + 176°09′51″	98°39′18″ + 176°09′51″
	94°49′09″ − 179°59′60″	274°49′09″ − 180°
Rat-Duck	S85°10′51″E	94°49′09″

Table 7-9 Computations for Example 7.3

Station	Corrected Bearing and Length	Cos, Sin	Unadjusted		Correction		Adjusted Coordinates	
			Lat.	Dep.	Lat.	Dep.	Lat.	Dep.
Fox	S68°00′36″W	0.37445					1059.89	1092.47
A	346.21	0.92725	− 129.64	− 321.02	− 0.03	− 0.07	− 129.67	− 321.09
A							930.22	771.38
B	S04°53′12″W	0.99637						
	448.62	0.08519	− 446.99	− 38.22	− 0.04	− 0.09	− 447.03	− 38.31
B							483.19	733.07
Rat	S81°20′42″E	0.15049						
	502.74	0.98861	− 75.65	+ 497.01	− 0.05	− 0.09	− 75.70	+ 496.92
Rat		0.08384					407.49	1229.99
Sums	1297.57		− 652.28	+ 137.77				
Coordinate diff.			− 652.40	+ 137.52				
Error			+ 0.12	+ 0.25				

7. Compute the corrections to the latitudes and departures; see Table 7-10.

8. Compute the adjusted latitudes and departures; see Table 7-9.

9. Compute the coordinates. Beginning with the fixed coordinates at the start of the traverse, compute the coordinates of each station by successive algebraic addition of the adjusted latitudes and departures. An arithmetic check is obtained when the computed coordinates of Rat agree with its fixed coordinates. See the right-most two columns of Table 7-9. The computed coordinates of Rat are 407.49, 1229.99. The fixed coordinates of Rat were given as N407.49, E1229.99, so this checks.

10. Plot the traverse. You could plot the traverse by tangent angles, by protractor and scale, or by coordinates. Since you have developed the coordinates, use them. Plotting coordinates is more accurate than plotting angles. See Fig. 7-9.

Table 7-10   Corrections to Latitudes and Departures (Example 7.3)

Course	Correction to Latitude	Correction to Departure
Fox-*A*	$\dfrac{-0.12}{1298}(346) = -0.03$	$\dfrac{-0.25}{1298}(346) = -0.07$
AB	$\dfrac{-0.12}{1298}(449) = -0.04$	$\dfrac{-0.25}{1298}(449) = -0.09$
B-Rat	$\dfrac{-0.12}{1298}(503) = -0.05$	$\dfrac{-0.25}{1298}(503) = -0.09$
	-0.12	-0.25

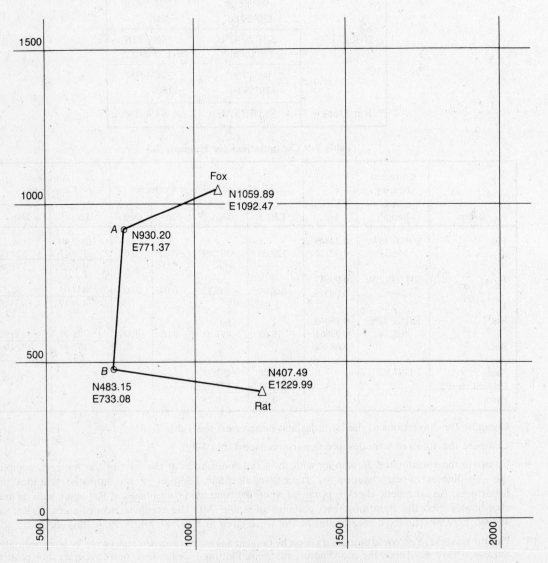

Fig. 7-9   Plotting coordinates.

Solved Problems

7.1   Given: A loop traverse *ABCDE* in which leg *DE* has a bearing of S88°50′30″E. Angles are as follows:

$$A = 46°20′20″ \qquad D = 100°30′30″$$
$$B = 231°30′10″ \qquad E = 100°39′40″$$
$$C = 61°00′10″$$

Work out all the bearing computations for traverse *ABCDE*.

Solution

Draw the traverse with all the angles to scale; see Fig. 7-10. First compute the angular error.

$$\text{Sum of the angles} = (n - 2)(180°) = 540°$$

$$
\begin{aligned}
A &= 46°20′20″ \\
B &= 231°30′10″ \\
C &= 61°00′10″ \\
D &= 100°30′30″ \\
E &= \underline{100°39′40″} \\
\text{Sum} &= 540°00′50″
\end{aligned}
$$

$$\text{Error per angle:} \ \frac{50 \text{ seconds}}{5 \text{ angles}} = 10 \text{ seconds per angle}$$

An angular error of 1 minute per angle is allowed, so this traverse is acceptable.

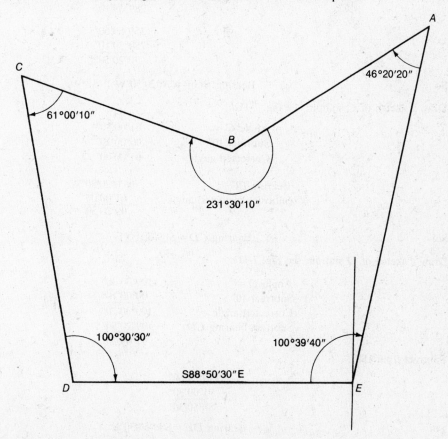

Fig. 7-10   Traverse computations.

Start with station E since the bearing is known to be S88°50′30″E. At point E we have a given angle of 100°39′40″.

Corrected angle = 100°39′40″ − 00°00′10″ = 100°39′30″

Draw a sketch of E station; see Fig. 7-11a.

$$DE = S88°50′30″E$$

Corrected angle =	100°39′30″E
Subtract DE	− 88°50′30″
	11°49′00″

So: Bearing EA = N11°49′00″E

Draw a sketch of A station; see Fig. 7-11b.

Angle A =	46°20′20″
Subtract 10″	− 00°00′10″
Corrected angle =	46°20′10″
Add bearing EA	+ 11°49′00″
	58°09′10″

So: Bearing AB = S58°09′10″W

Draw a sketch of B station; see Fig. 7-11c.

Angle B =	231°30′10″
Subtract 10″	− 00°00′10″
Corrected angle =	231°30′00″
Add bearing AB	+ 58°09′10″
	289°39′10″
	359°59′60″
	− 289°39′10″
	70°20′50″

So: Bearing BC = N70°20′50″W

Draw a sketch of C station; see Fig. 7-11d.

Angle C =	61°00′10″
Subtract 10″	− 00°00′10″
Corrected angle =	61°00′00″
Bearing BC	N70°20′50″W
Subtract corrected angle	− 61°00′00″
	09°20′50″

So: Bearing CD = S09°20′50″E

Draw a sketch of D station; see Fig. 7-11e.

Angle D =	100°30′30″
Subtract 10″	− 00°00′10″
Corrected angle =	100°30′20″
Subtract bearing CD	− 09°20′50″
	91°09′30″

Subtract from 180°.

	179°59′60″
	− 91°09′30″
	88°50′30″

So: Bearing DE = S88°50′30″E

This is also the bearing of DE at the start of the traverse, so the angles of the traverse check.

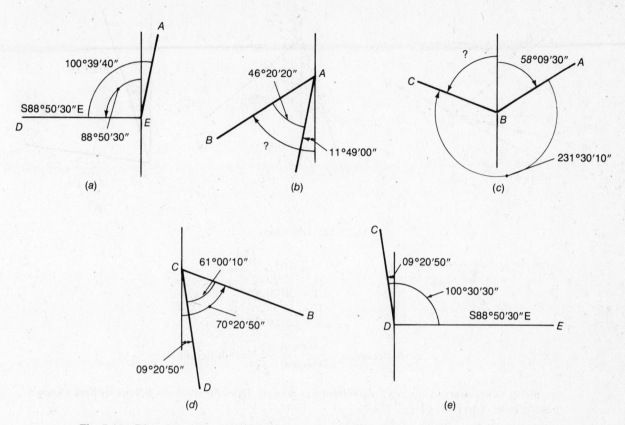

Fig. 7-11 Directions. (*From P. Kissam, Surveying Practice, McGraw-Hill, New York, 1978*)

7.2   Given: A triangular loop traverse *ABC* with a bearing on *AB* of N78°49′50″E. Angles are as
follows: *A* = 62°29′20″; *B* = 50°20′20″; *C* = 67°10′26″. Work out the bearings of the three courses
of the traverse.

Solution

   See Figs. 7-12 and 7-13. Compute the angular error.

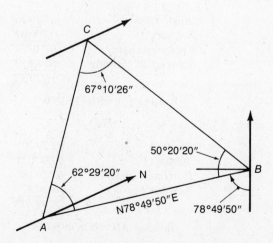

Fig. 7-12 Computing angular error.

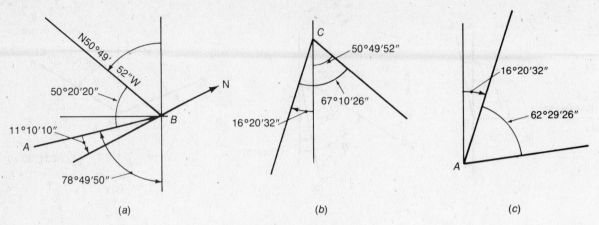

Fig. 7-13 Directions.

Since the figure is a triangle,   Sum of angles = 180°

$$A = 62°29'20''$$
$$B = 50°20'20''$$
$$C = \underline{67°10'26''}$$
$$Sum = 180°00'06''$$

$$\text{Error per angle} = \frac{6 \text{ seconds}}{3 \text{ angles}} = 2 \text{ seconds per angle}$$

It will thus be necessary to subtract 2 seconds from each angle. Draw the station at B, since we have a known bearing from A to B; see Fig. 7-13a.

Angle B =	50°20'20''
Subtract 02''	− 00°00'02''
Corrected angle =	50°20'18''
Add AB bearing	+ 78°49'50''
	129°10'08''

	179°59'60''
Subtract	− 129°10'08''
	50°49'52''

So: Bearing BC = N50°49'52''W

Draw station at C; see Fig. 7-13b.

Angle C =	67°10'26''
Subtract 02''	− 00°00'02''
Corrected angle =	67°10'24''
Subtract BC bearing	− 50°49'52''
	16°20'32''

So: Bearing CA = S16°20'32''W

Draw station at A; see Fig. 7-13c.

Angle A =	62°29'20''
Subtract 02''	− 00°00'02''
Corrected angle =	62°29'18''
Add CA bearing	+ 16°20'32''
	78°49'50''

So: Bearing AB = N78°49'50''E

This is the same as the starting bearing and confirms the traverse.

7.3   Given: Loop traverse *ABCDE* with a known bearing on course *DE* of S74°56′40″E. Angles
are as follows:

$$A = \ 50°40′30″ \qquad D = 126°30′02″$$
$$B = 220°10′10″ \qquad E = \ 93°58′48″$$
$$C = \ 48°40′50″$$

Compute the bearings of the traverse.

Solution

Draw a sketch of the traverse; see Fig. 7-14.

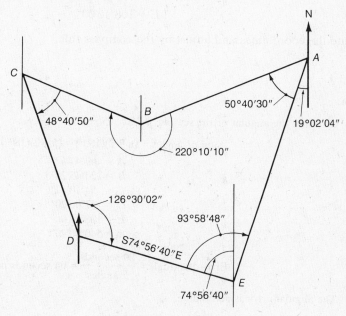

Fig. 7-14   Loop traverse.

1.   Compute the angular error.

$$\text{Sum of angles} = (n - 2)(180°) = (5 - 2)(180°) = 540°$$

$$
\begin{aligned}
A &= \ 50°40′30″ \\
B &= 220°10′10″ \\
C &= \ 48°40′50″ \\
D &= 126°30′02″ \\
E &= \ \underline{\ 93°58′48″} \\
\text{Sum} &= 540°00′20″
\end{aligned}
$$

$$\text{Error per angle} = \frac{20 \text{ seconds}}{5 \text{ angles}} = 4 \text{ seconds per angle}$$

An angular error of 1 minute per angle is allowed; thus this angular error is acceptable.

2.   Adjust the angles. Correct each angle by the same amount, as the chance for error is the same
for all angles.

$$
\begin{aligned}
A &= \ 50°40′30″ - 04″ = \ 50°40′26″ \\
B &= 220°10′10″ - 04″ = 220°10′06″ \\
C &= \ 48°40′50″ - 04″ = \ 48°40′46″ \\
D &= 126°30′02″ - 04″ = 126°29′58″ \\
E &= \ 93°58′48″ - 04″ = \ \underline{\ 93°58′44″} \\
& \qquad\qquad \text{Sum} = 540°00′00″ \qquad \text{Check}
\end{aligned}
$$

3. Compute the bearings. Start with a known bearing—DE = S74°56′40″E; compute the bearings by applying the corrected angles successively. See Table 7-11.

7.4 Given: A loop traverse $ABCDE$. The coordinates of point A are N692.00, E1100.00. Bearing of line EA = 11°29′20″E. Interior angles (see Figs. 7-15 and 7-16) are:

$$A = 69°47′42″$$
$$B = 220°43′26″$$
$$C = 54°10′52″$$
$$D = 87°06′20″$$
$$E = 108°12′00″$$

Compute the coordinates and adjust by the compass rule.

Solution

1. Compute the angular error; see Fig. 7-15.

Sum of angles = $(n - 2)(180°) = (5 - 2)(180°) = 540°$

$$
\begin{aligned}
A &= 69°47′42″ \\
B &= 220°43′26″ \\
C &= 54°10′52″ \\
D &= 87°06′20″ \\
E &= 108°12′00″ \\
\text{Sum} &= 540°00′20″
\end{aligned}
$$

$$\text{Error per angle} = \frac{20 \text{ seconds}}{5 \text{ angles}} = 04 \text{ seconds per angle}$$

The angular error is acceptable.

2. Adjust the angles.

$$
\begin{aligned}
A &= 69°47′42″ - 04″ = 69°47′38″ \\
B &= 220°43′26″ - 04″ = 220°43′22″ \\
C &= 54°10′52″ - 04″ = 54°10′48″ \\
D &= 87°06′20″ - 04″ = 87°06′16″ \\
E &= 108°12′00″ - 04″ = 108°11′56″ \\
\text{Corrected sum} &= 540°00′00″ \qquad \text{Check}
\end{aligned}
$$

3. Compute the bearings. Start with the known bearing EA = N11°29′20″E. Compute the bearings by applying corrected angles successively; see Figs. 7-15 and 7-16.

(a)

$$
\begin{aligned}
\text{Angle } A = &69°47′38″ \\
\text{Add } EA &\underline{+11°29′20″} \\
&81°16′58″
\end{aligned}
$$

Bearing AB = S81°16′58″W

(b)

$$
\begin{aligned}
&81°16′58″ \\
&\underline{+220°43′22″} \\
&302°00′2\upsilon \\[4pt]
&359°59′60″ \\
&\underline{-302°00′20″} \\
&57°59′40″
\end{aligned}
$$

Bearing BC = N57°59′40″W

Table 7-11   Computation of Bearings for Prob. 7.3

74°56′40″ ? A	DE = S74°56′40″E 93°58′44″
D 93°58′48″ E	EA = N19°02′04″E − 74°56′40″ ——————— 19°02′04″
A B 50°40′30″ 19°02′04″ E	AB = S69°42′30″W 50°40′26″ + 19°02′04″ ——————— 69°42′30″
69°42′30″ C A B 220°10′10″	BC = N70°07′24″W 69°42′30″ + 220°10′06″ ——————— 289°52′36″ 359°59′60″ − 289°52′36″ ——————— 70°07′24″
C 48°40′50″ B 21°26′38″ 70°07′24″ D	CD = S21°26′38″E 70°07′24″ − 48°40′46″ ——————— 21°26′38″
C 21°26′38″ 126°30′02″ 105°03′18″ D ? E	DE = S74°56′40″E 126°29′58″ − 21°26′38″ ——————— 105°03′20″ 179°59′60″ − 105°03′20″ ——————— 74°56′40″ Bearings check

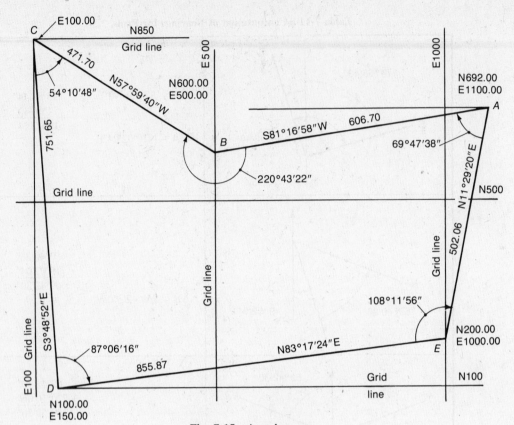

Fig. 7-15 Angular error.

(c)
$$57°59'40''$$
$$-\ 54°10'48''$$
$$\overline{03°48'52''}$$

Bearing $CD = S03°48'52''E$

(d)
$$87°06'16''$$
$$-\ 03°48'52''$$
$$\overline{83°17'24''}$$

Bearing $DE = N83°17'24''$

(e)
$$83°17'24''$$
$$+\ 108°11'56''$$
$$\overline{191°29'20''}$$
$$-\ 180°$$
$$\overline{11°29'20''}$$

Bearing $EA = N11°29'20''E$

4. Compute the latitudes and departures; see Table 7-12.

5. Compute the error; see Table 7-12.

Total error in latitude = 0.06   Total error in departure = 0.32

Total error = $\sqrt{(0.06)^2 + (0.32)^2} = \sqrt{0.0036 + 0.1024} = \sqrt{0.1060} = 0.326$

6. Compute the measure of accuracy (the ratio of the total error to the total length of the survey).

Sum of the lengths of the courses (see Table 7-12) = 3187.98

Accuracy = 0.326 : 3188 = 1 : 9780

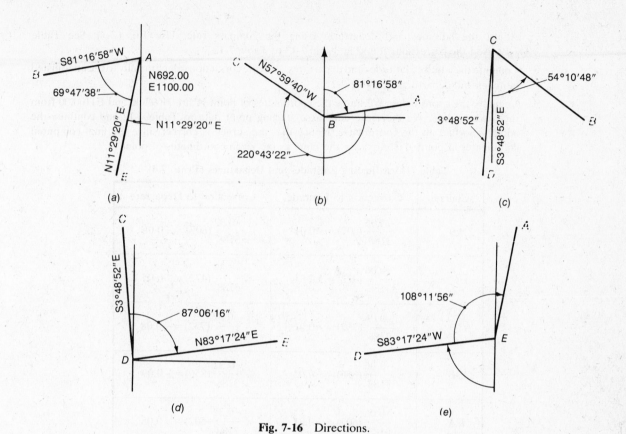

Fig. 7-16 Directions.

Table 7-12 Computations for Prob. 7.4

Station	Corrected Bearing and Length	Cos, Sin	Unadjusted		Correction		Adjusted Coordinates	
			N-S Lat.	E-W Dep.	Lat.	Dep.	Lat.	Dep.
A	S81°16′58″W	0.15156					692.00	1100.00
	606.70	0.98845	− 91.95	− 599.69	−0.01	−0.06	− 91.96	− 599.75
B							600.04	500.25
B	N57°59′40″W	0.53000						
	471.70	0.84800	+ 250.00	− 400.00	−0.01	−0.05	+ 249.99	− 400.05
C							850.03	100.20
C	S03°48′52″E	0.99779						
	751.65	0.06653	− 749.99	+ 50.00	−0.01	−0.07	− 750.00	+ 49.93
D							100.03	150.13
D	N83°17′24″E	0.11684						
	855.87	0.99315	+ 100.00	+ 850.01	−0.02	−0.09	+ 99.98	+ 849.92
E							200.01	1000.05
E	N11°29′20″E	0.97996						
	502.06	0.19918	+ 492.00	+ 100.00	−0.01	−0.05	+ 491.99	+ 99.95
Plus sums			+ 842.00	+ 1000.01			692.00	1100.00
Minus sums			− 841.94	− 999.69				
Sums	3187.98		+ 0.06	+ 0.32	−0.06	−0.32		

Error in dep. = +0.32

Total error = 0.326   Error in lat. = +0.06

7. Adjust the latitudes and departures using the compass rule. Use Eq. (*7-3*). See Table 7-13. Add the corrections found in Table 7-13 to Table 7-12.

8. Compute the adjusted latitudes and departures. Add the corrections algebraically to the unadjusted latitudes and departures in Table 7-12.

9. Compute the coordinates. We know the coordinates of point *A* are N692.00 and E1100.00 from the traverse data. We therefore add these starting point data to Table 7-12 and continue the addition to find all the coordinates. Notice that the survey is correct since the final computed coordinates of point *A* are exactly the same as the given coordinates of point *A*.

Table 7-13 Adjusting Latitudes and Departures (Prob. 7.4)

Course	Correction to Latitude	Correction to Departure
AB	$\dfrac{-0.06}{3188}(607) = -0.01$	$\dfrac{-0.32}{3188}(607) = -0.06$
BC	$\dfrac{-0.06}{3188}(472) = -0.01$	$\dfrac{-0.32}{3188}(472) = -0.05$
CD	$\dfrac{-0.06}{3188}(752) = -0.01$	$\dfrac{-0.32}{3188}(752) = -0.08$
DE	$\dfrac{-0.06}{3188}(856) = -0.02$	$\dfrac{-0.32}{3188}(856) = -0.09$
EA	$\dfrac{-0.06}{3188}(502) = -0.01$	$\dfrac{-0.32}{3188}(502) = -0.05$
	-0.06	-0.32

7.5 Given: A four-angle loop traverse *ABCD*. The bearing of line *BC* = N77°28′17″W. Coordinates of point *A* = N100.00, E1400.00. Interior angles are:

$$A = 59°29′30″$$
$$B = 136°13′11″$$
$$C = 52°16′17″$$
$$D = 112°01′22″$$

Compute the traverse, making corrections by the traverse rule.

Solution

Start by making a sketch to scale; see Fig. 7-17.

1. Compute the angular error.

$$\text{Sum of angles} = (n-2)(180°) = (4-2)(180°) = 360°$$

$$
\begin{aligned}
A &= 59°29′30″ \\
B &= 136°13′11″ \\
C &= 52°16′17″ \\
D &= \underline{112°01′22″} \\
\text{Sum} &= 360°00′20″
\end{aligned}
$$

$$\text{Error per angle} = \frac{20 \text{ seconds}}{4 \text{ angles}} = 5 \text{ seconds per angle}$$

This is a small error and is acceptable.

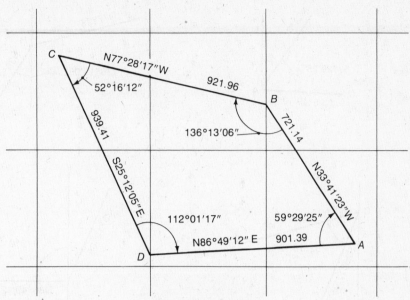

Fig. 7-17 Traverse.

2. Adjust the angles.

$$
\begin{aligned}
A &= 59°29'30'' - 05'' = 59°29'25'' \\
B &= 136°13'11'' - 05'' = 136°13'06'' \\
C &= 52°16'17'' - 05'' = 52°16'12'' \\
D &= 112°01'22'' - 05'' = \underline{112°01'17''} \\
&\qquad\qquad\text{Sum} = 360°00'00'' \qquad \text{Check}
\end{aligned}
$$

3. Compute the bearings; see Fig. 7-18. Starting with the given bearing of BC = N77°28'17''W, compute the bearings by applying the corrected angles successively.

Bearing DA =	N86°49'12''E
	+ 59°29'25''
	146°18'37''
	179°59'60''
	− 146°18'37''
	33°41'23''
Bearing AB =	N33°41'23''W
	136°13'06''
	− 33°41'23''
	102°31'43''
	179°59'60''
	− 102°31'43''
Bearing BC =	N77°28'17''W
	− 52°16'12''
Bearing CD =	S25°12'05''E
	+ 112°01'17''
Bearing DA =	N86°49'12''
	89°59'60''
	− 86°49'12''
	03°10'48''
	+ 90°00'00''
	93°10'48''
Subtract angle A =	− 59°29'25''
Bearing AB =	N33°41'23''W Check

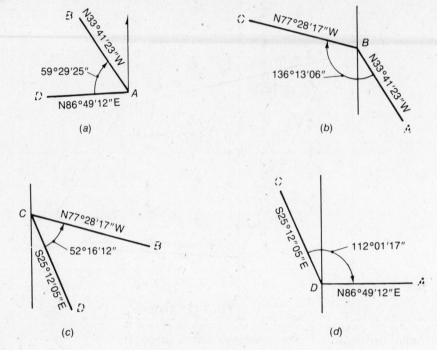

Fig. 7-18 Computing directions.

Table 7-14 Computations for Prob. 7.5

Station	Corrected Bearing and Length	Cos, Sin	Unadjusted		Correction		Adjusted Coordinates	
			N-S Lat.	E-W Dep.	Lat.	Dep.	Lat.	Dep.
A	N33°41'23"W	0.83205					100.00	1400.00
	721.14	0.55470	+ 600.03	− 400.01	−0.01	0.00	+ 600.02	− 400.01
B							700.02	999.99
B	N77°28'17"W	0.21693						
	921.96	0.97619	+ 200.00	− 900.01	−0.01	+0.01	+ 199.99	− 900.00
C							900.03	99.99
C	S25°12'05"E	0.90482						
	939.41	0.42580	− 849.99	+ 400.00	−0.02	0.00	− 850.01	+ 400.00
D							50.00	499.99
D	N86°49'12"E	0.05547						
	901.39	0.99846	+ 50.00	+ 900.00	0.00	+0.01	+ 50.00	+ 900.01
A							100.00	1400.00
Absolute sums			1700.02	2600.02	−0.04	+0.02		
Plus sums			850.03	1300.00				
Minus sums			− 849.99	− 1300.02				
Sums	3483.90		+ 0.04	− 0.02				

Error in dep. = 0.02

Total error = 0.045   Error in lat. = 0.04

4. Compute the latitudes and departures; see Table 7-14.

5. Compute the error; see Table 7-14.

$$\text{Total error} = \sqrt{(\text{departure error})^2 + (\text{latitude error})^2} = \sqrt{(-0.02)^2 + (0.04)^2} = \sqrt{0.0020} = 0.045$$

6. Compute the accuracy (ratio of the total error to the total length of the survey). The sum of the lengths of the courses is 3483.90; hence

$$\text{Accuracy} = 0.045 : 3483.90 = 1 : 77\,420$$

7. Adjust the latitudes and departures using the transit rule [use Eq. (7-4)]; see Table 7-15.

Table 7-15 Adjusting Latitudes and Departures (Prob. 7.5)

Course	Correction to Latitude	Correction to Departure
AB	$\dfrac{-0.04}{1700}(600) = -0.01$	$\dfrac{+0.02}{2600}(400) = 0.00$
BC	$\dfrac{-0.04}{1700}(200) = -0.01$	$\dfrac{+0.02}{2600}(900) = 0.01$
CD	$\dfrac{-0.04}{1700}(850) = -0.02$	$\dfrac{+0.02}{2600}(400) = 0.00$
DA	$\dfrac{-0.04}{1700}(50) = \quad 0.00$	$\dfrac{+0.02}{2600}(900) = 0.01$
	-0.04	$+0.02$

8. Transfer the values obtained in Table 7-15 to the adjustment of survey Table 7-14.

9. With the corrections logged into Table 7-14 you are now able to compute the coordinates. Since you have the coordinates of point A given as N100.00, E1400.00, put these coordinates into Table 7-14 and add and subtract as required to find all the coordinates of the station points.

7.6 Given: A connecting traverse with distance Mouse to Fox = 538.52, with bearing S21°48′05″E. Angle at Fox = 260°10′10″, distance Fox to A = 300.00. Angle at A = 120°20′10″; distance A to B = 500.00. Angle at B = 95°10′00″; distance B to C = 450.00. Angle at C = 170°30′10″; distance C to Rat = 300.00. Angle at Rat = 220°20′30″. Rat to Duck bearing = S52°41′05″E, distance = 450.00. Compute the angular error by bearings.

Solution

First draw the traverse; see Fig. 7-19. Compute the angular error. Compute the directions of the fixed lines upon which the traverse begins and closes.

$$\tan (\text{Mouse-Fox direction}) = \frac{800.00 - 1000.00}{1500.00 - 1000.00} = \frac{-200}{500} = -0.40000$$

$$\text{Mouse-Fox bearing} = \text{S21°48′05″E}$$

$$\tan (\text{Rat-Duck direction}) = \frac{1503.40 - 1861.21}{341.82 - 68.93} = \frac{-357.81}{272.89} = -1.31119$$

$$\text{Rat-Duck bearing} = \text{S52°40′06″E}$$

The angular error is computed in Table 7-16.

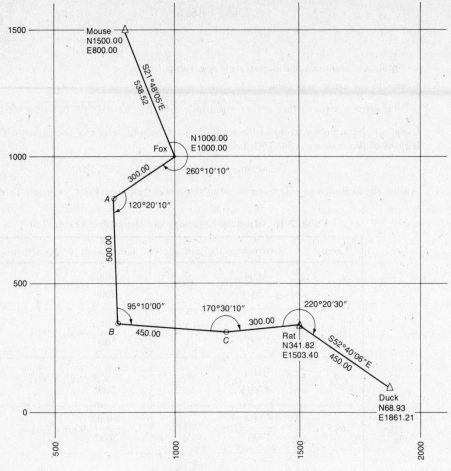

Fig. 7-19 Traverse.

**Table 7-16 Computation of Angular
Error by Bearings (Prob. 7.6)**

Course	Computation
Mouse-Fox	S21°48'05"E + 260°10'10"
	238°22'05" − 180°
Fox-A	S58°22'05"W + 120°20'10"
	178°42'15" − 180°
AB	S01°17'45"E + 95°10'00"
	96°27'45" − 179°59'60"
BC	S83°32'15"E + 170°30'10"
C-Rat	N86°57'55"E + 220°20'30"
	127°18'25" − 179°59'60"
	S52°41'35"E
C-Rat fixed	S52°40'06"E
Error	1°00'29"

Chapter 8

Topographic Surveys

8.1 DEFINITION OF TOPOGRAPHIC SURVEYING

Topographic surveying is the method of determining the positions, on the surface of the earth, of human-made and natural features. It also is used to determine the configuration of the terrain. The purpose of a topographic survey is to find the data necessary for the construction of a graphical portrayal of topographic features. The graphical portrayal from the gathered data forms a *topographic map*. A topographic map will show the character of the vegetation by conventional signs, as well as the horizontal distances between features and their elevations above a given datum.

Ground methods are needed in topographic surveying. The tools are the transit, level, plane table alidade, and tape. Hand levels are also often used in contouring. Aerial photography is now used for making most of the small-scale topographic maps; the process is known as *photogrammetry*. Even with the photogrammetric methods, however, a certain amount of the work must be done in the field. These methods are described in this chapter.

The first step for planning and designing an engineering project is the preparation of a topographic map and the necessary control surveys.

8.2 SCALES

A topographic map represents in a small area upon a drafting medium a portion of the surface of the earth. For this reason the distance between any two points on the map must have a known ratio to the distance between the same two points on the ground. The ratio of these points is the *scale* of the map. The scale is given in terms of map distance in a given number of units, which corresponds to a certain distance on the ground. Scales may be expressed by direct correspondence or by a ratio. For example, a typical scale is 1 in = 100 ft. Expressed as a ratio, this is 1:1200, or 1/1200. A scale of 1:2000, or 1/2000, indicates that 1 unit on the map corresponds to 2000 units on the ground. Some typical scales are as follows:

1 in = 200 ft for highway reconnaissance

1 in = 2000 ft for U.S. Geological Survey topographic maps

1 in = 50 ft for plot plan for buildings

8.3 TOPOGRAPHIC REPRESENTATION

Topography on the map may be represented by contour lines or by hill shading. Hill shading is accomplished by means of *hachures*, a series of short lines drawn in the direction of the slope. For a steep slope the lines are heavy and spaced close together. For a gentle slope the lines are widely spaced and of a light weight. The hachures give a general impression of the configuration of the ground. They are not used to give actual elevations of the ground.

A *contour line* is a line passing through points of equal elevation. A level plane that intersects the ground surface would show on the map as a contour line. In nature you may think of the shoreline of a still lake as a contour line. The contour interval for contour lines is the constant vertical distance between adjacent contour lines. Contour lines on a map are drawn in true horizontal positions with respect to the ground surface. Topographic maps with contour lines show the slopes of topographic features—hills, valleys, and ridges—and the lines give the elevations of these features.

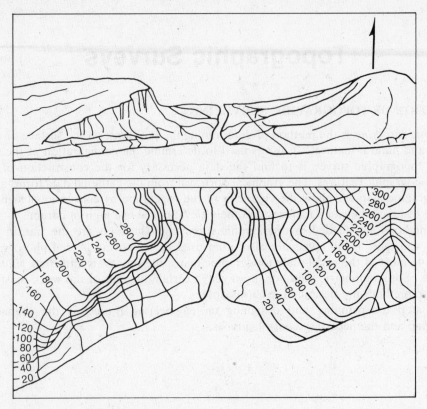

Fig. 8-1 Representing terrain by contour lines. (*Courtesy of U.S. Army, General Drafting, T M 5-230*)

EXAMPLE 8.1   Given: Profile and plan view of a certain piece of terrain (Fig. 8-1). Examine the two views and explain what the features show.

Solution: There is a stream meandering in a valley between a cliff on the left and a rounded hill on the right. The stream is most apparent in the plan view (view looking down from the top), as you can see it discharge into a small body of water. The plan view is the contour map, or topographic representation, of the terrain in the upper view of Fig. 8-1. The contour interval on this map is 20 units; this could represent 20 ft or 20 m. Where the lines are close together the cliffs are very steep. By looking at the plan (or contour) map, you can tell that the left-hand portion at the top of the cliff is very steep, as is the south edge of the right-hand hill. This is apparent because of the closeness of the spacing between contour lines. Contour lines spaced far apart indicate a slow rise or fall in the terrain. Close spacing indicates rapid rise or fall of the land. In the top view of Fig. 8-1 there are several ridges; all show up in the contour map. The contour lines run up to them and form a typical V shape.

8.4   CONTOUR LINES

The basic rules for drawing contour lines are as follows:

1.   Contour lines never end, meet, or cross, except in the unusual case of a vertical or overhanging cliff or a cave.

2.   Contour lines must be uniformly spaced, unless data to the contrary are available.

3.   Contour lines are drawn so that the ground higher than the contour line is always on the same side of the contour line.

4.   Since the earth is a continuous surface, all contours must close upon themselves. Although this closure may occur within the area being mapped, it often appears external to the map view and thus does not appear on the map itself.

5. Contour lines are perpendicular to the direction of maximum slope.

6. The spacing between contour lines indicates the steepness of the slope. Close spacing indicates a steep slope (see the spacing in Fig. 8-1 on the west side of the stream valley). Wide spacing indicates gentle slopes (note the wide spacing in Fig. 8-1 on the east side of the stream).

7. Concentric closed contours which increase in elevation indicate hills. See Fig. 8-2 for a graphic representation of a hill from three directions.

8. Contours forming a closed loop around lower ground are called *depression contours.* Hachures are put inside the lowest contour pointing to the bottom of a hole with no outlet to make it easier to interpret the map.

9. Smooth lines indicate gradual slopes. Irregular contour lines show rough country.

10. Contour lines never branch into two contours with the same elevation.

11. Valleys are shown as V-shaped contours, and ridges are U-shaped contours.

12. The V's formed by the contours crossing a stream always point upstream.

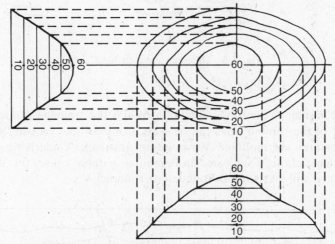

Fig. 8-2 Hill in three dimensions, showing contours. (*From R. C. Brinker and P. R. Wolf, Elementary Surveying, IEP Publishing, Dun-Donnelley, New York, 1977*)

Contour lines give the maximum amount of information of the map area, and they do not obscure other essential map detail. They also show elevation and configuration of the ground. The elevation of any point not falling on a contour line can be found by interpolating between the two contour lines which fall on either side of the point.

It must be emphasized that a contour cannot have an end within the map. It commences and ends at the edges of the map, or it closes on itself. Closing on itself will be shown by a series of contour lines which make circles or ovals on the map. They will indicate either a depression or a hill. You can identify a hill by the ascending elevations, which terminate with the innermost closed contour line being the highest elevation. In case of a depression the innermost closed contour line will be at the lowest elevation, and on the lowest contour line hachure marks pointing toward the depression will be shown. This makes it quite evident that you are looking at a depression because no hachure marks are used on hills.

When contour lines are equally spaced along a line normal to the contours, you know the slope is constant.

Straight, parallel, equally spaced contours represent human-made embankments or excavations.

To make it convenient to read the elevations from a topographic map every fifth contour is drawn as a heavier line. This line is called an *index contour.* If the interval between contours is 1 ft, contours whose elevations are multiples of 5 ft are shown as heavy lines. If the interval is 10 ft, the heavy contours have elevations that are multiples of 50 ft.

EXAMPLE 8.2 Given: Three different contour intervals, drawn to the same scale, as represented in Fig. 8-3. Find the steepness of the slopes from the contour intervals and horizontal spacing of the contour lines. Evaluate each representation.

Solution: The ground slopes are steepest in Fig. 8-3*a*, the contour interval is 10 ft. The ground slopes are flattest in Fig. 8-3*c*, where the contour interval is 1 ft.

Ascending elevations on contour lines indicate that Fig. 8-3*a* is a hill. Descending contour elevations and hachure marks located toward the lower elevation indicate that Fig. 8-3*b* represents a hole or sink.

Contour line elevations going from higher to lower indicate that Fig. 8-3*c* is a ravine. If the elevations were reversed, the same contours would represent a ridge.

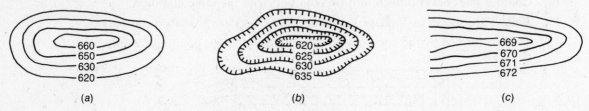

(a)   (b)   (c)

Fig. 8-3 Different contour lines.

Drainage

The primary agent in shaping the topography of terrain is the drainage. Notice in Fig. 8-4 how its influence on the shape of the contour lines can be seen. Contour lines crossing gullies or streams or any other drainage features are modified V's pointing upstream. Underlying material causes the determination of the shape of the V's. Coarse and granular material causes the V's to be sharp. Clay or soil that is fine-grained will give smooth rounded or rounded V's.

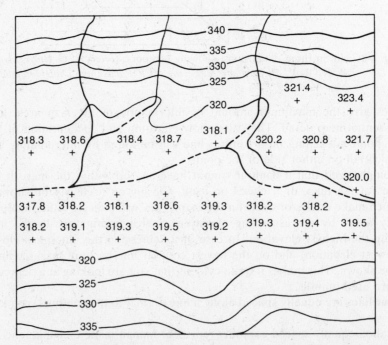

Fig. 8-4 Spot elevations supplementing contour lines. (*From P. Kissam, Surveying Practice, McGraw-Hill, New York, 1978*)

8.5   FIELD METHODS

Factors influencing the field method employed in compiling topographic maps are:

1. Scale of the map
2. Contour interval
3. Nature of the terrain
4. Type of project
5. Equipment available
6. Accuracy required
7. Type of existing control
8. Mapping area

The method of topography commonly employed for highway and railroad mapping is the *cross-section method* (see Sec. 8.6).

The *trace contour method* is used to make an engineering study involving drainage, water impounding, or irrigation, or to prepare a map of an area having little relief. Each contour line is accurately located in its correct horizontal position on the map by following it along the ground.

The *grid method* of obtaining topography is used when a small area is rolling and has many constant slopes; points forming a grid are located on the ground and the elevations of the grid points are determined.

The *controlling-point method* is used where the area to be mapped is extensive. This method utilizes contour lines located by determining the elevations of chosen points from which points on the contours are determined by interpolation. The technician making the map determines the shapes of the contours by experience and judgment.

8.6   CROSS-SECTION METHOD

The most common equipment for cross sectioning follows:

1. Transit and tape   3. Level and tape
2. Transit and stadia   4. Hand level and tape

Horizontal control is established by a theodolite-tape traverse, by an EDM (electronic distance measuring) traverse, or by a stadia traverse between fixed control points. Depending on the terrain, stakes are set at pertinent intervals along a centerline or traverse line. Generally the intervals are 50 or 100 ft, but occasionally a special interval must be included to pick up a special feature of the terrain.

Vertical control, which can be performed prior to or concurrently with the mapping, is obtained by profile leveling yielding centerline elevations. When the theodolite and tape are used, the instrumentperson stands at each station or plus station on the line. He or she determines the HI by holding the leveling rod alongside the instrument and observing the height of the horizontal axis. A right angle (90°) is turned off the line, and the rodperson, holding one end of the tape, proceeds along this crossline until a break occurs in the slope. If possible, the instrumentperson takes a level sight on the rod and records the reading and distance to the point.

The rodperson proceeds to the next break in the topography, or the 50- or 100-ft interval, and repeats the process.

EXAMPLE 8.3   Given: Data recorded in Table 8-1 for nine regular cross sections spaced at 50-ft intervals (see Fig. 8-5). The centerline elevations were obtained from a previous profile leveling of the line of stations. Describe the method of recording data for cross sectioning in the field book.

Table 8-1   Recorded Data for Cross Sections (Example 8.3)

Station	Centerline Elevation	Cross Section					
		L	C			R	
326 + 00	450.0		450.2/93	450.0/0	447.6/23	439.0/61	432.4/106
325 + 50	434.5		455.0/107	434.5/0	432.1/59	427.7/109	
325 + 00	439.3	451.3/103	449.1/31	439.3/0	437.3/10	421.7/100	
324 + 50	444.2	451.0/94	455.1/52	444.2/0	432.3/52	424.5/106	
324 + 00	451.0	450.0/101	456.3/65	451.0/0	429.1/100		
323 + 50	446.6	447.4/102	457.8/61	446.6/0	425.8/94		
323 + 00	450.0	452.0/108	459.8/69	450.0/0	447.6/20	432.0/109	
322 + 50	442.7	454.8/89	449.2/47	442.7/0	438.0/46	435.0/102	
322 + 00	439.3	445.5/61	441.4/19	439.3/0	439.7/32	428.6/102	

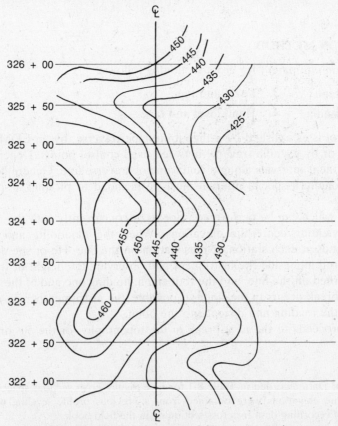

Fig. 8-5 Contour lines interpreted from cross-section notes.

Solution: If the leveling is carried along as the cross-sectional measurements are made, then profile-level notes are recorded on the left-hand side of the field book. This is done in combination with the cross-section notes on the right-hand side. The cross-section notes consist of the following:

1. Station number.

2. The horizontal distance to the right or left of the centerline to the point stated as the denominator of the fraction.

3. The elevation of the point stated as the numerator of the fraction. [Sometimes the rod reading (foresight) is entered as the numerator with the appropriate elevation entered directly above.]

4. Station numbers increase from the bottom of the right-hand field book page upward.

Note: Viewing the right-hand page of the field book (Table 8-1), imagine looking in the forward direction of the line. If you are standing at *C* and looking forward, the values to the left of *C* and to the right of *C* are entered as such. You merely have to remember which numbers constitute the numerator and the denominator of the fraction used for location and elevation of the point.

8.7 PLOTTING THE POINTS

The located points are plotted either in the field or in the office. The positions of the contour lines are obtained by interpolating between the elevations of the plotted points, as will be discussed in Sec. 8.8.

In compiling the topography by the cross-section method, positions of planimetric features, such as fences, buildings, property lines, and streams, are located with respect to the control line. Then they are plotted on the topographic map.

8.8 INTERPOLATING

Points of contours may be determined mathematically or mechanically.

EXAMPLE 8.4 See Fig. 8-6. Given: Two points *a* and *b* whose elevations are 673.4 and 696.2 ft, respectively. We know that contour lines spaced at 5-ft intervals will cause the following contour lines to cross between *a* and *b*: 675, 680, 685, 690, and 695. The horizontal distance between points *a* and *b* is 2.78 in. Find the points on the line where the contour lines are located.

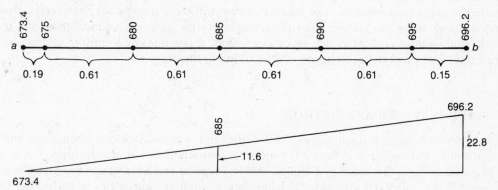

Fig. 8-6 Contour line intersection points.

Solution: The horizontal distance between *a* and *b* = 2.78 in. The corresponding vertical distance is 696.2 − 673.4 = 22.8 ft. The vertical distance from *a* to the 675-ft contour is 675.0 − 673.4 = 1.6 ft. The horizontal distance from *a* to the 675-ft contour is:

$$\frac{2.78 \text{ in}}{22.8 \text{ ft}}(1.6 \text{ ft}) = 0.19 \text{ in}$$

The distance between two adjacent 5-ft contours is:

$$\frac{2.78 \text{ in}}{22.8 \text{ ft}}(5 \text{ ft}) = 0.61 \text{ in}$$

The remaining points are found by successively adding 0.61 in to each contour point; the points are listed as follows:

Point a =   0.00 in
675-ft contour = 0.19 in
680-ft contour = 0.80 in
685-ft contour = 1.41 in
690-ft contour = 2.02 in
695-ft contour = 2.63 in
Point b =   2.78 in

8.9   OTHER METHODS OF OBTAINING TOPOGRAPHY

Several other commonly used methods of obtaining topography are as follows:

1.   Radiation method

2.   Stadia method

3.   Plane table method

4.   Coordinate square method

With the radiation method traverse hubs are covered by a transit or theodolite. The angles are measured to the desired point; then distances to that point are taped or taken by stadia. An EDM may also be used instead of the tape if the instrument is available. Corners of buildings, bridges, and other human-made features should be located. The lengths, widths, and projections are taped and sketched in the field book. This procedure is accurate, but it is the slowest method and usually too expensive for ordinary work.

With the stadia method a similar setup is used. The distances are obtained by stadia. This procedure is fast and fairly accurate for use in most topographic surveys. Lines that radiate from the transit to the required points have azimuths, vertical angles, and distances read by stadia.

An alidade, which is the upper part of the transit, is sighted on a rod held at the point to be located in the plane table method. The stadia distance and vertical angle are then read. Direction of the line is drawn along the alidade ruler. This eliminates the necessity of measuring and recording any horizontal angles. Vertical angles are also avoided, if possible, by using the alidade as a level. The instrumentperson sketches contours while looking at the area.

8.10   COORDINATE SQUARE METHOD

The coordinate square method, also called the *grid method*, is more often used and will be explained more thoroughly. It is better for locating contours than cultural features, but is used for both. Depending upon the accuracy required and the type of terrain, the surveyed area is staked in squares 10, 20, 50, and 100 ft on a side. The lines are laid out at right angles to each other by transit or theodolite; see Fig. 8-7. Note that lines AD and $D3$ are at right angles, as are all the other lines in the surveyed area which make up the grid squares.

Grid lengths are marked on AD and $D3$, and the other corners are staked by intersections of taped lines. A number and letter of intersecting lines identify the corners. In case no transit is available, the layout work can be done with a tape.

To get the corner elevations, a level is set up in the middle of the area, or in a convenient position from which it is possible to take level sights on each point. Contours are interpolated between the corner elevations (on the sides of the squares or on diagonals within the squares) by calculated proportional distances.

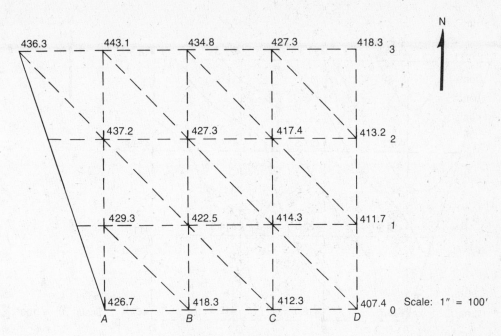

Fig. 8-7 Elevations on a grid.

In contour plotting by the grid method, elevations obtained by interpolation along diagonals will usually not agree with those found by interpolation along the four sides of the squares because of the warped surface of the ground.

EXAMPLE 8.5 Given: A surveyed plat of land as shown in Fig. 8-7. The elevations are all recorded on the stakes at the grid corners. The grid squares are 100 ft. Contour lines are at 5-ft intervals. Sketch in the contour lines on the grid.

Solution: See Fig. 8-8. To find the contour lines we are going to have to interpolate where they pass through the sides of the grid squares and the diagonals through the centers of the squares. Lightly draw in some diagonal lines; see Fig. 8-8 on a map drawn to the scale of 1 in = 100 ft. Contour lines are at 5-ft intervals, so we know the contours are at 410, 415, 420, 425, 430, 435, and 440 ft. The highest point on the map is $A3$, elevation 443.1, and the lowest is at $D0$, elevation 407.4. The contour lines listed above will be contained within the high and low elevations on the map.

Start at line $D0$–$D1$; total difference in elevation is $411.7 - 407.4 = 4.3$ ft. The distance the 410 contour line is from $D1$ is $411.7 - 410 = 1.7$ ft; $1.7/4.3 = 0.395$ or 0.40. This is 40 percent of the line length of $D0$–$D1$. Since this line is exactly 1 in long on the map, (1 in) (0.40) = 0.40 in. Measure south 0.40 in from point $D1$ and make a mark. *Note*: Since the map is at a scale of 1 in = 100 ft, you can make the readings directly from the 10 scale on the engineer's scale.

Start at point $D0$ and go northwest up the diagonal. Elevation differences are $414.3 - 407.4 = 6.9$ ft. 410 contour $= 414.3 - 410 = 4.3$ ft, so $4.3/6.9 = 0.62$. Thus the 410 contour is located 62 percent of the diagonal's length from point $C1$. Measure southeast down the diagonal 0.87 in. This is obtained because the diagonal measures 1.4 in in length and must be multiplied by the percentage figure on the line length: $1.4(0.62) = 0.87$ in.

Start at $C0$ running back to $D0$: elevations $412.3 - 407.4 = 4.9$ ft (total distance). $412.3 - 410 = 2.3$ ft from $C0$ to contour point. $2.3/4.9 = 0.47$, or 47 percent of distance of 1 in = 0.47 in from $C0$. Measure 0.47 in from C toward D and locate the contour point. Now the 410 contour can be drawn in on the map by connecting these points.

Locate contour 415 by interpolation. $B0 = 418.3 - C0 = 418.3 - 412.3 = 6.0$; $418.3 - 415 = 3.3$; $3.3/6.0 = 0.55$. So 55 percent of distance from B to $C = 0.55$ in east of point B. Mark this distance on the BC line.

Start up the diagonal $C0$ to point $B1$: $422.5 - 412.3 = 10.2$; $422.5 - 415 = 7.5$; $7.5/10.2 = 0.74$, or 74 percent of distance of the diagonal. Diagonal measures 1.4 in, so $1.4(0.74) = 1.03$ in. From point $B1$ measure southeast on the diagonal 1.03 in and mark a point.

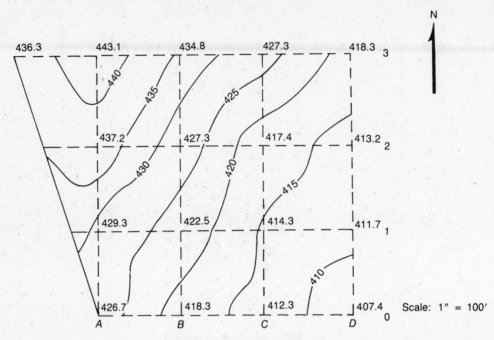

Fig. 8-8 Contour lines on a grid.

You can see from the point $C1$ that the contour line must pass northwest of $C1$ since $C1$ is 414.3 and the contour line is 415. We therefore must find the intersection between points $B1$ and $C1$: $422.5 - 414.3 = 8.2$; $422.5 - 415 = 7.5$; $7.5/8.2 = 0.91$, or 91 percent of distance $B1$ to $C1$ (which is 1 in). Thus $0.91(1) = 0.91$ in from point $B1$ so measure 0.91 in from point $B1$ east on the square line and mark the contour point. Start upward on diagonal from $D1$ northwest to point $C2$. $417.4 - 411.7 = 5.7$; $417.4 - 415 = 2.4$; $2.4/5.7 = 0.42$, or 42 percent of the diagonal length, which is 1.4 in. Thus $1.4(0.42) = 0.59$ in. Measure 0.59 in on diagonal from $C2$ and mark point.

The contour line is going northwest of point $D2$ since $D2$ is at elevation 413.2 and the contour line is 415. Develop the contour crossing point on line $C2$–$D2$: $417.4 - 413.2 = 4.2$; $417.4 - 415 = 2.4$; $2.4/4.2 = 0.57$, or 57 percent of 1 in = 0.57 in. Measure 0.57 in east on line $C2$–$D2$ to get the contour crossing point.

Since $D2$ is elevation 413.2, the 415 contour goes northwest of it; so go up diagonal $D2$–$C3$. Then $427.3 - 413.2 = 14.1$; $427.3 - 415 = 12.3$; $12.3/14.1 = 0.87$, or 87 percent of the diagonal's length. Diagonal measures 1.4 in: $1.4(0.87) = 1.22$ in. Measure 1.22 in down the diagonal from $C3$ toward $D2$ and mark a point.

The contour still has to pass between $D2$ and $D3$ as these two elevations are above and below 415: $418.3 - 413.2 = 5.1$; $418.3 - 415 = 3.3$; $3.3/5.1 = 0.65$, or 65 percent of 1 in = 0.65 in. Make a mark 0.65 in from point $D3$. This is where the contour line exits the map.

The above computations show the method of interpolating where contour lines pass through the grid sides and diagonals. Only two contour lines have been found, but the rest are included on Fig. 8-8. The method is used in the same way for proportioning line space in the grid squares to fix the location of contour lines.

8.11 GEOMETRIC LOCATION OF CONTOUR POINTS

A method for locating the points of contours will be illustrated by Example 8.6 which follows. This is an example of the geometric method of dividing a line into any number of equal parts.

EXAMPLE 8.6   See Figs. 8-9 and 8-10. Given: A line AB which is 2.78 in long. Elevation at station $A = 673.4$ ft and elevation at station $B = 696.2$ ft. Contour interval = 5 ft. Locate the contour points.

Solution: We know there will be five contour lines passing through between 673.4 and 696.2. They are 675, 680, 685, 690, and 695. A triangular engineer's scale (which has divisions in tenths, not eighths and sixteenths) and a triangle are required to perform the mechanical separation of the contours into their proper points. Using the last three figures of the elevations of points A and B, we take $A = 673.4$ as 7.34 in on the engineer's scale. Since $B = 696.2$, we use 9.62 in on the scale.

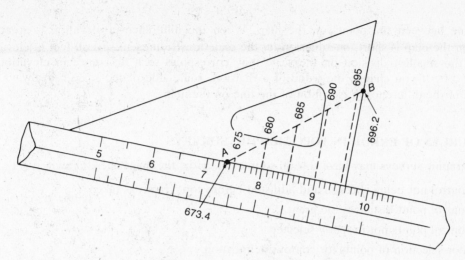

Fig. 8-9 Triangular scale combination for mechanical separation of contours into proper location. (*Courtesy of U.S. Army, General Drafting, T M 5-230*)

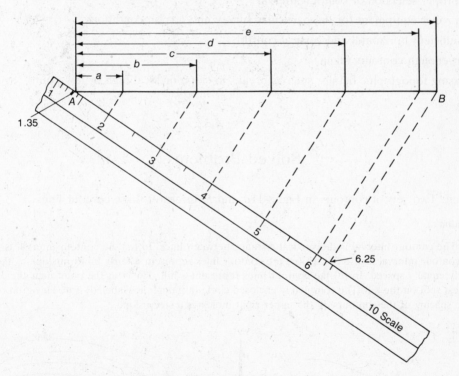

Fig. 8-10 Geometric method of separating a line into parts. (*Courtesy of U.S. Army, General Drafting, T M 5-230*)

Set the 7.34-in mark on the scale on point *A* as shown in Fig. 8-9. Place the corner of the triangle on the 9.62-in mark. Turn the scale and triangle, which must be held together, until the edge of the triangle passes through *B*, whose elevation is 696.2. The scale is then held in place while the edge of the triangle is moved successively to the 9.50-, 9.00-, 8.50-, 8.00-, and 7.50-in points on the scale. The point at which the edge of the triangle crosses line *AB* in these positions is the point where the contours cross the line.

This is a very accurate and fast method to locate contour lines, and it eliminates all mathematical computations. Any edge of the triangular scale can be used, but the length of the scale corresponding to the difference in elevation between the two plotted points must be shorter than the length of the straight line between the points on the map. When the difference in elevation is great and the distance on the map is short, one division on the scale must represent several feet in elevation. For example, the smallest division on the scale may correspond to a difference in elevation of 10 ft instead of 1 ft. If you change the value of a division, some side of the scale can always be used, regardless of the difference in elevation of the line on the map.

8.12 SOURCES OF ERROR IN TOPOGRAPHIC SURVEYS

Topographic surveys may have several sources of error; the main ones follow:

1. Control not being checked and adjusted before topography is taken
2. Control point distances too great
3. Control points not properly selected
4. Poor selection of points for contour delineation

Typical mistakes in topographic surveys follow:

1. Improper selection of contour interval
2. Improper equipment for the particular survey and terrain conditions
3. Insufficient horizontal and vertical control
4. Not enough contours taken
5. Missing topographic details, such as breaks in slope or local high or low point

Solved Problems

8.1   Given: Two sets of contours in Fig. 8-11a and b. Interpret these contour lines.

Solution

The contour interval is the vertical distance between lines. In (a) the contour interval is 5 ft. In (b) the contour interval is also 5 ft. In (a) the contour lines represent a fairly uniform slope as the lines are nearly equally spaced. In (b) the contour lines represent a hill. The × at the point near the 362.5 is the highest spot on the hill. The completely enclosed circle (although not labeled) is an elevation of 360. The close spacing of the contours on the upper right indicates a steep slope.

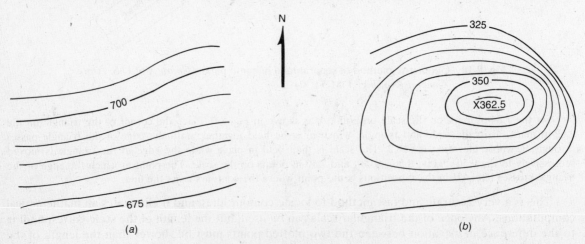

Fig. 8-11 Uniform slope and a hill.

8.2   Given: A contour and a hachure drawing in Fig. 8-12. Interpret these drawings.

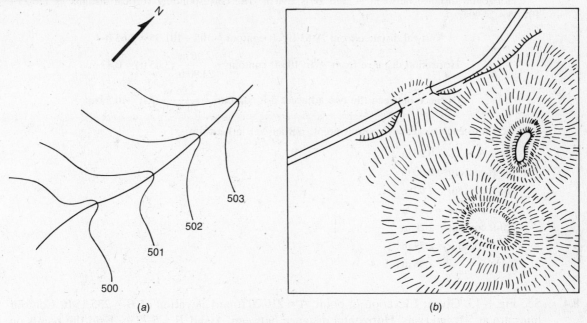

Fig. 8-12 Contour and hachures.

Solution

Figure 8-12a represents a flat slope indicated by the close intervals on the contour lines. They are only 1 ft apart. The higher elevation is to the north. The line running through the contours represents a stream. Since contour lines always bend upstream in crossing creeks or rivers this is a stream which has its origin in the north. The stream is flowing south—probably not rapidly since the elevations are dropping only 1 ft with each contour line.

Figure 8-12b is a drawing showing two low areas—probably filled with water since a road is shown with a bridge going over what is probably a low area containing either a swamp or possibly a small stream. It is made up of hachures which only indicate high or low elevations by their spacing or length. The hachures close together represent a steeper slope than is represented by the longer ones.

8.3   See Fig. 8-13. Given: Two points A and B whose elevations are 101.35 and 126.25 ft, respectively. Contour lines are spaced at 5-ft intervals causing contour lines to cross line AB at elevations 105, 110, 115, 120, and 125 ft. The horizontal distance between A and B is 2.96 in. Find the points on the line where the contour lines are located.

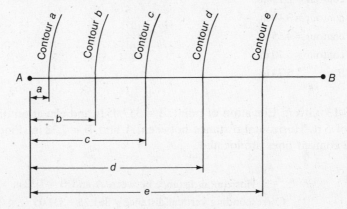

Fig. 8-13 Contour line intersection points for solved problems.

Solution

Horizontal distance between A and B is 2.96 in. The corresponding vertical distance is $126.25 - 101.35 = 24.90$ ft.

$$\text{Vertical distance from } A \text{ to 105-ft contour} = 105 - 101.35 = 3.65 \text{ ft}$$

$$\text{Horizontal distance from } A \text{ to 105-ft contour} = \frac{2.96 \text{ in}}{24.90 \text{ ft}} (3.65 \text{ ft}) = 0.43 \text{ in}$$

$$\text{Distance between the two adjacent 5-ft contours} = \frac{2.96 \text{ in}}{24.90 \text{ ft}} (5 \text{ ft}) = 0.59 \text{ in}$$

Add 0.59 in to each successive contour point; put answer in list form:

Point $A =$ 0.00 in
105-ft contour = 0.43 in
110-ft contour = 1.02 in
115-ft contour = 1.61 in
120-ft contour = 2.20 in
125-ft contour = 2.79 in
Point $B =$ 2.96 in

8.4   See Fig. 8-13. Given: Elevation of point $A = 210.20$ ft and elevation of $B = 235.15$ ft. Contour lines are at 5-ft intervals. Horizontal distance between A and $B = 5.73$ in. Find the points on the line where the contour lines are located.

Solution

$$\text{Horizontal distance between } A \text{ and } B = 5.73 \text{ in}$$
$$\text{Corresponding vertical distance} = 235.15 - 210.20 = 24.95 \text{ ft}$$

Contour lines will cross line AB at elevations 215, 220, 225, 230, and 235 ft.

$$\text{Vertical distance from } A \text{ to 215-ft contour} = 215.0 - 210.20 = 4.8 \text{ ft}$$

$$\text{Horizontal distance from } A \text{ to 215-ft contour} = \frac{5.73 \text{ in}}{24.95 \text{ ft}} (4.8 \text{ ft}) = 1.10 \text{ in}$$

$$\text{Distance between the two adjacent 5-ft contours} = \frac{5.73 \text{ in}}{24.95 \text{ ft}} (5 \text{ ft}) = 1.15 \text{ in}$$

Add 1.15 in to each successive contour point; put answer in list form:

Point $A =$ 0.00 in
215-ft contour = 1.10 in
220-ft contour = 2.25 in
225-ft contour = 3.40 in
230-ft contour = 4.55 in
235-ft contour = 5.70 in
Point $B =$ 5.73 in

8.5   See Fig. 8-13. Given: Elevation of point $A = 337.05$ ft and elevation of $B = 361.25$ ft. Contour intervals of 5 ft. Horizontal distance between A and $B = 7.32$ in. Find the points on the line where the contour lines are located.

Solution

$$\text{Horizontal distance between } A \text{ and } B = 7.32 \text{ in}$$
$$\text{Corresponding vertical distance} = 361.25 - 337.05 = 24.20 \text{ ft}$$

Contour lines will cross line AB at elevations 340, 345, 350, 355, and 360 ft.

$$\text{Vertical distance from } A \text{ to 340-ft contour} = 337.05 - 340.00 = 2.95 \text{ ft}$$

$$\text{Horizontal distance from } A \text{ to 340-ft contour} = \frac{7.32 \text{ in}}{24.20 \text{ ft}} (2.95 \text{ ft}) = 0.89 \text{ in}$$

$$\text{Distance between the two adjacent 5-ft contours} = \frac{7.32 \text{ in}}{24.20 \text{ ft}} (5 \text{ ft}) = 1.51 \text{ in}$$

Add 1.51 in to each successive contour point; put answer in list form:

Point A =   0.00 in

340-ft contour = 0.89 in

345-ft contour = 2.40 in

350-ft contour = 3.91 in

355-ft contour = 5.42 in

360-ft contour = 6.93 in

Point B =   7.32 in

8.6 Given: A surveyed plat of land as shown in Fig. 8-14. All elevations have been recorded on the stakes at the grid corners. The grid squares are 100 ft. Contour lines are at 5-ft intervals. Sketch in the contour lines.

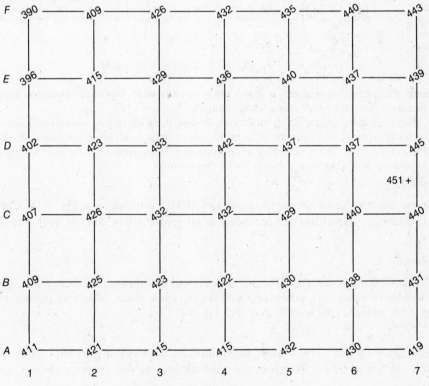

Fig. 8-14 Grid with elevations.

Solution

See Fig. 8-15. To find the contour lines we will estimate the points where the various contour lines will pass through the sides and the diagonals. This will not be as accurate as the mathematical method used in Example 8.6, but it is much faster. Figure 8-14 is just the plat with all the grids shown with elevations at the grid corners. Before looking at Fig. 8-15 where the contours are shown, try to estimate

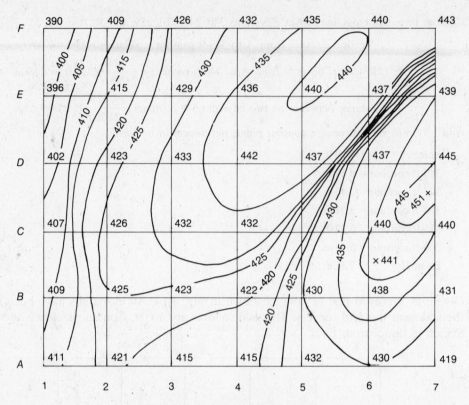

Fig. 8-15 Contours on a grid.

where these contours will be. A good place to start is at the 440-ft elevation (point *F*6). It can be estimated as an oval which closes within the map.

Next you have several 435-ft elevations. Follow these elevations around and you will notice that the contour line enters and exits the map in the upper right quadrant. Continue with the other elevations: 430 down to 400 ft. After sketching what appears to be a good contour map for this grid, refer to Fig. 8-15 and see how your map compares with this solution.

8.7 Given: Surveyed plat with grid squares of 100 ft each side; see Fig. 8-16. Contour lines are at 5-ft intervals. Elevations are recorded at all grid corners. Sketch in the contour lines on the grid.

Solution

Estimate the points where the 5-ft contours will pass through the sides of the grid. On Fig. 8-16 try to plot the contours you would expect with the elevations given. After your plot has been sketched in on Fig. 8-16, compare your contour map with Fig. 8-17.

8.8 See Figs. 8-18 and 8-19. Given: Elevations on a surveyed grid with sides of 100 ft. Contour lines are 5-ft intervals. Sketch in the contour lines as you estimate them.

Solution

After you have drawn in your contours on Fig. 8-18, compare your map with Fig. 8-19.

8.9 See Figs. 8-20 and 8-21. Given: A small grid, 100 ft on each side, with elevations of the grid points as given. The contours are at 5-ft intervals. Draw the contour lines in this grid from the elevations shown.

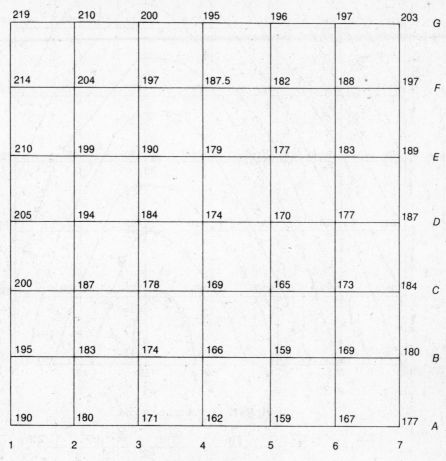

Fig. 8-16 Elevations on a grid.

Solution

Estimate your contour lines and draw in on the map. After drawing in the contours on Fig. 8-20, compare your map with Fig. 8-21.

8.10 See Figs. 8-22 and 8-23. Given: Elevations at the intersection points of a 100-ft-square grid. Contour intervals are 5 ft. Draw the contour map within this grid.

Solution

By estimating the intersections within the grid lines draw your contour map on Fig. 8-22. After finishing your map, compare it with Fig. 8-23.

8.11 See Fig. 8-13. Given: Same data as Prob. 8.3. Elevations at $A = 101.35$ and $B = 126.25$ ft. Contour lines are at 5-ft intervals, causing contour crossings at stations 105, 110, 115, 120, and 125. Horizontal distance between A and B is 2.96 in. By the mechanical method find the points on the line where the contours cross. Check your solution with that found by the mathematical method in Prob. 8.3.

Solution

We take the last three figures of the elevations of A and B to use on our scale. In this problem to attain more accuracy we should double the scale of line AB. So $2(2.96) = 5.92$; note we will divide the measurements on line AB by 2 after obtaining the points. This increases the accuracy of measuring when

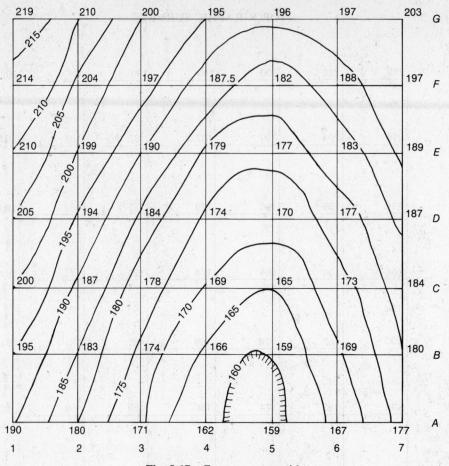

Fig. 8-17 Contours on a grid.

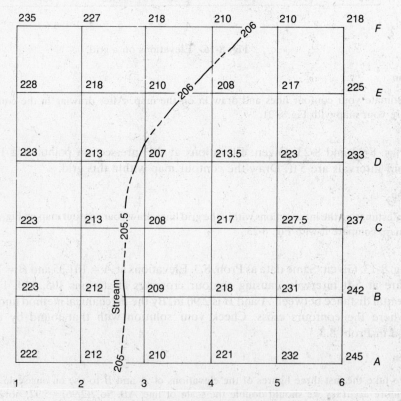

Fig. 8-18 Elevations on a grid.

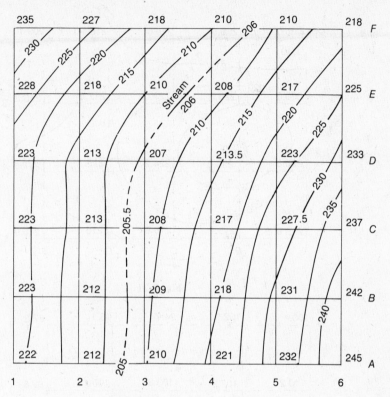

Fig. 8-19   Contours on a grid.

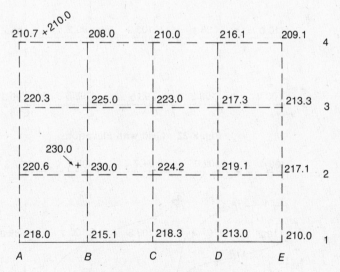

Fig. 8-20   Grid with elevations.

your line is short. Set your 1.35 figure on the tens of the engineer's scale with the triangle resting on the front edge of the scale; see Fig. 8-9. Now the 90° angle corner of the triangle must rest on the 6.25 on the scale; rotate scale and triangle together until the right-angle edge of the triangle intersects point *B*.

To get the contour intersections strike lines perpendicular to the scale, down the edge of the triangle as you slide the triangle along the edge of the scale. You will mark lines at these scale numbers: 6, 5, 4, 3, 2. These represent the 105-, 110-, 115-, 120-, and 125-ft contour lines.

Measure these intersection points on your line *AB*. Since we multiplied the line length by 2 to gain more accuracy in measuring, we will reduce the measured dimension by 2 to obtain the actual sizes.

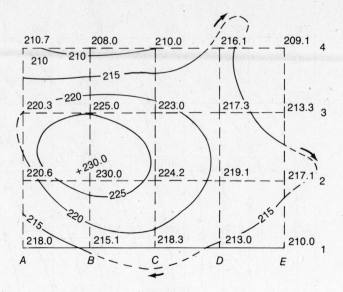

Fig. 8-21

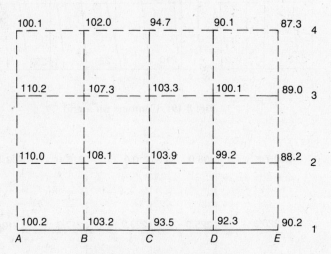

Fig. 8-22 Grid with elevations.

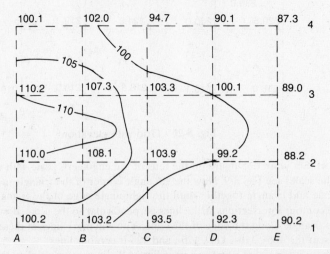

Fig. 8-23

First contour (station a) $105 = \dfrac{0.75}{2} = 0.38$ in

Second contour (station b) $110 = \dfrac{1.95}{2} = 0.97$ in

Third contour (station c) $115 = \dfrac{3.24}{2} = 1.62$ in

Fourth contour (station d) $120 = \dfrac{4.42}{2} = 2.21$ in

Fifth contour (station e) $125 = \dfrac{5.60}{2} = 2.80$ in

Point $B = \dfrac{5.92}{2} = 2.96$ in

Compare these figures with the tabulated measurements in Prob. 8.3; the above answers are close to the mathematical answers. The greatest inaccuracy occurs at contour 105.

8.12 See Fig. 8-13. Given: Same data as Prob. 8.4. Elevations at $A = 210.20$ and $B = 235.15$ ft. Contours at 5-ft intervals. Horizontal distance between A and $B = 5.73$ in. By the mechanical method find the points on the line where contours cross. Check your solution with that found by the mathematical method in Prob. 8.4.

Solution

$A = 210.20$; use 0.20 on scale. $B = 235.15$; use 5.15 on scale. Locate 0.20 number on the scale over point A, with triangle and scale positioned as shown in Fig. 8-9. Position the edge of the triangle over 5.15 on the scale and rotate the two together until the edge of the triangle which is perpendicular to the scale passes through point B. Make marks on line AB at scale points 5, 4, 3, 2, 1. These marks on line AB represent where the contour lines would pass through the line. *Note*: We did not double the scale to obtain more accuracy of reading as we did in the previous problem, so we read the dimension directly from the scale. The dimensions are: $a = 0.94$ in, $b = 2.18$ in, $c = 3.30$ in, $d = 4.42$ in, $e = 5.60$ in, and point $B = 5.73$ in. These figures are close to the mathematical solution obtained in Prob. 8.4.

Chapter 9

Construction Surveys

9.1 INTRODUCTION

In construction surveying the surveyor supplies all the reference marks so that each individual new improvement is placed at the correct location on the property. Once the map for a project is available, detailed plans for the construction of the project can be completed. The surveyor prepares a site plan which shows relationships between the land and all the improvements to be erected thereon, and marks all the horizontal positions and their elevations. This process is begun before the work is started and continues throughout the construction period.

The surveyor's work must proceed at a pace to provide the necessary marks just prior to the time they will be required by the builder for each day's operation. The surveyor must never get so far ahead of the work that the marks might be destroyed by the rough action of the building process.

The surveying process is called *construction surveying*, or *location surveying*. It requires special techniques. The construction surveying procedures for ordinary structures are covered in this chapter.

9.2 METHODS

Construction plans always give through scale, or by actual dimension, all positions and elevations of new work. These are given relative to survey control points or existing structures. The dimensions of the new construction, which are shown on the plans, give all the extra necessary data for the construction survey.

EXAMPLE 9.1 See Fig. 9-1. Given: A house to be constructed 30 ft from a monument; dimensions are as shown (total depth of 40 ft). The west property line starts at a monument which is 30 ft west of the westernmost side of the house; this monument has an elevation of 150.11 ft. The front property line extends 100 ft east to another monument whose elevation is 150.41 ft. The first floor of the new house is to be 2.00 ft higher than the monument whose elevation is 150.41. Required city setback from the property line is 25 ft. Find the setup of the stakes to mark position and elevation of the house.

Solution: Sighting from one monument to the other on the property line, set the first stake 30 ft east on the property line. Set the second stake on the same property line 14 ft farther. Then on the same property line set the third stake 16 ft farther. Go back to stake 1 and turn a 90° angle left from a backsight (north direction) and measure 25 ft. This gives the setback of 25 ft required by the city. This stake (number 4) is the southwest corner of the new house. Prolong this same sight 16 ft. This is the back of the house at the offset.

Come back to stake 2 on the property line and again set up and turn a 90° angle left from a backsight (north). Along this sighting first measure 41 ft and erect a stake. Prolong this sight and measure 24 ft more and set a stake. This gives the point of the northwest corner of the house.

Set up on stake 3 and turn a 90° angle left from a backsight (north), and set a stake 25 ft from stake 3. Prolong the sight and set another stake 40 ft beyond this stake. This last stake is the northeast corner of the house.

Finally, mark each stake with the elevation of 152.41, which is the required distance above the monument elevation of 150.41 ft.

Note: The stakes may be set at the corners or other points required; later they can be transferred to nearby marks which will not be disturbed during construction. These safe points can be located by short measurements with a carpenter's rule and the elevation transferred by carpenter's level.

9.3 TRANSFER FROM STAKES TO THE IMPROVEMENT

The equipment used to transfer line and grade from the stakes set by the construction survey are familiar to most persons. They are 6-ft folding rules, steel or reinforced cloth tapes, carpenter's level, plumb bob, and string lines.

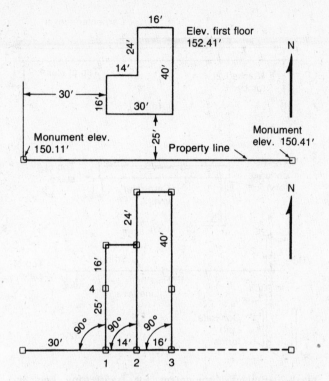

Fig. 9-1 Location dimensions for a house (top); one
method of staking out a house, showing stakes
set and angles and distances measured (bottom).

A carpenter's level, used with a level rod or a 6-ft folding rule as a rod, is a simple engineer's level, and the carpenter uses it in the same way. It has a very small field of vision, no magnification, and much less sensitivity in its level bubble. It is used to transfer grades for construction for distances up to 50 ft.

Carpenter's and mason's levels are made of long wooden or metal frames in which two level vials are fixed. One vial is perpendicular to the long edge of the level; the other is parallel to it. The use of levels is shown in Fig. 9-2.

Line levels are also used to transfer levels from the stakes to the construction; see Fig. 9-2. A line level consists of a level vial with hooks on both ends. These hooks allow the vial to be suspended from a string line to indicate whether the line is level. It is always placed at the midpoint of the string because its weight makes the line sag. At the center point a correct level reading can be obtained. The level line is accurate up to 30 ft.

9.4 BASELINE AND OFFSET CONTROL

Several reference systems are available to locate an improvement with respect to the land and any other buildings on the land. Example 9.1 and Probs. 9.1 to 9.3 are problems in which the corners of a new construction project are located by stakes directly on the point where the project is to be constructed. This is not usually done, as the construction would destroy the stake marking the building spot as well as the elevation of the point. The *baseline and offset method* eliminates the serious problem of the vulnerability of stakes at the location of construction. In this method the position and elevation are transferred to the construction point by carpenter's level and rule from stakes which have been offset from the actual construction point. Stakes are generally offset by 3 to 10 ft. The baseline method is often used where areas near the proposed structure are paved.

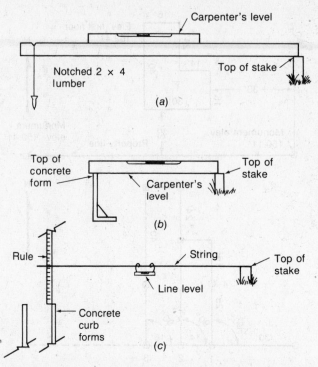

Fig. 9-2 Builder's equipment for transferring line and grade. (*From C. A. Herubin, Principles of Surveying, Reston Publishing, Reston, Va., 1982*)

Example 9.2 will demonstrate a building located with reference to two property lines defined by corner markers. The building will be constructed parallel to both property lines.

EXAMPLE 9.2   See Fig. 9-3. Given: Property of undisclosed length and width, but which has corner markers. Building is to be constructed parallel to both property lines. Stakes are to have 10-ft offsets. There must be a 25-ft offset from the front (east) and side (south) property lines. The building is 50 ft wide and 135 ft 4 in long. The front of the building faces east. Determine the setup of the stakes.

Solution: Set the transit over one property corner. Line is established by sighting to the other corner. Measure the distances shown in Fig. 9-3 along the property line from the southeast corner. Put nails in the ground at proper locations to measure offsets to the stakes. Backsights are set at each property corner unless corner markers can be seen with the transit when set up over the nails. The transit should then be set up over each nail, and a backsight taken on whichever property corner is farthest away. Stakes should be set at right angles at the correct distance. Mark finish floor elevation on all stakes.

9.5   ANGLE-AND-DISTANCE METHOD

Stake locations may be designated by angle and distance. This method will be illustrated by Example 9.3.

EXAMPLE 9.3   See Fig. 9-4. Given: A plot plan as shown in Fig. 9-4 with a building 50 ft wide and 135 ft 4 in long. Setbacks of 25 ft are required from the east side (front) of the property and 25 ft from the south property line. The stakes are to be offset 10 ft from the building corners so they will not be disturbed by the construction. Find the distance and angle of the stakes from the southeast and northeast control points.

Solution: Set a control point at the southeast property corner. From this point four offset stakes will be set. The formulas to use are as follows:

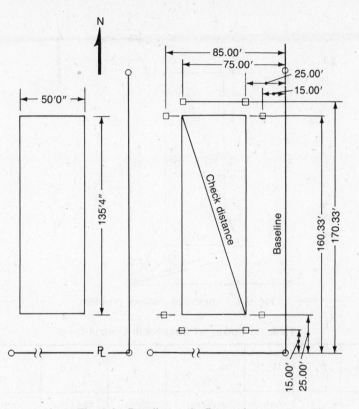

Fig. 9-3 Baseline and offset stakeout.

$$\tan a = \frac{W}{N} \qquad L = \frac{W}{\sin a}$$

where N = distance north of south property line (N can also be distance south of the north control point), ft

W = distance west of the east property line, ft

a = angle between property line and line from stake to control point, °

L = distance from control point to stake, ft

From Fig. 9-4 the distances from the given setback and offset information can be put into table form and easily worked out.

Stake 1 is north 25 ft and west 15 ft of the property lines. Use the formula to find tan a:

$$\tan a = \frac{W}{N} = \frac{15}{25} = 0.60000$$

Look up 0.60000 in a trigonometric function table or use a calculator to find

$$a = 30°58'$$

Find the sine of angle a (use a trigonometric function table or calculator):

$$\sin a = 0.51454$$

Now use the given formula:

$$L = \frac{W}{\sin a} = \frac{15}{0.51454} = 29.15 \text{ ft}$$

Make a table of values as shown in Fig. 9-4.

Stake 2 is north 15 ft and west 25 ft from the property lines. Set these figures into the formula:

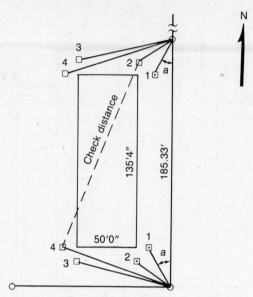

Fig. 9-4 Angle-and-distance problem.

Table for Stakes Set from South Control Point

Stake	N, ft	W, ft	$\tan a = \dfrac{W}{N}$	a	$\sin a$	$L = \dfrac{W}{\sin a}$, ft
1	25	15	0.60000	30°58′	0.51454	29.15
2	15	25	1.66667	59°02′	0.85747	29.15
3	15	75	5.00000	78°42′	0.98061	76.49
4	25	85	3.40000	73°37′	0.95940	88.60

Table for Stakes Set from North Control Point

Stake	S, ft	W, ft	$\tan a = \dfrac{W}{N}$	a	$\sin a$	$L = \dfrac{W}{\sin a}$, ft
1	25	15	0.60000	30°58′	0.51454	29.15
2	15	25	1.66667	59°02′	0.85747	29.15
3	15	75	5.00000	78°42′	0.98061	76.49
4	25	85	3.40000	73°37′	0.95940	88.60

$$\tan a = \frac{W}{N} = \frac{25}{15} = 1.66667 \quad \text{so}$$
$$a = 59°02' \quad \text{and} \quad \sin a = 0.85747$$

Now use the given formula: $\qquad L = \dfrac{W}{\sin a} = \dfrac{25}{0.85747} = 29.15\,\text{ft}$

Enter this value into the table.

Stake 3 is north 15 ft (25-ft setback − 10-ft offset) and west 75 ft (50-ft building width + 25-ft setback). Use the formula:

$$\tan a = \frac{W}{N} = \frac{75}{15} = 5.00000 \quad \text{so}$$
$$a = 78°42' \quad \text{and} \quad \sin a = 0.98061$$

Use the formula:

$$L = \frac{W}{\sin a} = \frac{75}{0.98061} = 76.49 \text{ ft}$$

Stake 4 is north of the property line by 25 ft and west of the property line by 85 ft (10-ft offset + 50-ft building width + 25-ft setback).

Use the formula:

$$\tan a = \frac{W}{N} = \frac{85}{25} = 3.40000 \quad \text{so}$$
$$a = 73°37' \quad \text{and} \quad \sin a = 0.95940$$

Use the formula:

$$L = \frac{W}{\sin a} = \frac{85}{0.95940} = 88.60 \text{ ft}$$

The above procedure has located the four offset stakes at the south of the building.

To get the north control point sight up the property line and measure a distance of 185.33 ft (the 25-ft setback on the south + 135.33-ft building length + 25 ft added to the north) to match figures on the south side of the building. This point is our control point on the north side of the building. Perform the same mathematical functions as for the south control point. The math is shown in the second chart of Fig. 9-4.

Since the same 25-ft distance is added in the north as the 25-ft setback required by specifications to building length + south 25-ft setback, the answers are going to be the same for both charts. The only difference is that the direction north is changed to south in the second chart.

Informational note: There is a possibility of roundoff error. For example, for stake 3 there is a roundoff of 0.40 minute when you round down to 41 minutes; 0.4 minute of angle affects the sine and hence the length. Using 41 minutes to find L yields 76.483, whereas using 41.4 minutes to find L yields 78.4857. In this case the difference is negligible, but for longer lengths it might cause error.

The angle-and-distance staking method saves field time because the distances to be measured are usually shorter, even if one additional long measurement between the north and south control points must be made.

One person in the office must do extra calculations for this method. Offsetting this work is the fact that the field party (who are several in number) will save time if several stakes are placed from one or two instrument setups.

The angle-and-distance method does have some limitations. Site conditions might limit visibility, making it impossible to use this method. The size of the project might require measuring much longer distances, in which case the method is subject to error. In most other instances, however, the angle-and-distance method is a good method to use, as it saves time and money.

The baseline and offset method (see Sec. 9.4) is easier to visualize in the field, which helps to prevent mistakes. With either method results can be checked by measuring diagonals. The correct diagonal length should be computed before going to the field, or two diagonals should be measured to see whether they are equal in length. The diagonals may be from nails set at the actual corner locations or from stake to stake.

9.6 COMPLEX PLANS: COORDINATE METHOD

If the design of a facility is complex, such as a chemical processing plant, the designers often provide coordinates at key construction points based on two mutually perpendicular baselines.

Monuments are used to establish baselines. The surveyor should use permanent monuments (iron rods set in a concrete base). These then will be available for control of repairs and alterations after the plant is built.

The surveyor should set construction stakes as needed throughout the construction process, just as would be done with the other methods. The baseline and offset or angle-and-distance method would be employed by using points on a baseline for the instrument setups.

9.7 ESTABLISHING LINE IN THE FIELD: MARKING POSITION

The process of giving line consists of establishing predetermined angles and distances and placing a series of marks in line, usually at given distances. The angles and alignment are established with a transit. The distances are measured with a steel tape.

Setting a Predetermined Angle

The procedure to establish an angle is as follows. First set the transit at the angle point, or vertex; then follow these steps:

1. Set the *A* vernier at zero, using the upper motion.
2. Point at the mark, using the lower motion.
3. Turn off the angle, using the upper motion, setting the *A* vernier accurately at the value of the angle.
4. Set a mark on the new line.

Such an angle can be set only to the nearest half minute. When greater accuracy is desired, the angle just established must be measured by repetition and the tack in the top of the stake adjusted accordingly. The distance the tack must be shifted is computed by trigonometry (see Fig. 9-5). If the error is less than 3 minutes of arc the following formula is accurate:

$$D = 0.00000485SR$$

where
D = distance to shift the tack
0.00000485 = number of radians in 1 second of arc
S = seconds of error
R = distance from transit to stake set

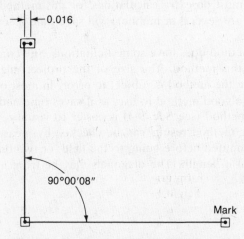

Fig. 9-5 Example of sighting 400 ft to tack.

Good procedure dictates rechecking the final angle by repetition. If more than one mark is available on a line of sight, always take the more distant mark as this will reduce the error of sighting.

EXAMPLE 9.4 See Fig. 9-5. Given: A sighting of 400 ft to a tack. The intended angle was 90°00′00″. The actual angle measured by repetition was 90°00′08″. Determine how far the tack in the top of the stake should be moved to adjust the angle.

Solution: 90°00′08″ − 90°00′00″ = 08 seconds error

Use the formula: $D = 0.00000485SR = 0.00000485(08)(400) = 0.016$ ft

So the tack should be moved 0.016 ft to the right.   *Ans.*

9.8 ESTABLISHING GRADE: MARKING ELEVATIONS

Marking elevations is called *giving grade* or *grade staking*. It consists of setting marks such as the tops of stakes, nails in vertical surfaces, and keel marks at required elevations. Marks are also set at random elevations with indications of the vertical heights at which future construction projects must be built above or below them. Marks for grade are placed near the work and transferred to the work by carpenter's levels and rules as noted earlier in this chapter.

Definitions

Surveyors use the word *grade* rather loosely. In this text *rate of grade* is used to mean steepness of slope. Grade is used here to mean the elevation of future construction. We will not use grade to mean slope.

Rate of Grade.   The rate of grade, usually referred to as the *gradient*, is the rate of change of elevation expressed as a ratio of the change in elevation divided by the horizontal distance.

EXAMPLE 9.5   Given: 100-ft distance; street sloping downward. Vertical drop in elevation = 1 ft. Find the rate of grade.

Solution:

$$\text{Gradient} = \frac{\text{change in vertical elevation}}{\text{horizontal distance}}$$

$$\text{Gradient} = \frac{-1}{100} = -0.01 \quad \text{or} \quad -1\%$$

Cut and Fill.   When the grade is above the grade mark, the notation "fill x ft and y in" is written at the mark, as F 3 ft 6 in. When the grade is below the grade mark, *cut* is used, as C 1 ft 10 in. The words *fill* and *cut* in this usage mean only up and down from a mark and have nothing to do with embankment or excavation.

9.9 METHODS OF GIVING GRADE

Two methods of grade staking will be discussed: (1) shooting in grade, (2) indicating cuts and fills.

Shooting in Grade

If marks are to be set for a uniform rate of grade, computation and field work can be saved by a process known as *shooting in grade*. This process is not independent. It is necessary to first set a grade stake or mark at each end of the uniform grade. A transit or level is then set up over the mark at one end (see Fig. 9-6). The difference in height between the instrument and the mark is measured (4.07), and the target on the rod is set at this value. The rod is held at the grade mark at the other end of the

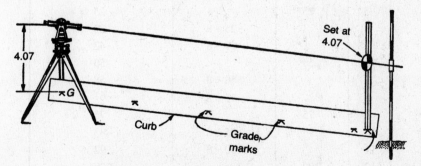

Fig. 9-6 Transit and rod for shooting in grade.

slope and the line of sight directed at the target. This places the line sight parallel to the grade line at a known height above it. With this arrangement, a grade mark can be set wherever desired by holding the rod at that point and raising or lowering the rod until the target is on the line of sight. The position of the bottom of the rod is marked; see Fig. 9-6.

A foresight on the line of sight should be established if many grades are to be set. When only a few are necessary, the slope of the line of sight should be checked when the work is completed by holding the rod at the original mark and making sure the line of sight strikes the target.

Indicating Cuts and Fills

The most rapid and, in most cases, the best method of giving grade is to indicate the cuts or fills measured from convenient objects near the work. Usually the objects are the tops of the line stakes or other line marks.

The elevations of the tops of the line stakes or the objects chosen are determined by profile leveling. The values of the cuts or fills are computed by comparing the elevation of each mark with the elevation at the proposed grade at that particular position. They are computed in hundredths of a foot, converted to inches, and marked on the stakes or near the marks (see Example 9.6). The tops of the stakes or other objects are usually covered with keel to indicate that grade should be measured from those points.

Reducing Hundredths to Inches. One inch equals $8\frac{1}{3}$ hundredths of a foot. Each $\frac{1}{4}$ ft (called a *quarter point*) can be expressed exactly in hundredths and in inches. By adding or subtracting 8 to or from the nearest quarter point, the inch values can be computed in hundredths of a foot to within one-third of a hundredth. Table 9-1 gives these values.

To reduce hundredths to inches, choose the nearest inch value and correct for the odd hundredths by calling them eighths of an inch. For example:

$$0.89\,\text{ft} = 0.92\,\text{ft} - 0.03\,\text{ft} = 11\,\text{in} - \tfrac{3}{8}\,\text{in} = 10\tfrac{5}{8}\,\text{in}$$
$$0.44\,\text{ft} = 0.42\,\text{ft} + 0.02\,\text{ft} = 5\,\text{in} + \tfrac{2}{8}\,\text{in} = 5\tfrac{1}{4}\,\text{in}$$
$$0.71\,\text{ft} = 0.75\,\text{ft} - 0.04\,\text{ft} = 9\,\text{in} - \tfrac{4}{8}\,\text{in} = 8\tfrac{1}{2}\,\text{in}$$

In the use of Table 9-1 the error is never greater than 0.005 ft.

Table 9-1   Converting Hundredths of Feet to Inches

In	Quarter Points	Computations	Inch Values in Hundredths of a Foot
0	0		0
1		0 + 8	8
2		25 − 8	17
3	25		25
4		25 + 8	33
5		50 − 8	42
6	50		50
7		50 + 8	58
8		75 − 8	67
9	75		75
10		75 + 8	83
11		100 − 8	92
12	100		100

Procedure for Setting Grade Stakes

Bench mark data, a list of required grades, and a sketch of the work to be done must be taken into the field. Example 9.6 illustrates the procedure for establishing grade.

EXAMPLE 9.6 See Fig. 9-7. Given: A concrete slab at elevation of 74.00 ft. To this will be added a platform. The elevation will be 76.17 ft. Stakes are set on line and profile leveled. The stations and elevations (in feet) follow: station $0 + 0$, elevation = 72.13; station $0 + 50$, elevation = 72.75; station $1 + 0$, elevation = 73.05; station $1 + 50$, elevation = 71.81; station $2 + 0$, elevation = 71.42; and station $2 + 50$, elevation = 71.02. Find the elevation grade and cut or fill for all stations.

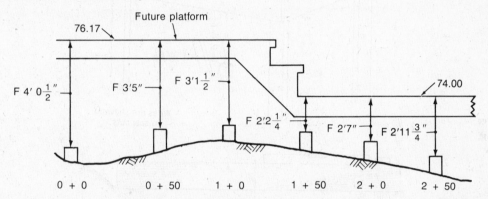

Fig. 9-7 Slab and platform for cut and fill.

Solution: Set up a schedule of station, elevation, grade, elevation grade, and cut or fill. By subtracting the grade at which the slab and future platform will be placed from the elevation at each station, you get the elevation grade. In other words, the vertical distance from the ground minus grade equals cut or fill measured in hundredths. Then convert this figure to inches for the cut or fill column. If the elevation grade is negative, fill is indicated; if it is positive, cut is indicated.

Station	Top of Stake Elevation, ft	Proposed Grade, ft	Elevation Grade	Cut or Fill
$0 + 0$	72.13	76.17	-4.04	F 4 ft $0\frac{1}{2}$ in
$0 + 50$	72.75	76.17	-3.42	F 3 ft 5 in
$1 + 0$	73.05	76.17	-3.12	F 3 ft $1\frac{1}{2}$ in
$1 + 50$	71.81	74.00	-2.19	F 2 ft $2\frac{1}{4}$ in
$2 + 0$	71.42	74.00	-2.58	F 2 ft 7 in
$2 + 50$	71.02	74.00	-2.98	F 2 ft $11\frac{3}{4}$ in

Computations are as follows:

$$\text{Elevation} - \text{grade} = 72.13 - 76.17 = -4.04 \text{ ft}$$

Convert -4.04 ft to inches, using Table 9-1:

$$-4.04 \text{ ft} = -4 \text{ ft } 0\tfrac{1}{2} \text{ in}$$

Since the elevation grade is negative, mark F for fill.

Station $0 + 50$:   Elevation $-$ grade $= 72.75 - 76.17 = -3.42 = -3$ ft 5 in (fill)
Station $1 + 0$:   Elevation $-$ grade $= 73.05 - 76.17 = -3.12 = -3$ ft $1\frac{1}{2}$ in (fill)
Station $1 + 50$:   Elevation $-$ grade $= 71.81 - 74.00 = -2.19 = -2$ ft $2\frac{1}{4}$ in (fill)
Station $2 + 0$:   Elevation $-$ grade $= 71.42 - 74.00 = -2.58 = -2$ ft 7 in (fill)
Station $2 + 50$:   Elevation $-$ grade $= 71.02 - 74.00 = -2.98 = -2$ ft $11\frac{3}{4}$ in (fill)

Note: After a surveyor has set marks for line and grade, it is usually necessary to use string or wire to guide the construction of the project. These strings are usually supported on pins, or *batter boards*; see Fig. 9-8.

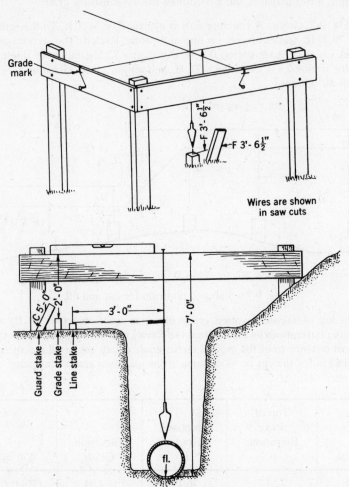

Fig. 9-8 Batter board. (*From P. Kissam, Surveying Practice, McGraw-Hill, New York, 1978*)

Solved Problems

9.1 See Fig. 9-9. Given: Plan for a house in which the front setback is 25 ft. The setback from the west lot line is 35 ft. Monument elevation is 78.06, and finished first floor must be 2 ft higher than the monument elevation. Front of house is 50 ft long. West side of the house is 34 ft, east side is 24 ft, and the offset in the rear of the house is 28 ft starting from the west wall. Find the distance and angles to be measured and the stakes to be placed to lay out this house.

Solution

Sight the transit from the property stake on the west of the lot to the monument at the east corner. Drive a stake 35 ft from property stake on the sighted line (stake 1). Measure an additional 28 ft east along the line and set stake 2. Measure an additional 22 ft east along the line and set stake 3. Come back

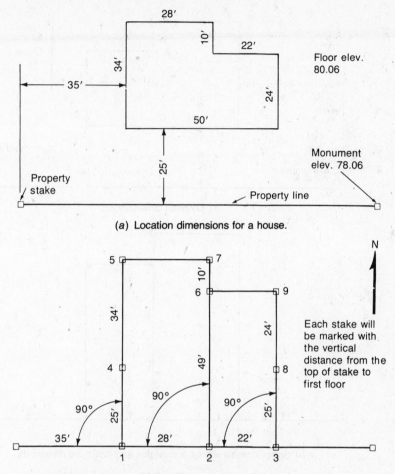

(a) Location dimensions for a house.

(b) Method of staking a house by setting stakes and angles and distances.

Fig. 9-9

to stake 1 and turn 90° clockwise (sighting from property stake 1 to monument at the west corner). Measure 25 ft north along the left (western) line and set stake 4 (the left-hand, or west, front of the house). Measure along the same line 34 ft more and set stake 5. Go to stake 2; there turn 90° clockwise from a backsight along the property line to the west corner and measure 49 ft to set stake 6. Measure an additional 10 ft on the same line and set stake 7. Come back to stake 3 on the property line and turn 90° as before. Measure 25 ft and set stake 8; continue on the same line 24 ft more to set stake 9. Mark the finish floor elevation of 80.06 on each stake.

9.2   See Fig. 9-10. Given: Plot plan (Fig. 9-10a) where $A = 27.5$ ft, $B = 25.0$ ft, $C = 46$ ft, $D = 20$ ft, $E = 26$ ft, $F = 40$ ft, $G = 28$ ft, $K = 12$ ft, $H = 50.03$. Find the stake locations and angles for setting up this house.

Solution

Sight the transit from the monument on the west, along the property line, to the monument on the east. Measure 27.5 ft from the west monument and set stake 1. Along the property line sight, set stake 2 at 20 ft east of stake 1. Again along the property line sight, measure 26 ft east from stake 2 and set stake 3.

Return to location of stake 1, and sighting east along the property line, turn 90° left for the lineup of stake 4, which is measured 25 ft from stake 1. Along the same sighting measure 40 ft from stake 4 and set stake 5. At stake 2 along the property line, sighting toward the monument with H elevation, turn left 90°. On this line measure 53 ft and set stake 6. On the same sighting measure 12 more ft and set stake 7.

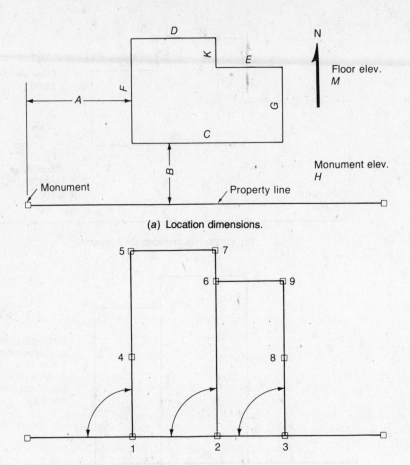

(a) Location dimensions.

(b) Setup to show stakes and angles and distances measured.

Fig. 9-10

Go back to the property line at stake 3, and sighting eastward to monument with elevation H, turn left 90°. Along this sighting measure 25 ft and set stake 8. On the same sighting set stake 9, 28 ft past stake 8. Mark the finish floor elevation of 52.03 on each stake.

9.3 See Fig. 9-10. Given: Plot plan for a house with the following dimensions: $A = 25$ ft, $B = 25$ ft, $C = 60$ ft, $D = 30$ ft, $E = 30$ ft, $F = 24$ ft, $G = 20$ ft, $K = 4$ ft, $H = 47.05$ ft, $M = 49.05$ ft. Find the angles and distances at which stakes are to be set for this house.

Solution

Set transit over west monument and sight toward east monument along property line. Measure 25 ft and set stake 1; measure on 30 ft and set stake 2; continue on 30 ft and set stake 3.

Go back to stake 1, and sighting along the property line toward the east, turn 90° counterclockwise. Along this line measure 25 ft and set stake 4; continuing on this sight, measure 24 ft and set stake 5. Come back to stake 2 on property line with transit pointing east. Turn 90° left, measure 45 ft, and set stake 6; on the same line measure 4 ft farther and set stake 7. Go back to stake 3 on the property line, transit pointed east, and turn a 90° angle left. Along this line measure 25 ft from stake 3 and set stake 8. On the same sighting measure an additional 20 ft and set stake 9. Mark the finish floor elevation of 49.05 on all stakes.

9.4 See Fig. 9-11. Given: Lot marked by corner monuments. SE monument has an elevation of 51.22 ft and the finish floor elevation should be 53.22 ft. The front and side (east and south) should be set back 25 ft from the property lines. The control stakes should be 10 ft from the

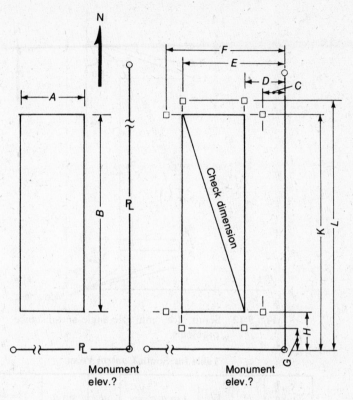

Fig. 9-11 Baseline and offset stakeout.

construction corners. Dimensions are as follows: $A = 50$ ft, $B = 170$ ft, $C = 15$ ft, $D = 25$ ft, $E = 75$ ft, $F = 85$ ft, $G = 15$ ft, $H = 25$ ft, $K = 195$ ft, $L = 205$ ft. Determine the setup of the stakes.

Solution

Measure the distances given down the property lines and set all stakes using the transit and steel tape. Mark all offset stakes, of which there are eight, as finish floor elevation 53.22.

9.5   Given: The building plot as shown in Fig. 9-12; the stake offsets are 8 ft. The setbacks are 25 ft from both lot lines. Building size: A dimension is 52 ft and B dimension is 140 ft. Find the angle and distance of the stakes from the control points. Find distance D between control points. If the south control point is a monument with elevation of 125.07 ft and the finished floor needs to be 0.5 ft higher, give proper finish floor elevations for each stake.

Solution

The monument at the south edge of the lot will be the south control point. Figure the N and W coordinates for the four stakes.

Stake 1: $N = 25$ ft, $W = 17$ ft (25-ft setback; 8-ft offset)

Stake 2: $N = 17$ ft (25-ft setback; 8-ft offset), $W = 25$ ft

Stake 3: $N = 17$ ft (25-ft setback; 8-ft offset), $W = 77$ ft (52-ft building width + 25-ft setback)

Stake 4: $N = 25$ ft, $W = 85$ ft (52-ft building width + 25-ft setback + 8-ft offset)

The second table would be the same except the north column would be labeled south.

Stake 1: $$\tan a = \frac{W}{N} = \frac{17}{25} = 0.68000 \quad \text{so}$$

$$a = 34°13' \quad \text{and} \quad \sin a = 0.56232$$

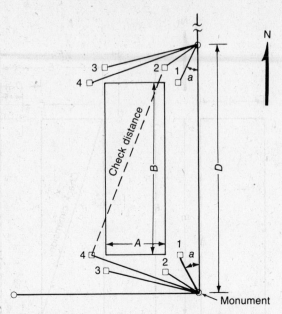

Fig. 9-12 Setup for multiple-angle-and-distance problems.

Table for South Control Point

Stake	N	W	$\tan a = \dfrac{W}{N}$	a	$\sin a$	$L = \dfrac{W}{\sin a}$
1						
2						
3						
4						

Table for North Control Point

Stake	S	W	$\tan a = \dfrac{W}{N}$	a	$\sin a$	$L = \dfrac{W}{\sin a}$
1						
2						
3						
4						

Formula:

$$L = \frac{W}{\sin a} = \frac{17}{0.56232} = 30.23 \text{ ft}$$

Stake 2:

$$\tan a = \frac{W}{N} = \frac{25}{17} = 1.47058 \quad \text{so}$$

$$a = 55°47' \quad \text{and} \quad \sin a = 0.82692$$

Formula:

$$L = \frac{W}{\sin a} = \frac{25}{0.82692} = 30.23 \text{ ft}$$

Stake 3:

$$\tan a = \frac{W}{N} = \frac{77}{17} = 4.52941 \quad \text{so}$$

$$a = 77°33' \quad \text{and} \quad \sin a = 0.97648$$

Formula:
$$L = \frac{W}{\sin a} = \frac{77}{0.97648} = 78.85 \text{ ft}$$

Stake 4:
$$\tan a = \frac{W}{N} = \frac{85}{25} = 3.40000 \quad \text{so}$$
$$a = 73°37' \quad \text{and} \quad \sin a = 0.95940$$

Formula:
$$L = \frac{W}{\sin a} = \frac{85}{0.95940} = 88.60 \text{ ft}$$

Put these figures in the top table; see Fig. 9-12.

$D = 25$-ft setback $+ 140$-ft length of building $+ 25$ ft (identical to south setback) $= 190$ ft

Monument elevation is 125.07 and finished floor is required to be 0.5 ft higher; so $125.07 + 0.5 = 125.57$ ft. Mark all offset stakes with finish floor elevation of 125.57.

9.6 See Fig. 9-12. Given: A factory building with A dimension of 200 ft and B dimension of 452 ft. The setbacks will be 30 ft from east and south property lines. The stakes will be offset 6 ft from building corners. Monument elevation is 632.08 ft, and finished floor will be 1.5 ft higher. Find D, stake distances and angles, and the finish floor elevation to be put on the stakes.

Solution

Stake 1: $N = 30$ ft, $W = 24$ ft (30-ft setback $-$ 6-ft offset)
Stake 2: $N = 24$ ft, $W = 30$ ft
Stake 3: $N = 24$ ft, $W = 230$ ft (building width $+$ setback)
Stake 4: $N = 30$ ft, $W = 236$ ft (setback $+$ building width $+$ offset)

The second table will be the same as the first since we will add the same dimension as the south setback to make it so.

$D = 30$-ft setback on south end of line $+ 452$-ft length of the building $+ 30$ ft on the north end of line
$ = 512$ ft

Monument elevation is 632.08. Add the required height above the monument elevation of 1.5 ft. This gives 633.58 as the finish floor elevation. This will be marked on all offset stakes.

Stake 1:
$$\tan a = \frac{W}{N} = \frac{24}{30} = 0.80000 \quad \text{so}$$
$$a = 38°40' \quad \text{and} \quad \sin a = 0.62479$$

Formula:
$$L = \frac{W}{\sin a} = \frac{24}{0.62479} = 38.42 \text{ ft}$$

Stake 2:
$$\tan a = \frac{W}{N} = \frac{30}{24} = 1.25000 \quad \text{so}$$
$$a = 51°20' \quad \text{and} \quad \sin a = 0.78079$$

Formula:
$$L = \frac{W}{\sin a} = \frac{30}{0.78079} = 38.42 \text{ ft}$$

Stake 3:
$$\tan a = \frac{W}{N} = \frac{230}{24} = 5.58333 \quad \text{so}$$
$$a = 84°2.57' \quad \text{and} \quad \sin a = 0.99460$$

Formula:
$$L = \frac{W}{\sin a} = \frac{230}{0.99460} = 231.25 \text{ ft}$$

Stake 4:
$$\tan a = \frac{W}{N} = \frac{236}{30} = 7.86666 \text{ ft} \quad \text{so}$$
$$a = 82°45' \quad \text{and} \quad \sin a = 0.99202$$

Formula:
$$L = \frac{W}{\sin a} = \frac{236}{0.99202} = 237.90 \text{ ft}$$

9.7 Given: Building shown in Fig. 9-12 with dimensions, offset stakes, and setbacks as follows: $A = 26$ ft 6 in; $B = 108$ ft 3 in. Setbacks all 25 ft; stakes to be offset 3 ft. Monument elevation 101.06; finished floor to be 1 ft higher than monument. Find all stake distances and angles from control points, distance D, and finish floor elevation.

Solution

Stake 1: $N = 25$ ft, $W = 22$ ft
Stake 2: $N = 22$ ft, $W = 25$ ft
Stake 3: $N = 22$ ft, $W = 51.5$ ft
Stake 4: $N = 25$ ft, $W = 54.5$ ft

$$D = 25\text{-ft setback} + 108.25\text{-ft building length} + 25 \text{ ft} = 158.25 \text{ ft}$$

Monument elevation $= 101.06 + 1.0 = 102.06$ to be marked on offset stakes as finish floor elevation.

Stake 1:
$$\tan a = \frac{W}{N} = \frac{22}{25} = 0.88000 \quad \text{so}$$
$$a = 41°21' \quad \text{and} \quad \sin a = 0.66067$$

Formula:
$$L = \frac{W}{\sin a} = \frac{22}{0.66067} = 33.30 \text{ ft}$$

Stake 2:
$$\tan a = \frac{W}{N} = \frac{25}{22} = 1.13636 \quad \text{so}$$
$$a = 48°39' \quad \text{and} \quad \sin a = 0.75069$$

Formula:
$$L = \frac{W}{\sin a} = \frac{25}{0.75069} = 33.30 \text{ ft}$$

Stake 3:
$$\tan a = \frac{W}{N} = \frac{51.5}{22} = 2.34090 \quad \text{so}$$
$$a = 66°52' \quad \text{and} \quad \sin a = 0.91959$$

Formula:
$$L = \frac{W}{\sin a} = \frac{51.5}{0.91959} = 56.00 \text{ ft}$$

Stake 4:
$$\tan a = \frac{W}{N} = \frac{54.5}{25} = 2.18000 \quad \text{so}$$
$$a = 65°21' \quad \text{and} \quad \sin a = 0.90887$$

Formula:
$$L = \frac{W}{\sin a} = \frac{54.5}{0.90887} = 59.96 \text{ ft}$$

9.8 Given: Building shown in Fig. 9-12 with dimensions, offset stakes, and setbacks as follows: $A = 46$ ft; $B = 118$ ft; all setbacks from east and south lot lines are 25 ft. Stakes to be offset 5 ft from building corners. Monument elevation is 65.00. Finish floor elevation to be 2 ft higher than monument. Find D, finish floor elevation, and the angle and distance to offset stakes from control points.

Solution

Stake 1: $N = 25$ ft, $W = 20$ ft

Stake 2: $N = 20$ ft, $W = 25$ ft

Stake 3: $N = 20$ ft, $W = 71$ ft

Stake 4: $N = 25$ ft, $W = 76$ ft

$$D = 25\text{-ft setback} + 118 \text{ ft} + 25 \text{ ft} = 168 \text{ ft}$$

The finish floor elevation to be marked on the offset stakes is monument elevation of $65.00 + 2.00 = 67.00$.

Stake 1:
$$\tan a = \frac{W}{N} = \frac{20}{25} = 0.80000 \quad \text{so}$$
$$a = 38°40' \quad \text{and} \quad \sin a = 0.62479$$

Formula:
$$L = \frac{W}{\sin a} = \frac{20}{0.62479} = 32.01 \text{ ft}$$

Stake 2:
$$\tan a = \frac{W}{N} = \frac{25}{20} = 1.25000 \quad \text{so}$$
$$a = 51°20' \quad \text{and} \quad \sin a = 0.78079$$

Formula:
$$L = \frac{W}{\sin a} = \frac{25}{0.78079} = 32.02 \text{ ft}$$

Stake 3:
$$\tan a = \frac{W}{N} = \frac{71}{20} = 3.55000 \quad \text{so}$$
$$a = 74°16' \quad \text{and} \quad \sin a = 0.96253$$

Formula:
$$L = \frac{W}{\sin a} = \frac{71}{0.96253} = 73.76 \text{ ft}$$

Stake 4:
$$\tan a = \frac{W}{N} = \frac{76}{25} = 3.04000 \quad \text{so}$$
$$a = 71°47.49' \quad \text{and} \quad \sin a = 0.94993$$

Formula:
$$L = \frac{W}{\sin a} = \frac{76}{0.94993} = 80.01 \text{ ft}$$

9.9 Given: Building shown in Fig. 9-12 with dimensions, offset stakes, and setbacks as follows: $A = 58$ ft, $B = 176$ ft. All setbacks from east and south lot lines are 25 ft. Stakes to be offset 10 ft from building corners. Monument elevation is 68.08. Finished floor will be 2 ft higher than the monument. Find D, finish floor elevation, and angle and distance to all the offset stakes from the control points.

Solution

Stake 1: $N = 25$ ft, $W = 15$ ft

Stake 2: $N = 15$ ft, $W = 25$ ft

Stake 3: $N = 15$ ft, $W = 83$ ft

Stake 4: $N = 25$ ft, $W = 93$ ft

$$D = 25 + 176 + 25 = 226 \text{ ft}$$

Finish floor elevation to be marked on stakes $= 68.08 + 2.00 = 70.08$

Stake 1: $\qquad$ $\tan a = \dfrac{W}{N} = \dfrac{15}{25} = 0.60000 \qquad$ so

$\qquad a = 30°58' \qquad$ and $\qquad \sin a = 0.51454$

Formula: $\qquad$ $L = \dfrac{W}{\sin a} = \dfrac{15}{0.51454} = 29.15 \text{ ft}$

Stake 2: $\qquad$ $\tan a = \dfrac{W}{N} = \dfrac{25}{15} = 1.66667 \qquad$ so

$\qquad a = 59°02' \qquad$ and $\qquad \sin a = 0.85747$

Formula: $\qquad$ $L = \dfrac{W}{\sin a} = \dfrac{25}{0.84805} = 29.15 \text{ ft}$

Stake 3: $\qquad$ $\tan a = \dfrac{W}{N} = \dfrac{83}{15} = 5.5333 \qquad$ so

$\qquad a = 79°45' \qquad$ and $\qquad \sin a = 0.98404$

Formula: $\qquad$ $L = \dfrac{W}{\sin a} = \dfrac{83}{0.98404} = 84.35 \text{ ft}$

Stake 4: $\qquad$ $\tan a = \dfrac{W}{N} = \dfrac{93}{25} = 3.72000 \qquad$ so

$\qquad a = 74°57' \qquad$ and $\qquad \sin a = 0.96570$

Formula: $\qquad$ $L = \dfrac{W}{\sin a} = \dfrac{93}{0.96570} = 96.30 \text{ ft}$

9.10 Given: Building shown in Fig. 9-12 with dimensions, offset stakes, and setbacks as follows: $A = 98$ ft, $B = 180$ ft. All setbacks from east and south lot lines are 25 ft. Stakes to be 10 ft from building corners. Monument elevation = 710.51 ft and finished floor is to be 2 ft higher. Find D, finish floor elevation, and angle and distance to all offset stakes from control points.

Solution

Stake 1: $N = 25$ ft, $W = 15$ ft

Stake 2: $N = 15$ ft, $W = 25$ ft

Stake 3: $N = 15$ ft, $W = 123$ ft

Stake 4: $N = 25$ ft, $W = 133$ ft

$$D = 25 + 180 + 25 = 230 \text{ ft}$$
$$\text{Finish floor elevation} = 710.51 + 2.00 = 712.51$$

Stake 1: $\qquad$ $\tan a = \dfrac{W}{N} = \dfrac{15}{25} = 0.60000 \qquad$ so

$\qquad a = 30°57.83' \qquad$ and $\qquad \sin a = 0.51454$

Formula: $\qquad$ $L = \dfrac{W}{\sin a} = \dfrac{15}{0.51454} = 29.15 \text{ ft}$

Stake 2: $\qquad$ $\tan a = \dfrac{W}{N} = \dfrac{25}{15} = 1.66667 \qquad$ so

$\qquad a = 59°02' \qquad$ and $\qquad \sin a = 0.85747$

Formula: $\qquad$ $L = \dfrac{W}{\sin a} = \dfrac{25}{0.85747} = 29.16 \text{ ft}$

Stake 3:
$$\tan a = \frac{W}{N} = \frac{123}{15} = 8.2000 \quad \text{so}$$
$$a = 83°03' \quad \text{and} \quad \sin a = 0.99265$$

Formula:
$$L = \frac{W}{\sin a} = \frac{123}{0.99265} = 123.91 \text{ ft}$$

Stake 4:
$$\tan a = \frac{W}{N} = \frac{133}{25} = 5.32000 \quad \text{so}$$
$$a = 79°21' \quad \text{and} \quad \sin a = 0.98279$$

Formula:
$$L = \frac{W}{\sin a} = \frac{133}{0.98279} = 135.33 \text{ ft}$$

9.11 See Fig. 9-13. Given: Actual measured angle $a = 89°59'53''$ and R (distance) $= 500$ ft. Find the distance and the direction to move the tack.

Solution

Subtract measured angle from desired angle:

$$90°00'00'' - 89°59'53'' = 07 \text{ seconds}$$

So　　　　　　　　　$S = 07 \text{ seconds} \quad R = 500 \text{ ft}$

Use the formula:

$$D = 0.00000485SR = 0.00000485(07)(500) = 0.017 \text{ ft}$$

Since the measured angle is less than 90°, move the tack 0.017 ft to the left.

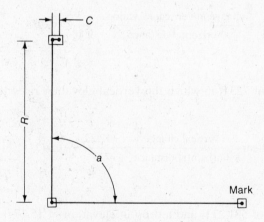

Fig. 9-13　Setup for tack movement.

9.12 See Fig. 9-13. Given: Measured angle $a = 90°00'08''$; distance $R = 600$ ft. Find the distance and the direction to move the tack.

Solution

$$S = 90°00'08'' - 90°00'00'' = 08 \text{ seconds}$$

Use the formula:

$$D = 0.00000485SR = 0.00000485(08)(600) = 0.023 \text{ ft}$$

The measured angle is greater than 90°, so move the tack 0.023 ft to the right.

9.13 See Fig. 9-13. Given: Measured angle $a = 89°59'54''$; distance $R = 462$ ft. Find the distance and the direction to move the tack.

Solution

$$S = 90°00'00'' - 89°59'54'' = 06 \text{ seconds}$$

Use the formula:

$$D = 0.00000485SR = 0.00000485(06)(462) = 0.013 \text{ ft}$$

Since the measured angle is less than 90°, move the tack 0.013 ft to the left.

9.14 See Fig. 9-13. Given: Measured angle $a = 90°00'05''$; distance $R = 673$ ft. Find the distance and the direction to move the tack.

Solution

$$S = 90°00'05'' - 90°00'00'' = 05 \text{ seconds}$$

Use the formula:

$$D = 0.00000485SR = 0.00000485(05)(673) = 0.016 \text{ ft}$$

Since the measured angle is greater than 90°, move the tack 0.016 ft to the right.

9.15 In a construction survey the distance of the street involved = 200 ft and the vertical drop = 3 ft. Find the rate of grade.

Solution

$$\text{Gradient} = \frac{\text{change in vertical elevation}}{\text{horizontal distance}} = \frac{-3}{200} = -0.015 \quad \text{or} \quad -1.5\% \quad Ans.$$

9.16 Given: Distance of 723 ft in which the vertical elevation rises 12.32 ft. Find the rate of grade.

Solution

$$\text{Gradient} = \frac{\text{vertical change}}{\text{horizontal distance}} = \frac{+12.32}{723} = +0.017 \quad \text{or} \quad +1.7\% \quad Ans.$$

9.17 Given: Distance of 531.23 ft and a drop in elevation of 15.17 ft. Find the rate of grade.

Solution

$$\text{Gradient} = \frac{\text{vertical change}}{\text{horizontal distance}} = \frac{-15.17}{531.23} = -0.029 \quad \text{or} \quad -2.9\% \quad Ans.$$

9.18 Given: Vertical drop of 30.08 ft in 650 ft. Find the gradient.

Solution

$$\text{Gradient} = \frac{\text{vertical change}}{\text{horizontal distance}} = \frac{-30.08}{650} = -0.04627 = -4.63\% \quad Ans.$$

9.19 Given: The following measures in decimal feet: 2.81, 4.78, 8.23, 7.91. Convert these measurements to feet and inches.

Solution

Use Table 9-1.

2.81: $0.81 \text{ ft} = 0.83 \text{ ft} - 0.02 \text{ ft} = 10 \text{ in} - \frac{2}{8} \text{ in} = 9\frac{3}{4} \text{ in}$ so
                    $2.81 \text{ ft} = 2 \text{ ft } 9\frac{3}{4} \text{ in}$ *Ans.*

4.78: $0.78 \text{ ft} = 0.75 \text{ ft} + 0.03 \text{ ft} = 9 \text{ in} + \frac{3}{8} \text{ in} = 9\frac{3}{8} \text{ in}$ so
                    $4.78 \text{ ft} = 4 \text{ ft } 9\frac{3}{8} \text{ in}$ *Ans.*

8.23: $0.23 \text{ ft} = 0.25 \text{ ft} - 0.02 \text{ ft} = 3 \text{ in} - \frac{2}{8} \text{ in} = 2\frac{3}{4} \text{ in}$ so
                    $8.23 \text{ ft} = 8 \text{ ft } 2\frac{3}{4} \text{ in}$ *Ans.*

7.91: $0.91 \text{ ft} = 0.92 \text{ ft} - 0.01 \text{ ft} = 11 \text{ in} - \frac{1}{8} \text{ in} = 10\frac{7}{8} \text{ in}$ so
                    $7.91 \text{ ft} = 7 \text{ ft } 10\frac{7}{8} \text{ in}$ *Ans.*

9.20 Given: The following measures in decimal feet: 5.64, 3.85, 1.92, 0.38, 9.25. Convert these measurements to feet and inches.

Solution

Use Table 9-1.

5.64: $0.64 \text{ ft} = 0.67 \text{ ft} - 0.03 \text{ ft} = 8 \text{ in} - \frac{3}{8} \text{ in} = 7\frac{5}{8} \text{ in}$ so
                    $5.64 \text{ ft} = 5 \text{ ft } 7\frac{5}{8} \text{ in}$ *Ans.*

3.85: $0.85 \text{ ft} = 0.83 \text{ ft} + 0.02 \text{ ft} = 10 \text{ in} + \frac{2}{8} \text{ in} = 10\frac{1}{4} \text{ in}$ so
                    $3.85 \text{ ft} = 3 \text{ ft } 10\frac{1}{4} \text{ in}$ *Ans.*

1.92: $0.92 \text{ ft} = 11 \text{ in}$ so $1.92 \text{ ft} = 1 \text{ ft } 11 \text{ in}$ *Ans.*

0.38: $0.38 \text{ ft} = 0.42 \text{ ft} - 0.04 \text{ ft} = 5 \text{ in} - \frac{4}{8} \text{ in} = 4\frac{1}{2} \text{ in}$ so
                    $0.38 \text{ ft} = 0 \text{ ft } 4\frac{1}{2} \text{ in}$ *Ans.*

9.25: $0.25 \text{ ft} = 3 \text{ in}$ so $9.25 \text{ ft} = 9 \text{ ft } 3 \text{ in}$ *Ans.*

9.21 See Fig. 9-14. Given: Building floor elevation is 102.31 ft and loading platform elevation is 100.81 ft. Elevations (in feet) at the tops of the stakes are as follows: station $0 + 0 = 97.20$; station $0 + 50 = 96.51$; station $1 + 0 = 96.03$; station $1 + 50 = 97.81$. Find the cut or fill for each station.

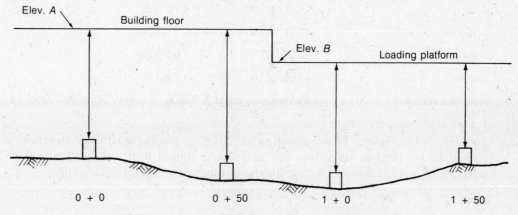

Fig. 9-14 Loading dock setup for cut and fill problems.

Solution

Set the information into tabular form.

Station	Elevation, ft	Grade, ft	Elevation Grade	Cut or Fill
0 + 0	97.20	102.31	−5.11	F 5 ft $1\frac{3}{8}$ in
0 + 50	96.51	102.31	−5.80	F 5 ft $9\frac{5}{8}$ in
1 + 0	96.03	100.81	−4.78	F 4 ft $9\frac{3}{8}$ in
1 + 50	97.81	100.81	−3.00	F 3 ft 0 in

Computations are as follows:

Station 0 + 0:   Elevation − grade = 97.20 − 102.31 = −5.11 ft

Convert −5.11 ft by Table 9-1 into −5 ft $1\frac{3}{8}$ in. Since elevation grade is negative, mark F for *fill*.

Station 0 + 50:   Elevation − grade = 96.51 − 102.31 = −5.80 ft = −5 ft $9\frac{5}{8}$ in

Station 1 + 0:   Elevation + grade = 96.03 − 100.81 = −4.78 = −4 ft $9\frac{3}{8}$ in

Station 1 + 50:   Elevation − grade = 97.81 − 100.81 = −3.00 = −3 ft 0 in

9.22 See Fig. 9-14. Given: Floor elevation is 45.00 ft; loading platform elevation is 43.00 ft. Elevations (in feet) at tops of stakes are as follows: station 0 + 0 = 43.25; station 0 + 50 = 42.51; station 1 + 0 = 42.26; station 1 + 50 = 43.01. Find the cut or fill for each station.

Solution

Set into tabular form.

Station	Elevation, ft	Grade, ft	Elevation Grade	Cut or Fill
0 + 0	43.25	45.00	−1.75	F 1 ft 9 in
0 + 50	42.51	45.00	−2.49	F 2 ft $5\frac{7}{8}$ in
1 + 0	42.26	43.00	−0.74	F 0 ft $8\frac{7}{8}$ in
1 + 50	43.01	43.00	+0.01	C 0 ft $\frac{1}{8}$ in

Computations are as follows:

Station 0 + 0:   43.25 − 45.00 = −1.75 = −1 ft 9 in

Station 0 + 50:   42.51 − 45.00 = −2.49 = −2 ft $5\frac{7}{8}$ in

Station 1 + 0:   42.26 − 43.00 = −0.74 = −0 ft $8\frac{7}{8}$ in

Station 1 + 50:   43.01 − 43.00 = +0.01 = +0 ft $\frac{1}{8}$ in

Note that station 1 + 50 has a positive elevation grade, so the final answer is cut instead of fill.

9.23 See Fig. 9-14. Given: Floor elevation is 10.23 ft; loading platform elevation is 9.71 ft. Elevations (in feet) at stake tops are as follows: station 0 + 0 = 9.65; station 0 + 50 = 9.90; station 1 + 0 = 9.75; station 1 + 50 = 10.00. Find the cut or fill for each station.

Solution

Set into tabular form.

Station	Elevation, ft	Grade, ft	Elevation Grade	Cut or Fill
0 + 0	9.65	10.23	−0.58	F 0 ft 7 in
0 + 50	9.90	10.23	−0.33	F 0 ft 4 in
1 + 0	9.75	9.71	+0.04	C 0 ft $0\frac{1}{2}$ in
1 + 50	10.00	9.71	+0.29	C 0 ft $3\frac{1}{2}$ in

Computations are as follows:

Station 0 + 0: $9.65 - 10.23 = -0.58 = -0$ ft 7 in

Station 0 + 50: $9.90 - 10.23 = -0.33 = -0$ ft 4 in

Station 1 + 0: $9.75 - 9.71 = +0.04 = +0$ ft $0\frac{1}{2}$ in

Station 1 + 50: $10.00 - 9.71 = +0.29 = +0$ ft $3\frac{1}{2}$ in

Supplementary Problems

9.24 See Fig. 9-11. Given: Same setback (25 ft) and offset (10 ft) as Prob. 9.4. Building dimensions are 52 ft 4 in wide and 152 ft 8 in long. Monument elevation (SE corner) = 87.02 ft; finished floor should be 2 ft higher. Figure out locations of all offset stakes.
Ans. $A = 52.33$ ft, $B = 152.67$ ft, $D = 25$ ft, $C = 15$ ft, $E = 77.33$ ft, $F = 87.33$ ft, $H = 25$ ft, $G = 15$ ft, $K = 177.67$ ft, $L = 187.67$ ft, finish floor elevation = 89.02

9.25 See Fig. 9-11. Given: The same information on setbacks and offsets as in Prob. 9.4. Monument height (SE corner) is 72.09, and finished floor is to be 2 ft higher. The building dimensions are: $A = 40$ ft 8 in, $B = 148$ ft 4 in. Find the positions of all offset stakes.
Ans. $A = 40.67$ ft, $B = 148.33$ ft, $D = 25$ ft, $C = 15$ ft, $E = 65.67$ ft, $F = 75.67$ ft, $H = 25$ ft, $G = 15$ ft, $K = 173.33$ ft, $L = 183.33$ ft, finish floor elevation = 74.09

9.26 See Fig. 9-11. Given: Setback and offset information as in Prob. 9.4. Monument on SE corner has an elevation of 103.51 ft, and finished floor should be 2 ft higher. $A = 48$ ft, and $B = 154$ ft. Find the positions of all offset stakes.
Ans. $A = 48$ ft, $B = 154$ ft, $D = 25$ ft, $C = 15$ ft, $E = 73$ ft, $F = 83$ ft, $H = 25$ ft, $G = 15$ ft, $K = 179$ ft, $L = 183.33$ ft, finish floor elevation = 105.51

9.27 See Fig. 9-13. Given: Measured angle $a = 90°00'07''$; distance $R = 434$ ft. Find the distance and the direction to move the tack.   *Ans.*   0.015 ft to the right

9.28 Given: Vertical rise = 25.07 ft in a horizontal distance of 452 ft. Find the gradient.
Ans. +0.0555 or 5.55%

9.29   Given: Vertical rise of 50.16 ft in distance of 751 ft. Find the gradient.   *Ans.*   +0.06679 or 6.68%

9.30   Given: The following measures in decimal feet: 3.53, 4.79, 8.23, 10.15, 7.56. Convert these measurements to feet and inches.
Ans.   3.53 ft = 3 ft $6\frac{3}{8}$ in, 4.79 ft = 4 ft $9\frac{1}{2}$ in, 8.23 ft = 8 ft $2\frac{3}{4}$ in, 10.15 ft = 10 ft $1\frac{3}{4}$ in, 7.56 ft = 7 ft $6\frac{3}{4}$ in

9.31   Given: The following measures in decimal feet: 6.25, 7.50, 2.75, 4.02. Convert these measurements to feet and inches.
Ans.   6.25 ft = 6 ft 3 in, 7.50 ft = 7 ft 6 in, 2.75 ft = 2 ft 9 in, 4.02 ft = 4 ft $\frac{1}{4}$ in

9.32   See Fig. 9-14. Given: Floor elevation is 52.23 ft; loading platform elevation is 49.21 ft. Elevations (in feet) at the tops of the stakes are as follows: station $0 + 0 = 48.01$; station $0 + 50 = 47.60$; station $1 + 0 = 47.21$; station $1 + 50 = 49.00$. Find the cut or fill for each station.
Ans.   Station $0 + 0$, F 4 ft $2\frac{5}{8}$ in; station $0 + 50$, F 4 ft $7\frac{1}{2}$ in; station $1 + 0$, F 2 ft 0 in; station $1 + 50$, F 0 ft $2\frac{1}{2}$ in

9.33   See Fig. 9-14. Given: Floor elevation of 26.75 ft, and loading platform elevation of 25.80 ft. Elevations (in feet) at the tops of the stakes are as follows: station $0 + 0 = 24.50$; station $0 + 50 = 24.25$; station $1 + 0 = 24.20$; station $1 + 50 = 26.00$. Find the cut or fill for each station.
Ans.   Station $0 + 0$, F 2 ft 3 in; station $0 + 50$, F 2 ft 6 in; station $1 + 0$, F 1 ft $7\frac{1}{4}$ in; station $1 + 50$, C 0 ft $2\frac{3}{8}$ in

9.34   See Fig. 9-14. Given: Floor elevation is 600.00 ft; loading platform elevation is 597.00 ft. Elevations (in feet) at tops of stakes are as follows: station $0 + 0 = 597.08$; station $0 + 50 = 596.75$; station $1 + 0 = 597.00$; station $1 + 50 = 597.50$. Find the cut or fill for each station.
Ans.   Station $0 + 0$, F 2 ft 11 in; station $0 + 50$, F 3 ft 3 in; station $1 + 0$, 0 ft 0 in; station $1 + 50$, C 0 ft 6 in

Chapter 10

Slope Staking

10.1 DEFINITION

Slope staking is the method of giving line and grade for the construction of sloping surfaces when such surfaces meet uneven ground. Its use is for staking out retaining walls and laying out earth excavations and embankments. Slope stakes are placed for every 50 ft of highway or railroad before construction can begin. Slope staking procedures are similar for all types of construction. When you have mastered the procedure for one type of construction, you can apply it to all other types. Slope staking is used mainly in highway construction for marking out cuts and fills.

10.2 METHOD

Stakes are placed on the existing ground surface, one on each side of the centerline stake. These stakes are placed where the edge of a cut or the toe of a fill will come when the earthwork is completed (see Fig. 10-1). In Fig. 10-1 the centerline stake is in place and the stakes on each side are *slope stakes*. They are placed where the limits of the earthwork must be located. Each slope stake must be marked with its horizontal distance to the right or left of the centerline and the vertical distance from the ground at the stake down to the elevation of the bottom of a cut or up to the elevation of the top of a fill, i.e., to the elevation of the line known as the *base* (see Fig. 10-2). All stakes *must* be marked with the station number.

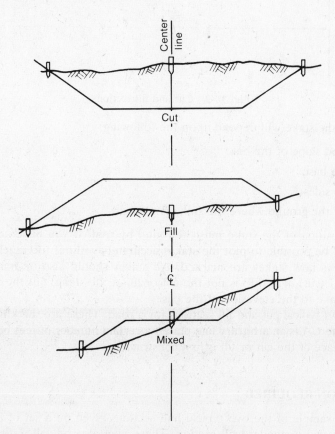

Fig. 10-1 Placing of stakes.

219

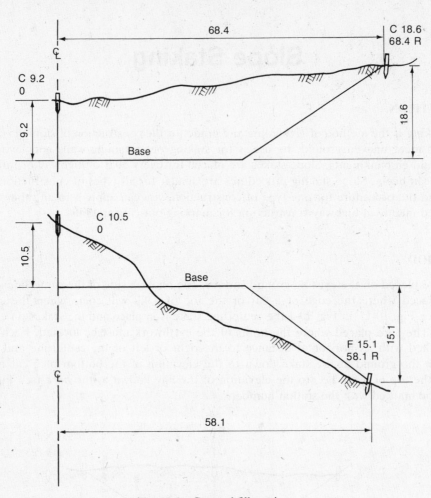

Fig. 10-2 Cut and fill sections.

The position of the stake will depend upon the following:

1.   Elevation and slope of the base
2.   Width of the base
3.   Slope of the sides
4.   Elevation of the ground where the stake is placed

Normally the position of the stake must be found by trial. If you have accurately plotted road cross sections, it will be possible to plot the stakes accurately without too much computation.

Figure 10-2 shows how stakes are marked. All stakes should also be marked with the station number. Note that C (cut) or F (fill) is not the cut or fill at the stake, but the vertical distance from the ground at the stake to the elevation of the base.

Figure 10-3 shows typical cut and fill sections for a road. These are the kinds of drawings shown on the plans. The line *CA* is an arbitrary line placed a certain number of feet below the profile grade, often where the surface of the cut or fill is first constructed.

10.3   INFORMATION REQUIRED

Figure 10-3 is a sample of the two typical half sections given on a set of road plans. There is a standard cut section and a standard fill section. These may also be called *template sections*. These

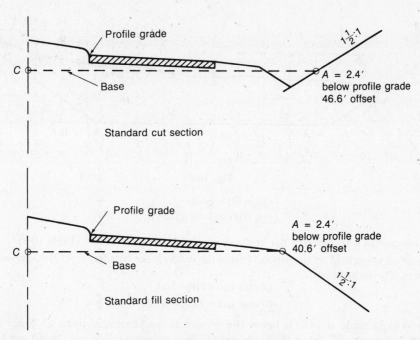

Fig. 10-3 Typical road sections.

sections are used for each section on the road unless the cross section of the road is changed or some unusual condition requires use of a special cross section. Generally, the base (line *CA*) is the line to which cut or fill is constructed. The line *CA* may be level or sloping and at different elevations below the profile grade. On curves, the whole half section may be tilted down or up by different amounts. The elevations and offset of the point *A* can always be determined from the set of road plans.

The procedure for slope staking is described by Examples 10.1 to 10.3, which illustrate the conditions generally encountered. In these examples, it is assumed that the base is level and that its elevation has been computed. Rod readings are as given; no units are given for dimensions—they could be meters or feet. Side slopes are given in terms of horizontal distance divided by vertical height (see Fig. 10-4). In Examples 10.1 to 10.3, the side slope is always $1\frac{1}{2}$: 1.

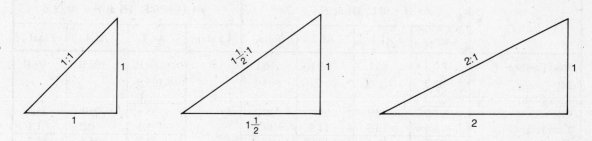

Fig. 10-4 Various side slopes.

EXAMPLE 10.1 In Fig. 10-5 the first HI (height of instrument) is located at *E*. From previous leveling its elevation is known to be 97.72. The calculations are then started. A rod shot is taken on the ground beside the center stake; therefore the notation 0 is used for the rod position. Determine the position of each stake.

Solution: The numbered steps below show the order of calculations. Computations are shown in Table 10-1.

1. Compute the *grade rod* (GR), or the theoretical rod reading that would occur if the rod were standing on the desired grade elevation when read from the HI. HI = 97.72 and the desired elevation is that of the base, shown on the plans as 70.00. The formula for grade rod is as follows:

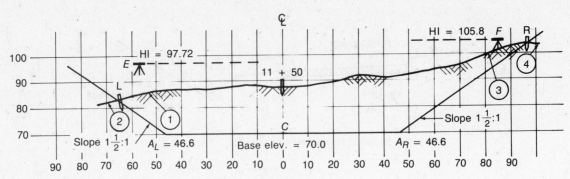

Fig. 10-5

$$GR = HI - \text{grade}$$
$$GR = 97.72 - 70.00 = 27.72$$

This computation is recorded in the first column of the calculation table (Table 10-1).

2. Read the rod when held on the ground beside the center stake (reading = 9.12). Compute the cut at the centerline. The formula is:

$$\text{Center cut} = GR - \text{rod}$$
$$\text{Center cut} = 27.72 - 9.12 = 18.60$$

So the finished grade is 18.60 ft below the ground at the centerline stake *C. Note*: *If this value is negative, it represents fill.* When the grade rod (27.7), and the centerline cut (18.60), are known, the first slope stake can be set.

3. Set the left grade stake *L*. Estimate the offset to the left grade stake *L*. This should be a guess based on experience. If the cross sections have been plotted, a very close estimate can be made by looking at the cross sections of the station you are concerned with. The following is a practical field method (see calculations in Table 10-1):

Compute the offset to *L* that would occur if the ground were level. This would be the offset to A_L (46.6) plus $1\frac{1}{2}$ times the center cut. For a $1\frac{1}{2}:1$ slope:

Center cut	18.6
Plus $\frac{1}{2}$ center cut	9.3
Plus *A* offset	46.6
Calculated offset stake *L* =	74.5

Table 10-1   Calculations for Example 10.1

	At 11 + 50 L, HI at $E = 97.72$				At 11 + 50 R, HI at $F = 105.8$			
	Center	At 1	At 2	Final	Center	At 3	At 4	Final
Rod position	0	55 L	69 L	66 L	0	88 R	102 R	99 R
HI	97.7					105.8		
Less base	− 70.0					− 70.0		
GR	27.7	27.7	27.7	27.7		35.8	35.8	35.8
Less rod	− 9.1	− 12.3	− 14.9	− 14.5		− 3.1	− 0.2	− 1.1
Cut	18.6	15.4	12.8	13.2		32.7	35.6	34.7
Plus $\frac{1}{2}$ cut	9.3	7.7	6.4	6.6		16.4	17.8	17.4
Plus A offset	46.6	46.6	46.6	46.6		46.6	46.6	46.6
Calculated offset	74.5	69.7	65.8	66.4	74.5	95.7	100.0	98.7
	Slopes opposite, move less				Slopes same, move more			
Try	55	69	66	OK	88	102	99	OK
Mark stakes:	$\dfrac{\text{C }13.2}{66.4\,\text{L}}$		$\dfrac{\text{C }18.6}{0}$	$\dfrac{\text{C }34.7}{98.7\,\text{R}}$				

But the ground slopes downward, so that the cut at stake L would be less than the center cut. So the offset would be somewhat less than 74.5 for the $1\frac{1}{2}:1$ slope.

Try 55. The rod is held at offset 55 shown on Fig. 10-5 at point 1. Offsets are usually measured with a woven tape. The offset for this rod reading is computed in the second column of Table 10-1. It has a value of 69.7.

4. The control procedure. It is now known that the cut measured at offset 55 should occur at 69.7. This indicates we should move the rod from 55 toward 69.7. Now the control procedure is needed. The control procedure is a method of determining which direction to move the rod. It is necessary to know the two slopes—the slope of the ground and the side slope of the earthwork. The control procedure rules are as follows:

    (*a*) When the slopes are opposite (i.e., one slopes up and the other down), *move* the rod *less* distance than called for.

    (*b*) When the slopes are the same (i.e., both slope up or both slope down), *move* the rod *more* distance than called for.

    In this example, the slopes are opposite, so move the rod less distance than called for. For example, try 69 on Fig. 10-5 and under point 2 in Table 10-1. At 69 the calculated offset is 65.8. The rod should be moved from 69 toward 65.8 but not all the way. Try 66. Here the calculated distance is 66.4. This is close enough to the actual rod position (difference $= 66.4 - 66 = 0.4$). A difference of 0.5 or less is close enough. Set the stake at the calculated offset (66.4) and assume the rod reading is the same as at 66. Mark the stake

$$\frac{\text{C } 13.2}{66.4 \text{ L}}$$

as shown in Table 10-1.

5. Set the right grade stake R. From the previous work the cut at the center is known to be 18.6. With level ground the calculated offset is 74.5 as before. In this case the slopes are the same, so move more. Try 88, shown at point 3 in Fig. 10-5. The calculated offset is 95.7. Move from 88 toward 95.7 and then more. Try 102. The calculated offset is 100.0. Move from 102 toward 100.0 but more. Try 99. The calculated offset is 98.7, which is close enough. Thus the stakes should be marked and positioned as follows:

$$\frac{\text{C } 13.2}{66.4 \text{ L}} \qquad \frac{\text{C } 18.6}{0} \qquad \frac{\text{C } 34.7}{98.7 \text{ R}}$$

The calculations are carried out in Table 10-1 and summarized in Table 10-2.

EXAMPLE 10.2 Figure 10-6 shows a roadway section requiring fill. Given: At station $23 + 50$ the centerline stake has been set out. The offsets for A_L (left) and A_R (right) are both 40.0. The side slopes both right and left are $1\frac{1}{2}:1$. Base elevation has been designated as 75.00. The instrument elevation at E is 68.12 and at F is 59.21. Find the location from the center stake for the left and right stakes.

Solution: Draw on graph paper a figure with the information given; see Fig. 10-6. Perform calculations. Use as a guide the numbered steps of Example 10.1. Note that the grade rod (GR) from the HI at E is negative (-6.9). The computed "cut" is -20.0, so it is a fill. The trials are summarized in Table 10-3. The calculations are shown in Table 10-4 and summarized in Table 10-2.

EXAMPLE 10.3 Figure 10.7 involves both cut and fill in the same road section. The station is $41 + 50$. Base elevation is 103.91. A_R (right) and A_L (left) both equal 50.6. On the right the slope is downward at $1\frac{1}{2}:1$ and on the left the slope is upward at $1\frac{1}{2}:1$. HI at $E = 121.06$, and HI at $F = 89.00$. Determine where the left and right stakes should be located.

Solution: Draw on graph paper a figure with the information given; see Fig. 10-7. The center cut had to be taken from an HI not shown. It turned out to be -13.9, which gives a calculated offset of 71.5. See Table 10-5 for the calculations; they are summarized in Table 10-2.

Note: On the right side of the center stake the figure obtained from subtracting the rod reading of 6.3 from the GR of -14.9 gives -21.2; since the figure is negative, it is fill.

Table 10-2   Calculations for Examples 10.1 to 10.3

Station	+	HI	−	Rod	Elev.	Grade	Cut or Fill	L	C	R
Example 10.1										
		97.72								
11 + 50 C				9.1	88.6	70.0	C 18.6		$\dfrac{\text{C 18.6}}{0}$	
66.4 L				14.5	83.2	70.0	C 13.2	$\dfrac{\text{C 13.2}}{\text{66.4 L}}$		
98.7 R		105.8		1.1	104.7	70.0	C 34.7			$\dfrac{\text{C 34.7}}{\text{98.7 R}}$
Example 10.2										
		68.12								
23 + 50 C				13.1	55.0	75.0	F 20.0		$\dfrac{\text{F 20.0}}{0}$	
55.0 L				3.1	65.0	75.0	F 10.0	$\dfrac{\text{F 10.0}}{\text{55.0 L}}$		
		59.21								
C				4.2	55.0	75.0	F 20.0		$\dfrac{\text{F 20.0}}{0}$	
77.5 R				9.2	50.0	75.0	F 25.0			$\dfrac{\text{F 25.0}}{\text{77.5 R}}$
Example 10.3										
		121.06								
41 + 50 C									$\dfrac{\text{F 13.9}}{0}$	
73.1 L				2.2	118.9	103.9	C 15.0	$\dfrac{\text{C 15.0}}{\text{73.1 L}}$		
83.5 R				7.0	82.0	103.9	F 21.9			$\dfrac{\text{F 21.9}}{\text{83.5 R}}$

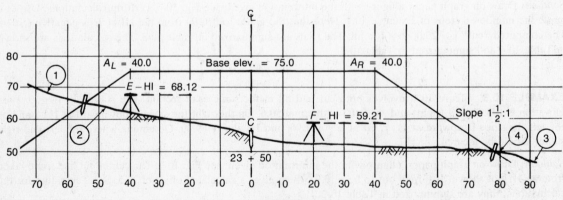

Fig. 10-6

Table 10-3 Trial Calculations for Example 10.2

Rod Position	Calculated Offset	Move	Try
0	70.0	Less	65 L
65 L	50.5	Less	51 L
51 L	56.5	Less	55 L
55 L	55.0	OK	
0	70.0	More	90 R
90 R	82.0	More	80 R
80 R	78.3	More	78 R
78 R	77.5	OK	

Table 10-4 Calculations for Example 10.2

	At $23+50$ L, HI at $E = 68.12$				At $23+50$ R, HI at $F = 59.21$			
	Center	At 1	At 2	Final	Center	At 3	At 4	Final
Rod position	0	65 L	51 L	55 L	0	90 R	80 R	78 R
HI*	68.12				39.2			
Less base	− 75.0				− 75.0			
GR	− 6.9	− 6.9	− 6.9	− 6.9	− 15.8	− 15.8	− 15.8	− 15.8
Less rod	− 13.1	− 0.1	− 4.1	− 3.1	− 4.2	− 12.2	− 9.7	− 9.2
	− 20.0	− 7.0	− 11.0	− 10.0	− 20.0	− 28.0	− 25.5	− 25.0
Fill	20.0	7.0	11.0	10.0	20.0	28.0	25.5	25.0
Plus $\frac{1}{2}$ fill	10.0	3.5	5.5	5.0	10.0	14.0	12.8	12.5
Plus A offset	40.0	40.0	40.0	40.0	40.0	40.0	40.0	40.0
Calculated offset	70.0	50.5	56.5	55.0	70.0	82.0	78.3	77.5
	Slopes opposite, move less				Slopes same, move more			
Try	65	51	55	OK	90	80	78	OK
Mark stakes:	$\dfrac{\text{F } 10.0}{55.0 \text{ L}}$	$\dfrac{\text{F } 20.0}{0}$	$\dfrac{\text{F } 25.0}{77.5 \text{ R}}$					

* Notice that when we use the instrument height F the center shot fill of 20.0 on the left is equal to the same center shot on the right. This checks that the two instrument heights were correct.

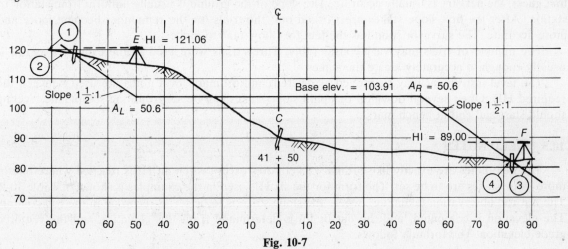

Fig. 10-7

Table 10-5 Calculations for Example 10.3

	At 41 + 50 L, HI at $E = 121.06$				At 41 + 50 R, HI at $F = 89.00$			
	Center	At 1	At 2	Final	Center	At 3	At 4	Final
Rod position	0	80 L	75 L	74 L	0	90 R	83 R	84 R
HI	From previous deter- mination	121.1				89.0		
Less base		− 103.9				− 103.9		
GR		17.2	17.2	17.2		− 14.9	− 14.9	− 14.9
Less rod		− 1.3	− 2.1	− 2.2		− 6.3	− 7.0	− 7.0
	− 13.9	15.9	15.1	15.0		− 21.2	− 21.9	− 21.9
Cut or fill	F 13.9	C 15.9	C 15.1	C 15.0		F 21.2	F 21.9	F 21.9
Plus $\frac{1}{2}$ cut or fill	7.0	8.0	7.6	7.5		10.6	11.0	11.0
Plus A offset	50.6	50.6	50.6	50.6		50.6	50.6	50.6
Calculated offse	71.5	74.5	73.3	73.1	71.5	82.4	83.5	83.5
	Slopes same, move more				Slopes same, move more			
Try	80	75	74	OK	90	83	84	OK

| | | Mark stakes: | C 15.0 / 73.1 L | F 13.9 / 0 | F 21.9 / 83.5 R | |

10.4 SUMMING UP SLOPE STAKING PROCEDURES

Memorize these rules:

1. GR = HI − grade
2. Cut or fill = GR − rod reading
3. Calculated offset = cut or fill + $\frac{1}{2}$ cut or fill + offset
4. Move "toward" calculated offset
5. Slopes opposite, move less
6. Slopes same, move more

The first estimate is usually a guess based on the elevation difference between the elevation at the centerline and the elevation at the left or right stake. After you find the calculated offset for this first guess, the next try is usually accurate. The slope of the ground is usually uniform from station to station. After the first slope stakes are worked out, the trials for the remainder become more and more accurate. The surveyor soon has the feel for "how far" to move.

The number of trials may vary considerably. Three trials were used in the examples as this is usually enough to accurately locate the stakes.

The level instrument may be moved as desired, and ordinary level notes are used between the required HIs. Often several downhill or uphill slope stakes can be set from one setup. The leveling should start and end at bench marks.

10.5 FIELD NOTES

The field notes are exactly like ordinary level notes except when an HI is reached where one or more slope stakes are to be set. The form for the six HIs used in the examples is shown in Table 10-2. The last rod observation taken, i.e., where the slope stake is to be set, is recorded in the rod column. The elevation is computed, and the cut or fill is determined from this elevation and the required grade elevation. The formula follows:

$$\text{Cut or fill} = \text{ground elevation} - \text{grade elevation}$$

A positive answer indicates cut; a negative answer indicates fill. These values should check with the cuts or fills computed by the grade-minus-rod method. On the right-hand side of Table 10-2 is the record of the marks placed on the slope stakes.

Solved Problems

10.1 Figure 10-8 shows a station on a highway with a ground slope from the left to the right side. The center stake, at station $11 + 50$, has already been set. Base elevation is 60.0. Toe of slope at left is 30.0 and toe of slope right is 30.0. Height of the surveying instrument at E is 81.10 and at point F, 90.0. Side slopes are $1\frac{1}{2}$: 1. Determine the position at which the stakes should be set to the right and left of centerline.

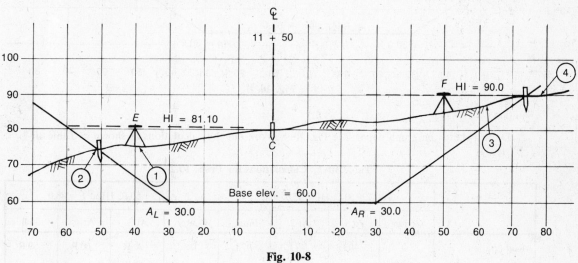

Fig. 10-8

Solution

Draw a sketch; see Fig. 10-8. Work out the calculations as shown in Table 10-6.

Table 10-6   Calculations for Prob. 10.1

	At 11 + 50 L, HI at E = 81.10				At 11 + 50 R, HI at F = 90.0			
	Center	At 1	At 2	Final	Center	At 3	At 4	Final
Rod position	0	40 L	52 L	51 L	0	65 R	80 R	74 R
HI	81.1					90.0		
Less base	− 60.0					− 60.0		
GR	21.1	21.1	21.1	21.1		30.0	30.0	30.0
Less rod	− 1.1	− 6.1	− 7.1	− 6.7		− 2.4	− 0.0	− 0.3
Cut	20.0	15.0	14.0	14.4		27.6	30.0	29.7
Plus $\frac{1}{2}$ cut	10.0	7.5	7.0	7.2		13.8	15.0	14.9
Plus A offset	30.0	30.0	30.0	30.0		30.0	30.0	30.0
Calculated offset	60.0	52.5	51.0	51.6	60.0	71.4	75.0	74.6
	Slopes opposite, move less				Slopes same, move more			
Try	40	52	51	OK	65	80	74	OK
Mark stakes:	$\dfrac{\text{C } 14.4}{51.6\,\text{L}}$	$\dfrac{\text{C } 20.0}{0}$	$\dfrac{\text{C } 29.7}{74.6\,\text{L}}$					

10.2   Figure 10-9 shows a road cross section at station $12 + 50$. The ground is on an incline from the low spot on the left to the high spot on the right. Information available: Base elevation 63.0; $A_L = 30.0$; $A_R = 30.0$. Survey instrument elevation at $E = 86.0$ and at $F = 88.0$. Side slopes are $1\frac{1}{2} : 1$. Determine the proper locations for left and right stakes.

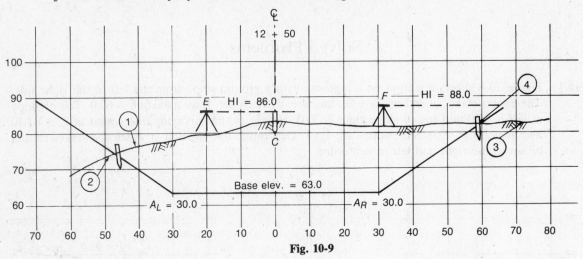

Fig. 10-9

Solution

Draw a sketch on graph paper; see Fig. 10-9. Work out the calculations as shown in Table 10-7.

Table 10-7   Calculations for Prob. 10.2

	At 12 + 50 L, HI at E = 86.0				At 12 + 50 R, HI at F = 88.0			
	Center	At 1	At 2	Final	Center	At 3	At 4	Final
Rod position	0	40 L	49 L	47 L	0	70 R	60 R	59 R
HI	86.0					88.0		
Less base	− 63.0					− 63.0		
GR	23.0	23.0	23.0	23.0		25.0	25.0	25.0
Less rod	− 3.0	− 9.7	− 13.0	− 12.0		− 5.5	− 5.7	− 5.8
Cut	20.0	13.3	10.0	11.0		19.5	19.3	19.2
Plus $\frac{1}{2}$ cut	10.0	6.7	5.0	5.5		9.8	9.7	9.6
Plus A offset	30.0	30.0	30.0	30.0		30.0	30.0	30.0
Calculated offset	60.0	50.0	45.0	46.5	60.0	59.3	59.0	58.8
	Slopes opposite, move less				Slopes same, move more			
Try	40.0	49.0	47.0	OK	70.0	60	59	OK
Mark stakes:		$\dfrac{C\,11.0}{46.5\,L}$	$\dfrac{C\,20.0}{0}$	$\dfrac{C\,19.2}{58.8\,R}$				

10.3   Given: A roadway cross section at station $8 + 00$. Ground sloping from low on the left to high on the right. Elevation of the instrument at $E = 73.0$ and at $F = 83.2$. Base elevation is 50.0; $A_L = 20.0$; $A_R = 30.0$. Side slopes are $1\frac{1}{2} : 1$. The center stake is already positioned. Determine where the two side stakes should be set from the center stake.

Solution

Draw a sketch on graph paper; see Fig. 10-10. Work out the calculations as shown in Table 10-8.

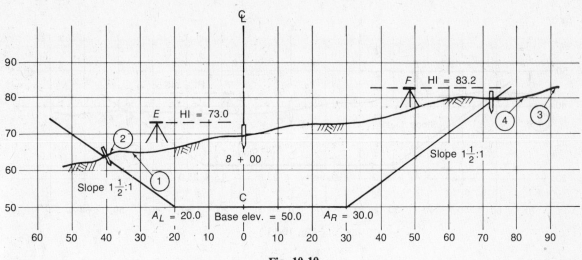

Fig. 10-10

Table 10-8   Calculations for Prob. 10.3

	At 8 + 00 L, HI at E = 73.0				At 8 + 00 R, HI at F = 83.2			
	Center	At 1	At 2	Final	Center	At 3	At 4	Final
Rod position	0	30 L	38 L	41 L	0	90 R	80 R	75 R
HI	73.0					83.2		
Less base	− 50.0					− 50.0		
GR	23.0	23.0	23.0	23.0		32.2	32.2	32.2
Less rod	− 3.5	− 8.0	− 8.4	− 9.1		− 1.1	− 3.4	− 4.2
Cut	19.5	15.0	14.6	13.9		32.1	29.8	29.0
Plus $\frac{1}{2}$ cut	9.8	7.5	7.3	7.0		16.1	14.9	14.5
Plus A offset	20.0	20.0	20.0			30.0	30.0	30.0
Calculated offset	49.3	42.5	41.9	40.9	49.3	78.2	74.7	73.5
	Slopes opposite, move less				Slopes same, move more			
Try	30	38	41	OK	90	80	75	OK
Mark stakes:		$\dfrac{\text{C 13.9}}{\text{40.9 L}}$	$\dfrac{\text{C 19.5}}{0}$	$\dfrac{\text{C 14.5}}{\text{73.5 R}}$				

10.4 Given: A roadway cross section located at station 7 + 50. The center stake has already been set on the baseline of survey for the road. The side slopes are $1\frac{1}{2}$: 1. Base elevation = 23.6; A_L = 30.0; A_R = 25.0. Height of the instrument at E = 49.7 and HI at F = 43.5. Determine where the two side stakes should be set from the center stake.

Solution

Draw a sketch on graph paper; see Fig. 10-11. Work out the calculations as shown in Table 10-9.

10.5 Given: A roadway cross section located at station 27 + 00. This section is on a level piece of ground with a base elevation of 75.8. Side slopes are $1\frac{1}{2}$: 1. A_L = 35.0 and A_R = 20.0. Height of instrument at E = 105.9 and HI at F = 106.0. Determine where the left and right stakes should be set.

Solution

Draw a sketch on graph paper; see Fig. 10-12. Work out the calculations as shown in Table 10-10.

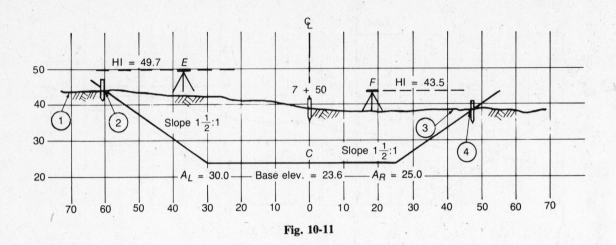

Fig. 10-11

Table 10-9   Calculations for Prob. 10.4

	At 7 + 50 L, HI at E = 49.7				At 7 + 50 R, HI at F = 43.5			
	Center	At 1	At 2	Final	Center	At 3	At 4	Final
Rod position	0	70 L	60 L	61 L	0	35 R	46 R	46.6 R
HI	49.7					43.5		
Less base	− 23.6					− 23.6		
GR	26.1	26.1	26.1	26.1		19.9	19.9	
Less rod	− 10.8	− 5.7	− 5.9	− 5.9		− 5.6	− 5.5	
Cut	15.3	20.4	20.2	20.2		14.3	14.4	
Plus $\frac{1}{2}$ cut	7.7	10.2	10.1	10.1		7.2	7.2	
Plus A offset	30.0	30.0	30.0	30.0		25.0	25.0	
Calculated offset	53.0	60.6	60.3	60.3	53.0	46.5	46.6	
	Slopes same, move more				Slopes opposite, move less			
Try	70	60	61	OK	35	46	OK	
Mark stakes:	$\dfrac{\text{C }20.2}{60.3\text{ L}}$		$\dfrac{\text{C }15.3}{0}$		$\dfrac{\text{C }14.4}{46.6\text{ R}}$			

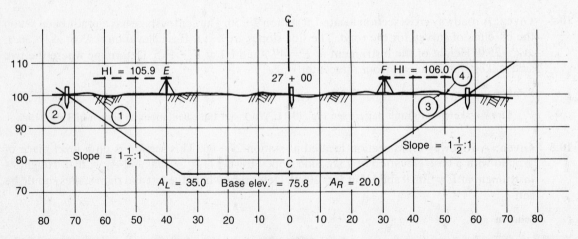

Fig. 10-12

Table 10-10   Calculations for Prob. 10.5

	At 27+00 L, HI at E = 105.9				At 27+00 R, HI at F = 106.0			
	Center	At 1	At 2	Final	Center	At 3	At 4	Final
Rod position	0	60 L	75 L	71.3 L	0	50 R	51 R	57 R
HI	105.9					106.0		
Less base	− 75.8					− 75.8		
GR	30.1	30.1	30.1	30.1		30.2	30.2	30.2
Less rod	− 5.9	− 5.9	− 5.9	− 5.6		− 6.0	− 6.0	− 6.0
Cut	24.2	24.2	24.2	24.5		24.2	24.2	24.2
Plus ½ cut	12.1	12.1	12.1	12.3		12.1	12.1	12.1
Plus A offset	35.0	35.0	35.0	35.0		20.0	20.0	20.0
Calculated offset	71.3	71.3	71.3	71.8		56.3	56.3	56.3
	Slopes opposite, move less				Slopes opposite, move less			
Try	60	75	71.3	OK	50	51	57	OK
Mark stakes:		C 24.5 ⎯⎯⎯ 71.8 L	C 24.2 ⎯⎯⎯ 0	C 24.2 ⎯⎯⎯ 56.3 R				

10.6   Given: A roadway cross section that is designed to be filled to a base elevation of 83.91. Side slopes are $1\frac{1}{2}$:1. Station is 21+00 with A_L = 35.0 and A_R = 35.0. Height of instrument at E = 68.3 and HI at F = 63.6. Determine where the slope stakes should be set to the left and right of the center stake.

Solution

Draw the figure on graph paper from the given information; see Fig. 10-13.

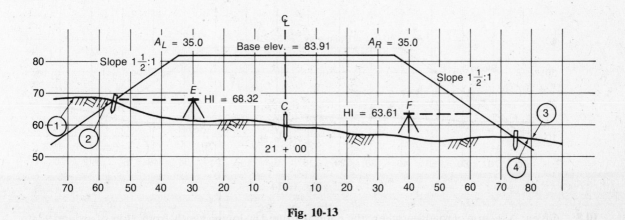

Fig. 10-13

Note: You can add the natural ground line after finding elevations from your calculations. Work out the calculations as shown in Table 10-11. Note that when using both instrument elevations at E and F the fill of 23.9 at the center checks. This means that we have the proper instrument readings at these locations.

Table 10-11   Calculations for Prob. 10.6

	At 21+00 L, HI at $E = 68.3$				At 21+00 R, HI at $F = 63.6$			
	Center	At 1	At 2	Final	Center	At 3	At 4	Final
Rod position	0	69 L	58 L	55 L	0	80 R	78 R	77 R
HI	68.3				63.6			
Less base	− 83.9				− 83.9			
GR	− 15.6	− 15.6	− 15.6	− 15.6	− 20.3	− 20.3	− 20.3	− 20.3
Less rod	− 8.3	− 0.0	− 0.1	− 0.6	− 3.6	− 8.1	− 8.0	− 7.6
	− 23.9	− 15.6	− 15.7	− 15.0	− 23.9	− 28.4	− 28.3	− 27.9
Fill	23.9	15.6	15.7	15.0	23.9	28.4	28.3	27.9
Plus $\frac{1}{2}$ fill	12.0	7.8	7.9	7.6	12.0	14.2	14.2	14.0
Plus A offset	35.0	35.0	35.0	35.0	35.0	35.0	35.0	35.0
Calculated offset	70.9	58.4	58.6	57.6	70.9	77.6	77.5	76.9
	Slopes opposite, move less				Slopes same, move more			
Try	69	58	55	OK	80	78	77	OK
Mark stakes:		$\dfrac{F\ 15.0}{57.6\ L}$	$\dfrac{F\ 24.3}{0}$	$\dfrac{F\ 27.9}{76.9\ R}$				

10.7   Given: A roadway cross section located at station $16+50$. The base elevation to which this section has been designed is 37.23. There are uneven A stations on this cross section. A_L is at 40.9, while A_R is located at 23.2. The side slopes are the usual $1\frac{1}{2}:1$ on both slopes. Height of the surveying instrument at E is 21.17. At F there is no new instrument reading since the natural ground at this location is relatively level. Determine the locations of the stakes.

Solution

We will use the E instrument reading for both right and left sides of the section; see Fig. 10-14. Work out the calculations as shown in Table 10-12. The trials made, taken from the calculations, are shown in Table 10-13.

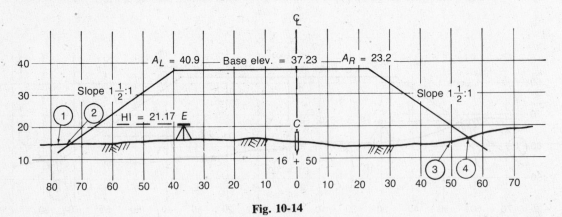

Fig. 10-14

10.8   Given: A section of roadway where the natural ground is sloping gently from a low of around 92 ft to a high of about 102 ft where the slope stakes will intersect on the right and left sides of the road section. Base elevation to which the section has been designed = 116.13. A_L (on the left) is 34.0 ft left of the center stake, and A_R (on the right) is 26.1 ft right of the center stake. Side slopes are the usual $1\frac{1}{2}:1$. Height of the surveying instrument at $E = 100.00$ and at $F = 102.91$. Determine the distance from the center stakes for both side stakes.

Table 10-12   Calculations for Prob. 10.7

	At 16 + 50 L, HI at E = 21.17				At 16 + 50 R, HI at E = 21.17			
	Center	At 1	At 2	Final	Center	At 3	At 4	Final
Rod position	0	80 L	75 L	74.2 L	0	45 R	55 R	54 R
HI	21.2				21.2			
Less base	− 37.2				− 37.2			
GR	− 16.0	− 16.0	− 16.0	− 16.0	− 16.0	− 16.0	− 16.0	− 16.0
Less rod	− 6.1	− 6.2	− 6.2	− 6.2	− 6.1	− 6.2	− 4.9	− 5.0
	− 22.1	− 22.2	− 22.2	− 22.2	− 22.1	− 22.2	− 20.9	− 21.0
Fill	22.1	22.2	22.2	22.2	22.1	22.2	20.9	21.0
Plus $\frac{1}{2}$ fill	11.1	11.1	11.1	11.1	11.1	11.1	10.5	10.5
Plus A offset	40.9	40.9	40.9	40.9	23.2	23.2	23.2	23.2
Calculated offset	74.1	74.2	74.2	74.2	56.4	56.5	54.6	54.7
	Slopes same, move more				Slopes opposite, move less			
Try	80	75	74.2	OK	45	55	54	OK
Mark stakes:	$\dfrac{\text{F 22.2}}{\text{74.2 L}}$		$\dfrac{\text{F 22.1}}{0}$		$\dfrac{\text{F 21.0}}{\text{54.7 R}}$			

Table 10-13   Trials Made for Calculations for Prob. 10.7

Rod Position	Calculated Offset	Move	Try
0	74.1	More	80
80 L	74.2	More	75
75 L	74.2	More	74.2
74.2 L	74.2		OK
0	56.4	Less	45
45 R	56.5	Less	55
55 R	54.6	Less	54
54 R	54.7		OK

Solution

First draw the figure from the given information; see Fig. 10-15. You can complete the figure on your graph paper after your calculations are completed. At that time you will have the location of the first try, second try, center stake, and fourth try. This is enough to put the natural ground line in. Work out the calculations; see Table 10-14.

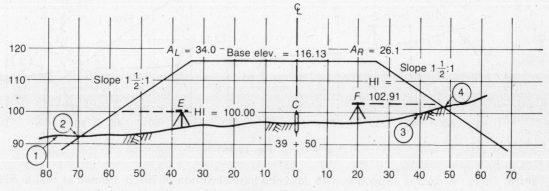

Fig. 10-15

Table 10-14   Calculations for Prob. 10.8

	At 39 + 50 L, HI at E = 100.00				At 39 + 50 R, HI at F = 102.91			
	Center	At 1	At 2	Final	Center	At 3	At 4	Final
Rod position	0	80 L	71 L	69 L	0	40 R	49 R	46 R
HI	100.00				102.91			
Less base	− 116.13				− 116.13			
GR	− 16.1	− 16.1	− 16.1	− 16.1	− 13.2	− 13.2	− 13.2	− 13.2
Less rod	− 3.0	− 8.0	− 7.2	− 7.5	− 5.9	− 2.9	− 0.7	− 1.0
	− 19.1	− 24.1	− 23.3	− 23.6	− 19.1	− 16.1	− 13.9	− 14.2
Fill	19.1	24.1	23.3	23.6	19.1	16.1	13.9	14.2
Plus $\frac{1}{2}$ fill	9.6	12.1	11.6	11.8	9.6	8.1	7.0	7.1
Plus A offset	34.0	34.0	34.0	34.0	26.1	26.1	26.1	26.1
Calculated offset	62.7	70.2	68.9	69.4	54.8	50.3	47.0	47.4
	Slopes same, move more				Slopes opposite, move less			
Try	80	71	69	OK	40	49	46	OK

Mark stakes: $\dfrac{F\,23.6}{69.4\,L}$ $\dfrac{F\,19.1}{0}$ $\dfrac{F\,14.2}{47.4\,R}$

10.9   Given: A road section at station 31 + 50 which was designed to a base elevation of 34.03 ft. A_L = 42.1 and A_R = 22.2. Natural ground is sloping from left to right. Height of surveying instrument at E = 25.00 and at F = 22.12. Slope on the left is 1 : 1 and on the right it is $1\frac{1}{2}$: 1. Find the points at which to drive the left and right slope stakes.

   Note: In this problem the left side slope is an unusual one. The slope is nearly always $1\frac{1}{2}$: 1, as this makes a stable ground condition. However, here the left slope is 1 : 1. Thus when we compute "plus $\frac{1}{2}$ fill," there is no $\frac{1}{2}$ fill. Enter 0.0 as values for this row.

Solution

   Draw the figure on graph paper from the given information; see Fig. 10-16. Work out the calculations; see Table 10-15.

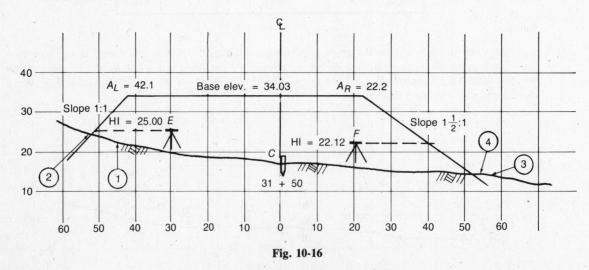

Fig. 10-16

10.10   Given: A road cross section with designed base elevation of 708.10 located at station 511 + 00. The side slopes are the usual $1\frac{1}{2}$: 1 ratio. A dimensions, both right and left, are 26.0. HI

Table 10-15   Calculations for Prob. 10.9

	At 31 + 50 L, HI at E = 25.00				At 31 + 50 R, HI at F = 22.12			
	Center	At 1	At 2	Final	Center	At 3	At 4	Final
Rod position	0	40 L	54 L	53 L	0	60 R	55 R	53 R
HI	25.0				22.1			
Less base	− 34.0				− 34.0			
GR	− 9.0	− 9.0	− 9.0	− 9.0	− 11.9	− 11.9	− 11.9	− 11.9
Less rod	− 7.8	− 3.6	− 2.0	− 1.4	− 4.9	− 9.1	− 8.1	− 8.6
	− 16.8	− 12.6	− 11.0	− 10.4	− 16.8	− 21.0	− 20.0	− 20.5
Fill	16.8	12.6	11.0	10.4	16.8	21.0	20.0	20.5
Plus $\frac{1}{2}$ fill	0.0	0.0	0.0	0.0	8.4	10.5	10.0	10.3
Plus A offset	42.1	42.1	42.1	42.1	22.2	22.2	22.2	22.2
Calculated offset	58.9	54.7	53.1	52.5	47.4	53.7	52.2	53.0
	Slopes opposite, move less				Slopes same, move more			
Try	40	54	53	OK	60	55	53	OK
Mark stakes:		$\dfrac{\text{F 10.4}}{\text{52.5 L}}$	$\dfrac{\text{F 16.8}}{0}$	$\dfrac{\text{F 20.5}}{\text{53 R}}$				

at E = 692.00; HI at F = 699.10. Determine where to place the slope stakes in relation to the center stake.

Solution

First draw the figure on graph paper from the given information; see Fig. 10-17. The calculations will give enough elevations to draw the natural ground. Perform the calculations; see Table 10-16.

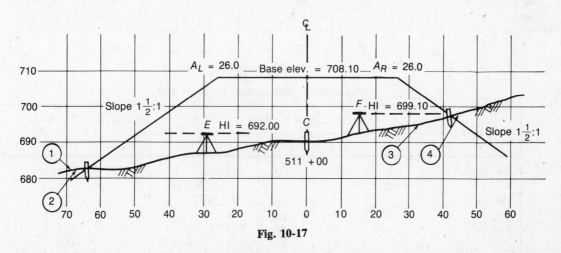

Fig. 10-17

10.11   Given: Station 28 + 00 of a road section where the natural ground is sloping gently from a high area on the left to a lower area on the right. The side slopes are $1\frac{1}{2}$: 1. A_L = 30.1 and A_R = 40.6. Base elevation = 57.61. Height of the surveying instrument at E is 66.91, and at F is 54.41. Find the distance from the center stake (station 28 + 00) of the right and left slope stakes.

Solution

First draw the figure on graph paper—natural ground may be put in after you do the calculations and find several important elevations—see Fig. 10-18. Perform the calculations; see Table 10-17.

Table 10-16 Calculations for Prob. 10.10

	At 511 + 00 L, HI at E = 692.00				At 511 + 00 R, HI at F = 699.10			
	Center	At 1	At 2	Final	Center	At 3	At 4	Final
Rod position	0	70 L	68 L	66 L	0	30 R	43 R	42 R
HI	692.0				699.1			
Less base	− 708.1				− 708.1			
GR	− 16.1	− 16.1	− 16.1	− 16.1	− 9.0	− 9.0	− 9.0	− 9.0
Less rod	− 2.0	− 11.2	− 10.5	− 10.0	− 9.1	− 4.5	− 1.6	− 1.3
	− 18.1	− 27.3	− 26.6	− 26.1	− 18.1	− 13.5	− 10.6	− 10.3
Fill	18.1	27.3	26.6	26.1	18.1	13.5	10.6	10.3
Plus ½ fill	9.1	13.7	13.3	13.1	9.1	6.8	5.3	5.2
Plus A offset	26.0	26.0	26.0	26.0	26.0	26.0	26.0	26.0
Calculated offset	53.2	67.0	65.9	65.2	53.2	46.3	41.9	41.5
	Slopes same, move more				Slopes opposite, move less			
Try	70	68	66	OK	30	43	42	OK
Mark stakes:	$\dfrac{\text{F 26.1}}{\text{65.2 R}}$		$\dfrac{\text{F 18.1}}{0}$		$\dfrac{\text{F 10.3}}{\text{41.5 R}}$			

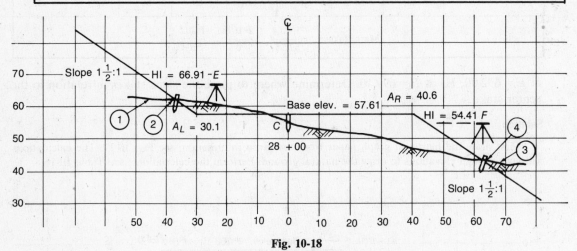

Fig. 10-18

Table 10-17 Calculations for Prob. 10.11

	At 28 + 00 L, HI at E = 66.91				At 28 + 00 R, HI at F = 54.41			
	Center	At 1	At 2	Final	Center	At 3	At 4	Final
Rod position	0	50 L	39 L	37 L	0	70 R	64 R	63 R
HI	66.9				54.4			
Less base	− 57.6				− 57.6			
GR	9.3	9.3	9.3	9.3	− 3.2	− 3.2	− 3.2	− 3.2
Less rod	− 11.4	− 3.9	− 4.9	− 4.9	+ 1.1	− 12.2	− 11.4	− 11.4
	− 2.1	5.4	4.4	4.4	− 2.1	− 15.4	− 14.6	− 14.6
Fill or cut	F 2.1	C 5.4	C 4.4	C 4.4	F 2.1	F 15.4	F 14.6	F 14.6
Plus ½ fill or cut	1.1	2.7	2.2	2.2	1.1	7.7	7.3	7.3
Plus A offset	30.1	30.1	30.1	30.1	40.6	40.6	40.6	40.6
Calculated offset	33.3	38.2	36.7	36.7	43.8	63.7	62.5	62.5
	Slopes same, move more				Slopes same, move more			
Try	50	39	37	OK	70	64	63	OK
Mark stakes:	$\dfrac{\text{C 4.4}}{\text{36.7 R}}$		$\dfrac{\text{F 2.1}}{0}$		$\dfrac{\text{F 14.6}}{\text{62.5 R}}$			

10.12 In Fig. 10-19 station $29+00$ has the center stake placed on the centerline of the proposed roadway project. The terrain at this location slopes from a high on the left to lower on the right. The proposed base elevation is 113.62 ft. $A_L = 30.8$ ft; $A_R = 24.9$ ft. Side slopes are the usual $1\frac{1}{2}:1$ on both right and left sides. The instrument at E is set at 125.10 and the instrument at F is at 106.21. Determine where to place the slope stakes in relation to the center stake.

Solution

First draw a sketch of all the given information on graph paper; see Fig. 10-19. Enough information will be obtained from your calculations to put in the natural ground line. Perform the calculations; see Table 10-18.

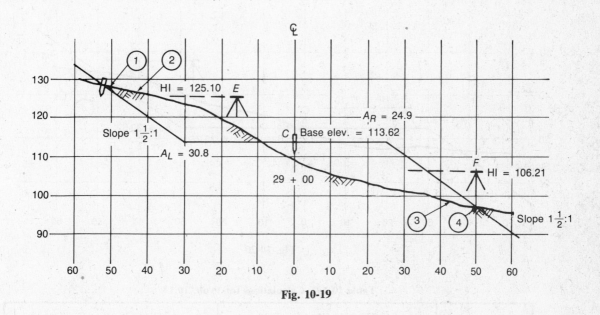

Fig. 10-19

Table 10-18   Calculations for Prob. 10.12

	At $29+00$ L, HI at $E = 125.10$				At $29+00$ R, HI at $F = 106.21$			
	Center	At 1	At 2	Final	Center	At 3	At 4	Final
Rod position	0	50 L	40 L	52 L	0	40 R	50 R	49.8
HI	125.1				106.2			
Less base	− 113.6				− 113.6			
GR	11.5	11.5	11.5	11.5	− 7.4	− 7.4	− 7.4	
Less rod	− 16.1	− 2.7	− 0.0	+ 2.9	+ 2.8	− 7.2	− 9.2	
	− 4.6	8.8	11.5	14.4	− 4.6	− 14.6	− 16.6	
Fill or cut	F 4.6	C 8.8	C 11.5	C 14.4	F 4.6	F 14.6	F 16.6	
Plus $\frac{1}{2}$ cut or fill	2.3	4.4	5.8	7.2	2.3	7.3	8.3	
Plus A offset	30.8	30.8	30.8	30.8	24.9	24.9	24.9	
Calculated offset	37.7	44.0	48.1	52.4	31.8	46.8	49.8	
	Slopes same, move more				Slopes same, move more			
Try	50	40	52	OK	40	50	OK	
Mark stakes:	$\dfrac{\text{C }14.4}{52.4\,\text{L}}$	$\dfrac{\text{F }4.6}{0}$	$\dfrac{\text{F }16.6}{49.8\,\text{R}}$					

10.13 Given: Station $23 + 50$ where natural ground slopes from a low position on the left to a higher position on the right. $A_L = 30.1$ ft and $A_R = 30.3$ ft. The side slopes on the designed section are at the usual $1\frac{1}{2}:1$ on both the left and right sides. At E the instrument setup shows the HI to be 37.10, and at F the HI = 48.21. Find the distance both slope stakes must be from the previously set center stake.

Solution

First draw as much of the section as you can from the given material; see Fig. 10-20. The calculations will give the additional information required to draw the natural ground line. Perform the calculations; see Table 10-19.

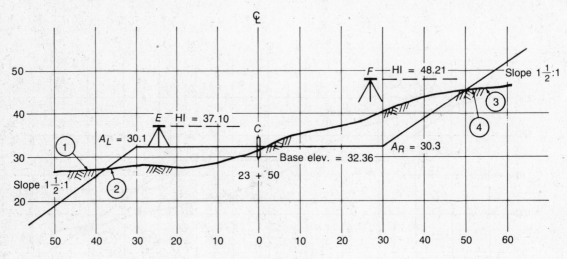

Fig. 10-20

Table 10-19   Calculations for Prob. 10.13

	At $23 + 50$ L, HI at $E = 37.10$				At $23 + 50$ R, HI at $F = 48.21$			
	Center	At 1	At 2	Final	Center	At 3	At 4	Final
Rod position	0	45 L	39 L	38 L	0	60 R	52 R	50 R
HI	37.1				48.2			
Less base	− 32.4				− 32.4			
GR	4.7	4.7	4.7	4.7	15.8	15.8	15.8	15.8
Less rod	− 5.9	− 10.1	− 9.6	− 9.5	− 17.0	− 1.7	− 2.6	− 3.2
	− 1.2	− 5.4	− 4.9	− 4.8	− 1.2	14.1	13.2	12.6
Fill or cut	F 1.2	F 5.4	F 4.9	F 4.8	F 1.2	C 14.1	C 13.2	C 12.6
Plus $\frac{1}{2}$ cut or fill	0.6	2.7	2.5	2.4	0.6	7.1	6.6	6.3
Plus A offset	30.1	30.1	30.1	30.1	30.3	30.3	30.3	30.3
Calculated offset	31.9	38.2	37.5	37.3	32.1	51.5	50.1	49.2
	Slopes same, move more				Slopes same, move more			
Try	45	39	38	OK	60	52	50	OK

Mark stakes: $\dfrac{\text{F 4.8}}{\text{37.3 L}}$ $\dfrac{\text{F 1.2}}{0}$ $\dfrac{\text{C 12.6}}{\text{49.2 R}}$

Chapter 11

Earthwork

11.1 IMPORTANCE OF EARTHWORK

In road, railroad, and airfield construction, the movement of large volumes of earth (*earthwork*) is one of the most important construction operations. From the standpoint of both workers and equipment, it requires a great amount of engineering effort.

The planning, scheduling, and supervision of earthwork operations are of major importance in obtaining an efficiently operated construction project. In order to plan a schedule, the quantities of clearing, grubbing, and stripping, as well as the quantities and position of cuts and fills, must be known so that the most efficient type and number of pieces of earthmoving equipment can be chosen, the proper number of personnel assigned, and the appropriate time allotted for each construction phase.

Earthwork computations involve the calculation of volumes or quantities, the determination of final grades, the balancing of cuts and fills, and the planning of the most economical haul of material. Field notes and established grades are used to plot the cross sections at regular stations and at any plus stations which may have been established at critical points. The line representing the existing ground surface, and those lines representing the proposed cut or fill, enclose cross-sectioned areas. These areas and the measured distance along the centerline are used to compute earthwork volumes.

11.2 CROSS SECTIONS

The cross section used in earthwork computations is a vertical section, perpendicular to the centerline at full and plus stations, which represents the boundaries of a proposed or existing cut or fill. Typical cross sections for a roadbed are illustrated in Fig. 11-1. Determination of cross-sectional areas is simplified when the sections are plotted on cross-sectional graph paper. They are usually plotted to the same vertical and horizontal scale, standard practice being 1 in = 10 ft. However, if the vertical cut or fill is small in comparison with the width, an exaggerated vertical scale may be used to gain additional precision in plotting such sections. However, when computing areas of this type of plotted section, care must be taken that the proper area is obtained. For example, a 1 in = 10 ft scale, both vertical and horizontal, yields 100 ft^2 per grid square, but 1 in = 10 ft horizontal and 1 in = 2 ft vertical yields only 20 ft^2.

The side slope is usually determined by the design specifications based on the stability of the soil in cut or fill. However, the need for economy in construction operations is often a consideration. For example, cut slopes may be flattened more than is required by soil characteristics solely to produce enough material for a nearby fill rather than to operate a "borrow" pit to obtain this material.

11.3 AREAS

Cross-sectional areas for construction earthwork volumes are usually determined by one of the following methods: by counting the squares, by the geometry of trapezoids and triangles, by the stripper method, by the double-meridian-distance method, or by using a planimeter. The stripper and counting-the-squares methods are simple and give approximate results, while the other methods give results as accurate as the cross-sectional field data will permit. Standard practice requires that cut and fill areas of a cross section, where both occur simultaneously, be determined separately.

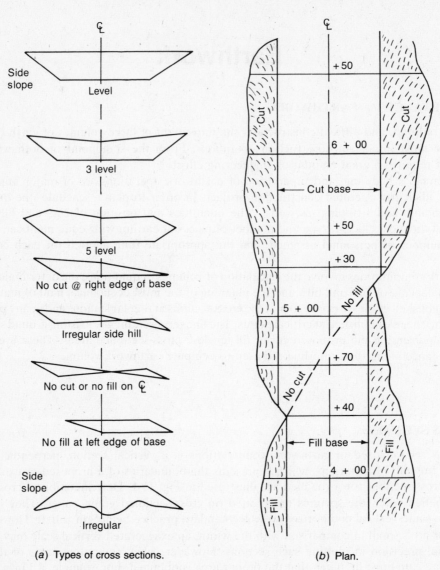

(a) Types of cross sections.   (b) Plan.

Fig. 11-1 Typical roadbed cross sections. (*Courtesy of U.S. Army, General Drafting,*
T M 5-230)

11.4 COUNTING-THE-SQUARES METHOD

To make a quick approximation of a cross-sectional area plotted on cross-sectional graph paper, count the number of squares enclosed by the boundary lines of the section. Then multiply the total number of counted squares by the number of square feet represented by a single square.

EXAMPLE 11.1 Figure 11-2 shows a cut section with both vertical and horizontal scales of 1 in = 10 ft. Determine the area by counting the squares.

Solution: The $\frac{1}{2}$-in square blocks will be easiest to count. So count the $\frac{1}{2}$-in squares; approximately 24 squares are obtained from the count. Each $\frac{1}{2}$-in square is 5 ft on a side, so $5 \times 5 = 25$ ft^2 for the area of each square. Multiply number of square feet in a single square times number of counted squares:

$$25(24) = 600 \text{ ft}^2 \qquad Ans.$$

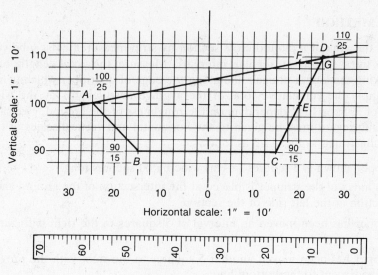

Fig. 11-2 Cut section.

11.5 GEOMETRIC METHOD

To compute the area of a cross section by the geometric method (sometimes called the trapezoidal method), subdivide the area into simple geometric figures, calculate each area according to its geometry, and total the results. There is no set rule for working out the subdivisions; the person performing the computations exercises judgment in selecting those subdivisions which will produce the most direct and accurate results. Figure 11-12 illustrates division of a typical three-level section into three triangles and a trapezoid. The formulas used are as follows:

Area of a triangle: $A = \frac{1}{2}bh$

where A = area
             b = base
             h = height

Area of a trapezoid: $A = \dfrac{h(AE + BC)}{2}$

where A = area
                  h = height of trapezoid
       AE, BC = the lengths of the base and top of the trapezoid, respectively

EXAMPLE 11.2 Figure 11-2, which was used to show the counting-the-squares method, is divided into three triangles—*AEF*, *EFG*, and *FGD*—and trapezoid *AECB*. Determine the area by the geometric method.

Solution: Find the area of each triangle:

AEF: $A = \frac{1}{2}bh = \frac{1}{2}(9)(45) = 202.50 \text{ ft}^2$

EFG: $A = \frac{1}{2}bh = \frac{1}{2}(4.5)(9) = 20.25 \text{ ft}^2$

FGD: $A = \frac{1}{2}bh = \frac{1}{2}(1)(4.5) = 2.25 \text{ ft}^2$

Find the area of the trapezoid:

$$A = \frac{h(AE + BC)}{2} = \frac{10(30 + 45)}{2} = 375.00 \text{ ft}^2$$

Add to find the total area:

$$\text{Area} = 202.50 + 20.25 + 2.25 + 375.00 = 600.00 \text{ ft}^2 \qquad Ans.$$

11.6 STRIPPER METHOD

To determine the area of a plotted cross section by strip measurements, subdivide the area into strips by vertical lines spaced at regular intervals. Measure the total length of these lines by cumulatively marking the length of each line along the edge of a stripper which is made of paper or plastic. Then multiply the cumulative total of the average base lengths by the width of the strip. Regular intervals of 3, 5, or 10 ft, depending on the roughness of the ground, give satisfactory results for strip widths. Due regard must be given to the horizontal and vertical scales of the cross section. The procedure is illustrated in Fig. 11-3 as follows:

- (a) The stripper shown is 5 squares ($=\frac{1}{2}$ in) wide by 60 squares long.

- (b) The zero index of the stripper is placed at the intersection of the ground and side-slope line of the section at the left side of the section.

- (c) The stripper has been moved an interval of 5 squares to the right with zero reading at the bottom.

- (d) The stripper has been moved another 5 squares to the right with the previous top reading (2.5) now adjacent to the bottom line.

- (e) The stripper has again been moved 5 squares to the right for another interval with the previous top reading (6.0) adjacent to the bottom line.

This process of moving 5 squares to the right and bringing the top reading to the bottom line is continued until the stripper reaches the right edge of the cross section with a final reading of 53.0. Multiply this last reading (53.0) by the strip width in number of squares (5) to get 265.0, the number of squares in the section. Multiply the number of squares by the area in square feet of one square to find the area of the cross section in square feet.

EXAMPLE 11.3   Using the stripper method, find the area of the geometric figure in Fig. 11-2, and compare your answer with the 600 ft^2 obtained from the counting-the-squares and geometric methods.

Solution: Prepare a stripper by laying a piece of notebook paper on the graph paper in Fig. 11-2. We must do this since we cannot be sure what will constitute 1 in in the printed textbook. Thus a stripper which conforms to the graph paper as printed will be made. The stripper should be cut to a width of 5 squares. We now have a paper stripper of a width of 5 squares and a length of 70 or 80 squares.

Start the zero point of the stripper at the bottom of the first 5-square line (on the horizontal scale this is at point 20 and vertically it is at 95). Measure to the top line. First stripper figure = 6; second = 18; third = 31; fourth = 45; fifth = 60. Now return to zero on your stripper and start the next amount. First = 16; second = 33; third = 51.2; fourth = 60.3. Add the stripper totals:

$$
\begin{array}{lr}
\text{First total} = & 60 \\
\text{Second total} = & +\ \ 60.3 \\
\hline
\text{Total} = & 120.3 \\
\end{array}
$$

Multiply the total by the stripper width of 5: 5(120.3) = 601.5. The squares equaled 1 ft^2 each, so 601.5(1) = 601.5 ft^2. This answer is close to the answer of 600 ft^2 found by the geometric and counting-the-squares methods.

The stripper method is fast and most often used in road design by consulting engineering firms and state highway departments.

11.7 DOUBLE-MERIDIAN-DISTANCE METHOD

The double-meridian-distance (DMD) method gives a more precise value for a cross-sectional area than the stripper method. It does, however, involve more effort and time. It is essential that the elevations (latitudes) and the distance from the centerline (departures) of all points on the cross section be known.

The method is based on the fact that the area of a right triangle equals one-half of the product of the two sides. Since latitudes and departures are at right angles to each other, the area bounded by the distance, the latitude, and the departure is a right triangle. This area can be determined by taking one-half of the product of the latitude and the departure. However, the triangle may add to or subtract from the total area of the irregular figure depending on its location.

To avoid determining a plus or minus area for each triangle, a slight refinement is made. The departure is added twice; first in determining the DMD of the course and then when the next

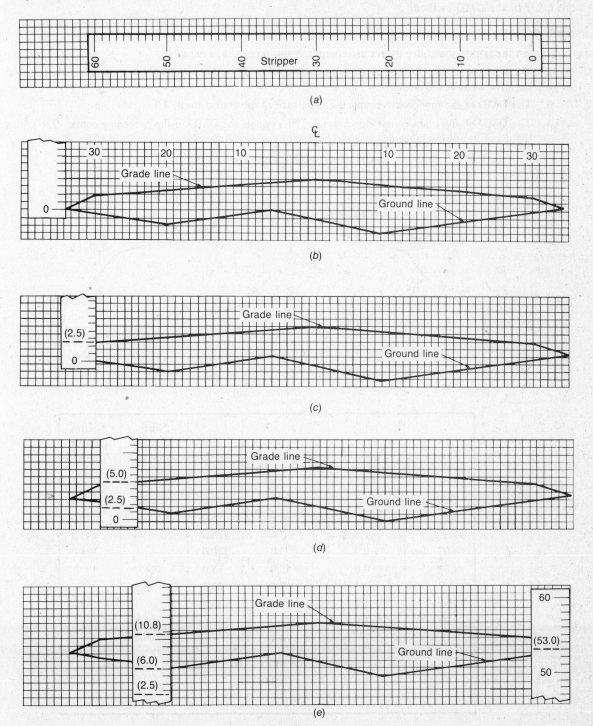

Fig. 11-3 Stripper method of calculating area. (*Courtesy of U.S. Army, General Drafting, T M 5-230*)

course's DMD is determined. Multiplying the DMD of each course by its latitude results in twice the area, but the sign of this product illustrates whether the area adds to or subtracts from the figure area.

EXAMPLE 11.4   Given: The area and table shown in Fig. 11-4. Follow a step-by-step procedure to find the area of this figure by the DMD method.

Solution:

1. All the latitudes and departures are computed and recorded in the table.

2. The leftmost station (*D*) is selected as the first point and line *DE* is selected as the first course to avoid negative areas in the DMD.

3. The DMD of the first course equals the departure of the course itself, 4.0.

4. The DMD of any other course (for example, *EF*) equals the DMD of the preceding course (*DE*), plus the departure of the preceding course (*DE*), plus the departure of the course itself (*EF*). Thus

$$\text{DMD of } EF = 4.0 + 4.0 + 30.0 = 38.0$$

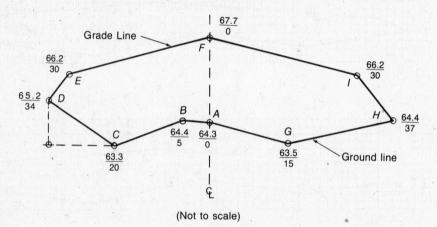

(Not to scale)

Course	Latitude	Departure	DMD	North Double Area (+)	South Double Area (−)
DE	+1.0	+4.0	4.0	4.0	
EF	+1.5	+30.0	38.0	57.0	
FI	−1.5	+30.0	98.0		147.0
IH	−1.8	+7.0	135.0		243.0
HG	−0.9	−22.0	120.0		108.0
GA	+0.8	−15.0	83.0	66.4	
AB	+0.1	−5.0	63.0	6.3	
BC	−1.1	−15.0	43.0		47.3
CD	+1.9	−14.0	14.0	26.6	
				+160.3	−545.3

$$\text{Difference} = 545.3 - 160.3 = 385.0$$
$$\text{Area} = \frac{385.0}{2} = 192.5 \text{ ft}^2$$

Fig. 11-4   Double meridian distance method for calculating area.

For the next course, the same procedure is followed. Thus

$$\text{DMD of } FI = \text{DMD of preceding course}$$
$$+ \text{ departure of preceding course} + \text{departure of the course itself}$$
$$= 38 + 30.0 + 30.0 = 98.0$$

5. The DMD of the last course is numerically equal to its departure but with opposite sign (+14.0).

6. Each DMD value is multiplied by its latitude, and positive products are entered under north double areas and negative products under south double areas.

7. The sum of all the south double areas minus the sum of all the north double areas, *disregarding sign*, equals twice the cross-sectional area. Dividing by 2 gives the true cross-sectional area.

All the computations have been worked out in the chart which accompanies Fig. 11-4.

11.8 AREA BY COORDINATES

Using the coordinates of the points of departure of a figure, we may compute the area as follows:

1. Draw an ordinate (N-S line) through the westernmost point of the figure of which we want to determine the area.

2. Draw an abscissa (E-W line) through the southernmost point of the figure.

3. List the xy coordinates.

4. Draw a series of solid lines starting at the westernmost clockwise point as the lines progress around the figure. *Note*: In Fig. 11-4 the solid lines are D at the x coordinate to E on the y coordinate; E at the x coordinate to F on the y coordinate; etc.

5. Draw a series of dotted lines starting at the westernmost point in a counterclockwise direction progressing around the figure. *Note*: In Fig. 11-14 the dotted lines run from D at the x coordinate to C on the y coordinate, etc.; see Table 11-1.

6. Carry out the multiplication for solid-line products and dotted-line products.

7. Sum up each set of products.

8. Divide the difference of the sums by 2. The result is the area in square units.

Table 11-1　Area by Coordinates (Example 11.5)

Point	x	y	Solid-Line Product	Dotted-Line Product
D	0	1.9	$0 \times 2.9 = 0$	$4 \times 1.9 = 7.6$
E	4	2.9	$4 \times 4.4 = 17.6$	$34 \times 2.9 = 98.6$
F	34	4.4	$34 \times 2.9 = 98.6$	$64 \times 4.4 = 281.6$
I	64	2.9	$64 \times 1.1 = 70.4$	$71 \times 2.9 = 205.9$
H	71	1.1	$71 \times 0.2 = 14.2$	$49 \times 1.1 = 53.9$
G	49	0.2	$49 \times 1 = 49.0$	$34 \times 0.2 = 6.8$
A	34	1.0	$34 \times 1.1 = 37.4$	$29 \times 1.0 = 29.0$
B	29	1.1	$29 \times 0 = 0.0$	$14 \times 1.1 = 15.4$
C	14	0	$14 \times 1.9 = 26.6$	$0 \times 0 = 0.0$
D	0	1.9		
			313.8	698.8

$$\text{Area} = \frac{698.8 - 313.8}{2} = \frac{385.0}{2} = 192.5 \text{ ft}^2$$

EXAMPLE 11.5   Given: Fig. 11-4 (top). Find the area by the coordinate method. Compare your answer with that obtained by the DMD method.

Solution: Draw an ordinate (N-S line) through the westernmost point. Draw an abscissa (E-W line) through the southernmost point. List the xy coordinates. Carry out the solid- and dotted-line procedures as listed in Sec. 11.8; see Table 11-1. The answer agrees with that found by the DMD method.

11.9   VOLUMES

The computation of earthwork volumes is basically a problem in solid geometry. Earthwork volumes are determined, primarily, by one of three methods:

1.   The average-end-area method
2.   The prismoidal formula
3.   The contour line method

Of these methods, the average-end-area method and the prismoidal formula method are more common and are considered to be more accurate; so only these two methods will be discussed.

11.10   VOLUME BY AVERAGE END AREAS

The average-end-area method is most commonly used to determine the volume between two cross sections or end areas.

The formulas for end areas shown in the following section give good results when the two end areas are of approximately the same shape and size. However, the greater the difference in shape between the two areas, the greater the error in volume, and when one area approaches a point (zero value), the error approaches a maximum.

If two adjacent end areas are identical in size and shape, the solid geometric figure is a prism. The volume V of a prism is given by the following formula:

$$V = \frac{A_1 + A_2}{2} L$$

where V = volume in cubic units
          A_1, A_2 = the respective end areas in square units
          L = the perpendicular distance in linear units between the end areas

Note: If $A_1 = A_2$, $V = AL$.

When one end area is zero, the geometric figure is a pyramid whose volume equals one-third the area of the base multiplied by the length ($V = \frac{1}{3}AL$).

The formula method assumes that the volume contained between successive end areas is a product of the average of the two end areas and the perpendicular distance between them. This is expressed by the formula given above: $V = (A_1 + A_2)L/2$. If V is desired in cubic yards (yd³) and A_1 and A_2 are expressed in square feet and L in feet, the formula becomes:

$$V = \frac{(A_1 + A_2)L}{54} \qquad yd^3$$

For $L = 100$ ft, the normal spacing of cross sections, and with A_1 and A_2 in square feet,

$$V = 1.852(A_1 + A_2) \qquad yd^3$$

EXAMPLE 11.6   Given: Two end areas 100 ft apart. A_1 = the area shown in Fig. 11-2; A_2 = the area shown in Fig. 11-5. Calculate the volume of material included between these two end areas.

Solution: Find the area of Fig. 11-2 geometrically. The calculations are carried out in Example 11.2; $A_1 = 600$ ft².

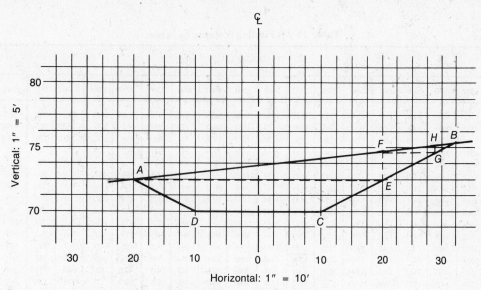

Fig. 11-5 Cut section (geometric).

Next find the area of Fig. 11-5. Use the formula for the area of a triangle, $A = \frac{1}{2}bh$. Find the area of the trapezoid by the formula $A = [h(AE + BC)]/2$.

AEF: $A = \frac{1}{2}(40)(2.2)$ $=$ 44.0 ft^2

EFG: $A = \frac{1}{2}(2.2)(9)$ $=$ 9.9 ft^2

FGH: $A = \frac{1}{2}(0.50)(9)$ $=$ 2.25 ft^2

GHB: $A = \frac{1}{2}(0.50)(3)$ $=$ 0.75 ft^2

Trapezoid $AECD$: $A = \dfrac{2.5(20 + 40)}{2} =$ 75.0 ft^2

Total = 131.9 ft^2

Use the formula for cubic yards of material within a length of 100 ft:

$$V = \frac{(A_1 + A_2)L}{54} = \frac{(600 + 131.9)100}{54} = 1355 \text{ yd}^3 \qquad Ans.$$

11.11 VOLUME BY END-AREAS TABLE

Another method of finding the volume by the use of end areas is by adding together the two end areas and using Table 11-2. Note that the table is to be used with the *sum* of end areas, and that it is not necessary to calculate the average end areas. Example 11.7 illustrates the use of Table 11-2.

EXAMPLE 11.7 Two cross sections 100 ft apart have areas of 121.1 and 316.2 ft^2, respectively. Find the volume of the section, using Table 11-2.

Solution:

Sum of end areas = 121.1 + 316.2 = 437.3 ft^2

Look at Table 11.2. The figure 437 is found to the left of the heavy line under the column headed 800. On the same line with 437 and under the column headed 0.3 is found the figure 9.81. Add these two figures to find the volume:

800 + 9.81 = 809.81 = 810 yd^3 *Ans.*

This is the volume between stations.

Note: For distances between stations other than 100 ft, multiply the result obtained by the actual distance between stations and divide by 100. Should the sum of the areas be greater than 1673.9 ft^2, break the sum into portions that will fit the table and add the results to get the total volume.

Table 11-2 Finding Volume by Areas

Volume, yd^3

3000	2900	2800	2700	2600	2500	2400	2300	2200	2100	2000	1900	1800	1700	1600	1500	1400	1300	1200	1100	1000
1620	1566	1512	1458	1404	1350	1296	1242	1188	1134	1080	1026	972	918	864	810	756	702	648	594	540
1621	1567	1513	1459	1405	1351	1297	1243	1189	1135	1081	1027	973	919	865	811	757	703	649	595	541
1622	1568	1514	1460	1406	1352	1298	1244	1190	1136	1082	1028	974	920	866	812	758	704	650	596	542
1623	1569	1515	1461	1407	1353	1299	1245	1191	1137	1083	1029	975	921	867	813	759	705	651	597	543
1624	1570	1516	1462	1408	1354	1300	1246	1192	1138	1084	1030	976	922	868	814	760	706	652	598	544
1625	1571	1517	1463	1409	1355	1301	1247	1193	1139	1085	1031	977	923	869	815	761	707	653	599	545
1626	1572	1518	1464	1410	1356	1302	1248	1194	1140	1086	1032	978	924	870	816	762	708	654	600	546
1627	1573	1519	1465	1411	1357	1303	1249	1195	1141	1087	1033	979	925	871	817	763	709	655	601	547
1628	1574	1520	1466	1412	1358	1304	1250	1196	1142	1088	1034	980	926	872	818	764	710	656	602	548
1629	1575	1521	1467	1413	1359	1305	1251	1197	1143	1089	1035	981	927	873	819	765	711	657	603	549
1630	1576	1522	1468	1414	1360	1306	1252	1198	1144	1090	1036	982	928	874	820	766	712	658	604	550
1631	1577	1523	1469	1415	1361	1307	1253	1199	1145	1091	1037	983	929	875	821	767	713	659	605	551
1632	1578	1524	1470	1416	1362	1308	1254	1200	1146	1092	1038	984	930	876	822	768	714	660	606	552
1633	1579	1525	1471	1417	1363	1309	1255	1201	1147	1093	1039	985	931	877	823	769	715	661	607	553
1634	1580	1526	1472	1418	1364	1310	1256	1202	1148	1094	1040	986	932	878	824	770	716	662	608	554
1635	1581	1527	1473	1419	1365	1311	1257	1203	1149	1095	1041	987	933	879	825	771	717	663	609	555
1636	1582	1528	1474	1420	1366	1312	1258	1204	1150	1096	1042	988	934	880	826	772	718	664	610	556
1637	1583	1529	1475	1421	1367	1313	1259	1205	1151	1097	1043	989	935	881	827	773	719	665	611	557
1638	1584	1530	1476	1422	1368	1314	1260	1206	1152	1098	1044	990	936	882	828	774	720	666	612	558
1639	1585	1531	1477	1423	1369	1315	1261	1207	1153	1099	1045	991	937	883	829	775	721	667	613	559
1640	1586	1532	1478	1424	1370	1316	1262	1208	1154	1100	1046	992	938	884	830	776	722	668	614	560
1641	1587	1533	1479	1425	1371	1317	1263	1209	1155	1101	1047	993	939	885	831	777	723	669	615	561
1642	1588	1534	1480	1426	1372	1318	1264	1210	1156	1102	1048	994	940	886	832	778	724	670	616	562
1643	1589	1535	1481	1427	1373	1319	1265	1211	1157	1103	1049	995	941	887	833	779	725	671	617	563
1644	1590	1536	1482	1428	1374	1320	1266	1212	1158	1104	1050	996	942	888	834	780	726	672	618	564
1645	1591	1537	1483	1429	1375	1321	1267	1213	1159	1105	1051	997	943	889	835	781	727	673	619	565
1646	1592	1538	1484	1430	1376	1322	1268	1214	1160	1106	1052	998	944	890	836	782	728	674	620	566
1647	1593	1539	1485	1431	1377	1323	1269	1215	1161	1107	1053	999	945	891	837	783	729	675	621	567
1648	1594	1540	1486	1432	1378	1324	1270	1216	1162	1108	1054	1000	946	892	838	784	730	676	622	568
1649	1595	1541	1487	1433	1379	1325	1271	1217	1163	1109	1055	1001	947	893	839	785	731	677	623	569
1650	1596	1542	1488	1434	1380	1326	1272	1218	1164	1110	1056	1002	948	894	840	786	732	678	624	570
1651	1597	1543	1489	1435	1381	1327	1273	1219	1165	1111	1057	1003	949	895	841	787	733	679	625	571
1652	1598	1544	1490	1436	1382	1328	1274	1220	1166	1112	1058	1004	950	896	842	788	734	680	626	572
1653	1599	1545	1491	1437	1383	1329	1275	1221	1167	1113	1059	1005	951	897	843	789	735	681	627	573
1654	1600	1546	1492	1438	1384	1330	1276	1222	1168	1114	1060	1006	952	898	844	790	736	682	628	574
1655	1601	1547	1493	1439	1385	1331	1277	1223	1169	1115	1061	1007	953	899	845	791	737	683	629	575
1656	1602	1548	1494	1440	1386	1332	1278	1224	1170	1116	1062	1008	954	900	846	792	738	684	630	576
1657	1603	1549	1495	1441	1387	1333	1279	1225	1171	1117	1063	1009	955	901	847	793	739	685	631	577
1658	1604	1550	1496	1442	1388	1334	1280	1226	1172	1118	1064	1010	956	902	848	794	740	686	632	578
1659	1605	1551	1497	1443	1389	1335	1281	1227	1173	1119	1065	1011	957	903	849	795	741	687	633	579
1660	1606	1552	1498	1444	1390	1336	1282	1228	1174	1120	1066	1012	958	904	850	796	742	688	634	580
1661	1607	1553	1499	1445	1391	1337	1283	1229	1175	1121	1067	1013	959	905	851	797	743	689	635	581
1662	1608	1554	1500	1446	1392	1338	1284	1230	1176	1122	1068	1014	960	906	852	798	744	690	636	582
1663	1609	1555	1501	1447	1393	1339	1285	1231	1177	1123	1069	1015	961	907	853	799	745	691	637	583
1664	1610	1556	1502	1448	1394	1340	1286	1232	1178	1124	1070	1016	962	908	854	800	746	692	638	584
1665	1611	1557	1503	1449	1395	1341	1287	1233	1179	1125	1071	1017	963	909	855	801	747	693	639	585
1666	1612	1558	1504	1450	1396	1342	1288	1234	1180	1126	1072	1018	964	910	856	802	748	694	640	586
1667	1613	1559	1505	1451	1397	1343	1289	1235	1181	1127	1073	1019	965	911	857	803	749	695	641	587
1668	1614	1560	1506	1452	1398	1344	1290	1236	1182	1128	1074	1020	966	912	858	804	750	696	642	588
1669	1615	1561	1507	1453	1399	1345	1291	1237	1183	1129	1075	1021	967	913	859	805	751	697	643	589
1670	1616	1562	1508	1454	1400	1346	1292	1238	1184	1130	1076	1022	968	914	860	806	752	698	644	590
1671	1617	1563	1509	1455	1401	1347	1293	1239	1185	1131	1077	1023	969	915	861	807	753	699	645	591
1672	1618	1564	1510	1456	1402	1348	1294	1240	1186	1132	1078	1024	970	916	862	808	754	700	646	592
1673	1619	1565	1511	1457	1403	1349	1295	1241	1187	1133	1079	1025	971	917	863	809	755	701	647	593

Source: P. Kissam, *Surveying Practice*, McGraw-Hill, New York, 1978.

Table 11-2 (cont.)

Volume, yd³										Area, ft²									
900	800	700	600	500	400	300	200	100	0	0.0	0.1	0.2	0.3	0.4	0.5	0.6	0.7	0.8	0.9
486	432	378	324	270	216	162	108	54	0	0.00	0.18	0.37	0.56	0.74	0.93	1.11	1.30	1.48	1.67
487	433	379	325	271	217	163	109	55	1	1.85	2.04	2.22	2.41	2.59	2.78	2.96	3.15	3.33	3.52
488	434	380	326	272	218	164	110	56	2	3.70	3.89	4.07	4.26	4.44	4.63	4.82	5.00	5.18	5.37
489	435	381	327	273	219	165	111	57	3	5.56	5.74	5.93	6.11	6.30	6.48	6.67	6.85	7.04	7.22
490	436	382	328	274	220	166	112	58	4	7.41	7.59	7.78	7.96	8.15	8.33	8.52	8.70	8.89	9.07
491	437	383	329	275	221	167	113	59	5	9.26	9.44	9.63	9.81	10.00	10.19	10.37	10.66	10.74	10.95
492	438	384	330	276	222	168	114	60	6	11.11	11.30	11.48	11.67	11.85	12.04	12.22	12.41	12.59	12.78
493	439	385	331	277	223	169	115	61	7	12.96	13.15	13.33	13.52	13.70	13.89	14.07	14.26	14.44	14.68
494	440	386	332	278	224	170	116	62	8	14.82	15.00	15.19	15.37	15.56	15.74	15.93	16.11	16.30	16.48
495	441	387	333	279	225	171	117	63	9	16.67	16.85	17.04	17.22	17.41	17.59	17.78	17.96	18.15	13.33
496	442	388	334	280	226	172	118	64	10	18.52	18.70	18.89	19.07	19.26	19.44	19.63	19.82	20.00	20.18
497	443	389	335	281	227	173	119	65	11	20.37	20.56	20.74	20.93	21.11	21.30	21.48	21.67	21.85	22.04
498	444	390	336	282	228	174	120	66	12	22.22	22.41	22.59	22.78	22.96	23.15	23.33	23.52	23.70	23.89
499	445	391	337	283	229	175	121	67	13	24.07	24.26	24.44	24.63	24.82	25.00	25.18	25.37	25.56	25.74
500	446	392	338	284	230	176	122	68	14	25.93	26.11	26.30	26.48	26.67	26.85	27.04	27.22	27.41	27.59
501	447	393	339	285	231	177	123	69	15	27.78	27.96	28.15	28.33	28.52	28.70	28.89	29.07	29.26	29.44
502	448	394	340	286	232	178	124	70	16	29.63	29.82	30.00	30.18	30.37	30.56	30.74	30.93	31.11	31.30
503	449	395	341	287	233	179	125	71	17	31.48	31.67	31.85	32.04	32.22	32.41	32.59	32.78	32.96	33.15
504	450	396	342	288	234	180	126	72	18	33.33	33.52	33.70	33.89	34.07	34.26	34.44	34.63	34.82	35.00
505	451	397	343	289	235	181	127	73	19	35.18	35.37	35.56	35.74	35.93	36.11	36.30	36.48	36.67	36.85
506	452	398	344	299	236	182	128	74	20	37.04	37.22	37.41	37.59	37.78	37.96	38.15	38.33	38.52	38.70
507	453	399	345	291	237	183	129	75	21	38.89	39.07	39.26	39.44	39.63	39.82	40.00	40.19	40.37	40.56
508	454	400	346	292	238	184	130	76	22	40.71	40.93	41.11	41.30	41.48	41.67	41.85	42.04	42.22	42.41
509	455	401	347	293	239	185	131	77	23	42.59	42.78	42.96	43.15	43.33	43.52	43.70	43.89	44.07	44.26
510	456	402	348	294	240	186	132	78	24	44.44	44.63	44.81	45.00	45.19	45.37	45.56	45.74	45.93	46.11
511	457	403	349	295	241	187	133	79	25	46.30	46.48	46.67	46.85	47.04	47.22	47.41	47.59	47.78	47.98
512	458	404	350	296	242	188	134	80	26	48.15	48.33	48.52	48.70	48.89	49.07	49.26	49.44	49.63	49.82
513	459	405	351	297	243	189	135	81	27	50.00	50.19	50.37	50.56	50.74	50.93	51.11	51.30	51.48	51.67
514	460	406	352	298	244	190	136	82	28	51.84	52.04	52.22	52.41	52.69	52.78	52.96	53.15	53.33	53.52
515	461	407	353	299	245	191	137	83	29	53.70	53.89	54.07	54.26	54.44	54.63	54.81	55.00	55.18	55.37
516	462	408	354	300	246	192	138	84	30	55.56	55.74	55.93	56.11	56.30	56.48	56.67	56.85	57.04	57.22
517	463	409	355	301	247	193	139	85	31	57.41	57.59	57.78	57.96	58.15	58.33	58.52	58.70	58.89	59.07
518	464	410	356	302	248	194	140	86	32	69.26	59.44	59.63	59.82	60.00	60.18	60.37	60.56	60.74	60.93
519	465	411	357	303	249	195	141	87	33	61.11	61.30	61.48	61.67	61.85	62.04	62.22	62.41	62.59	62.73
520	466	412	358	304	250	196	142	88	34	62.96	63.15	63.33	63.52	63.70	63.89	64.07	64.26	64.44	64.62
521	467	413	359	305	251	197	143	89	35	64.82	65.00	65.18	65.37	65.56	65.74	65.93	66.11	66.30	66.43
522	468	414	360	306	252	198	144	90	36	66.67	66.85	67.04	67.22	67.41	67.59	67.78	67.96	68.15	68.33
523	469	415	361	307	253	199	145	91	37	68.52	68.70	68.89	69.07	69.26	69.44	69.63	69.82	70.00	70.18
524	470	416	362	308	254	200	146	92	38	70.37	70.56	70.74	70.93	71.11	71.30	71.48	71.67	71.85	72.04
525	471	417	363	309	255	201	147	93	39	72.22	72.41	72.59	72.78	72.96	73.15	73.33	73.52	73.70	73.89
526	472	418	364	310	256	202	148	94	40	74.07	74.26	74.44	74.63	74.82	75.00	75.18	75.37	75.56	75.74
527	473	419	365	311	257	203	149	95	41	75.93	76.11	76.30	76.48	76.67	76.85	77.04	77.22	77.41	77.59
528	474	420	366	312	258	204	150	96	42	77.78	77.96	78.15	78.33	78.52	78.70	78.89	79.07	79.25	79.44
529	475	421	367	313	259	205	151	97	43	79.63	79.82	80.00	80.18	80.37	80.56	80.74	80.93	81.11	81.30
530	476	422	368	314	260	206	152	98	44	81.84	81.67	81.85	82.04	82.22	82.41	82.59	82.78	82.96	83.15
531	477	423	369	315	261	207	153	99	45	83.33	83.52	83.70	83.89	84.07	84.26	84.44	84.63	84.82	85.00
532	478	424	370	316	262	208	154	100	46	85.19	85.37	85.56	85.74	85.93	86.11	86.30	86.48	86.67	86.85
533	479	425	371	317	263	209	155	101	47	87.04	87.22	87.41	87.59	87.78	87.96	88.15	88.33	88.52	88.70
534	480	425	372	318	264	210	156	102	48	88.89	89.07	89.26	89.44	89.63	89.82	90.00	90.18	90.37	90.56
535	481	427	373	319	265	211	157	103	49	90.74	90.93	91.11	91.30	91.48	91.67	91.85	92.04	92.22	92.41
536	482	428	374	320	266	212	158	104	50	92.59	92.78	92.96	93.15	93.33	93.52	93.70	93.89	94.07	94.26
537	483	429	375	321	267	213	159	105	51	94.44	94.63	94.82	95.00	95.18	95.37	95.55	95.74	95.93	96.11
538	484	430	376	322	268	214	160	106	52	96.30	96.48	96.67	96.85	97.04	97.22	97.41	97.59	97.78	97.95
539	485	431	377	323	269	215	161	107	53	98.15	98.33	98.52	98.70	98.89	99.07	99.26	99.44	99.63	99.82

11.12 VOLUME BY PRISMOIDAL FORMULA

When more exact values for earthwork volumes are required, the prismoidal formula is used. This formula is often used to determine volumes of costly construction materials in complex shapes, such as concrete in place.

The volume of a prismoid is expressed by the following formula:

$$V = \frac{L}{6}(A_1 + 4A_m + A_2)$$

where

V = the volume in cubic units
A_1, A_2 = the end areas in square units
A_m = the area in square units of the section midway between A_1 and A_2
L = the perpendicular distance in linear units between the end areas

A_m is obtained by first averaging the corresponding dimensions of A_1 and A_2, then determining the area A_m—*not* by averaging the areas A_1 and A_2.

EXAMPLE 11.8 In Fig. 11-6, $A_1 = 175.2\,\text{ft}^2$ and $A_2 = 116.8\,\text{ft}^2$ are the computed cross-sectional areas at stations $110 + 00$ and $111 + 00$, respectively. Area $A_m = 144.4\,\text{ft}^2$ at station $110 + 50$ is determined by averaging the corresponding linear dimensions of A_1 and A_2. Find the volume between end areas A_1 and A_2, using the prismoidal formula.

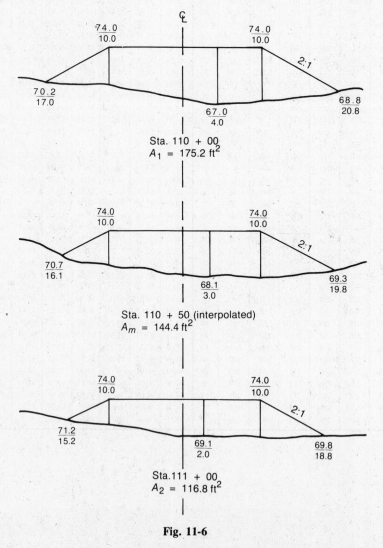

Fig. 11-6

Solution: The mean values 70.7 and 16.1 at the left slope and the ground intersection in A_m equal one-half the sum of the values at the corresponding locations in A_1 and A_2, respectively, or

$$\tfrac{1}{2}(70.2 + 71.2) = 70.7 \qquad \text{and} \qquad \tfrac{1}{2}(17.0 + 15.2) = 16.1$$

Values 68.1 and 3.0, and 69.3 and 19.8 are calculated by the same procedure. The values 74.0 and 10.0 are constant for all three sections. Substituting in the prismoidal formula, the volume of earthwork between stations $110 + 00$ and $111 + 00$ is

$$V = \frac{100}{6}[175.2 + 4(144.4) + 116.8] = 14\,493 \text{ ft}^3$$

Divide this number by 27 (the number of cubic feet in a cubic yard):

$$\frac{14\,493}{27} = 537 \text{ yd}^3 \qquad Ans.$$

The average-end-area method gives 541 yd^3 for this same volume, a 0.99 percent greater value, an error which is obviously negligible.

Solved Problems

11.1 Given: A figure with an area of cut enclosed; see Fig. 11-5. The vertical scale is 1 in = 5 ft. The horizontal scale is 1 in = 10 ft. Find the area enclosed within the figure *ABCD* by the method of counting the squares.

   Solution

   Use blocks $\tfrac{1}{4}$-in square. Find the area of one square block:

   Horizontal distance: $(\tfrac{1}{4} \text{ in})(10 \text{ ft}) = 2.5 \text{ ft}$

   Vertical distance: $(\tfrac{1}{4} \text{ in})(5 \text{ ft}) = 1.25 \text{ ft}$

   So   Area of a $\tfrac{1}{4}$-in square block $= 2.5(1.25) = 3.125 \text{ ft}^2$

   Count the blocks in Fig. 11-5. Estimate where the blocks are not complete. This gives about 42 squares. Thus

   $$\text{Area of } ABCD = 42(3.125) = 131.25 \text{ ft}^2 \simeq 131 \text{ ft}^2 \qquad Ans.$$

11.2 Given: An area bounded by points *RSTUV*; see Fig. 11-7. Estimate the area by counting the squares.

   Solution

   Use blocks $\tfrac{1}{4}$-in square. Find the area of one square block:

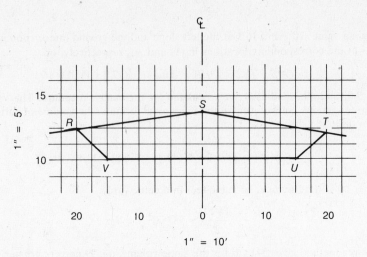

Fig. 11-7 Cut section (counting the squares).

Horizontal distance: $(\frac{1}{4}\,\text{in})(10\,\text{ft}) = 2.5\,\text{ft}$

Vertical distance: $(\frac{1}{4}\,\text{in})(5\,\text{ft}) = 1.25\,\text{ft}$

So   Area of a $\frac{1}{4}$-in square block $= 2.5(1.25) = 3.125\,\text{ft}^2$

Count the blocks in the figure; they add up to approximately 35. So

$$\text{Area of } RSTUV = 35(3.125) = 109.38\,\text{ft}^2 \approx 109\,\text{ft}^2 \qquad Ans.$$

11.3   Given: A cut area $ADCB$ with a vertical scale of $1\,\text{in} = 5\,\text{ft}$ and a horizontal scale of $1\,\text{in} = 10\,\text{ft}$; see Fig. 11-5. Find the area of $ADCB$ by geometrics.

Solution

Divide the area into four triangles AEF, EFG, FGH, and GHB, and trapezoid $ADCE$, and solve. Use the formulas for the area of a triangle, $A = \frac{1}{2}bh$, and for the area of a trapezoid, $A = h(DC + AE)/2$.

AEF: $A = \frac{1}{2}(40)(2.2)\qquad = 44.00\,\text{ft}^2$

EFG: $A = \frac{1}{2}(2.2)(9)\qquad = 9.90\,\text{ft}^2$

FGH: $A = \frac{1}{2}(0.50)(9)\qquad = 2.25\,\text{ft}^2$

GHB: $A = \frac{1}{2}(0.50)(3.0)\qquad = 0.75\,\text{ft}^2$

$ADCE$: $A = \dfrac{h(DC + AE)}{2} = 75.00\,\text{ft}^2$

$$\text{Total area} = 131.9 \ \text{ft}^2 \qquad Ans.$$

11.4   Given: An area bounded by $ABCDEF$; see Fig. 11-8. Find the area by geometrics.

Solution

Divide the area into triangles AIB, AFG, and CDE, plus a square $FGIJ$ and a rectangle $JECB$. Find the area of each figure and then add. Use the formula for the area of a triangle, $A = \frac{1}{2}bh$, and the formula for the area of a square or rectangle, $A = bh$.

AIB: $A = \frac{1}{2}(15)(5)\quad = 37.5\,\text{ft}^2$

AFG: $A = \frac{1}{2}(10)(2.5) = 12.5\,\text{ft}^2$

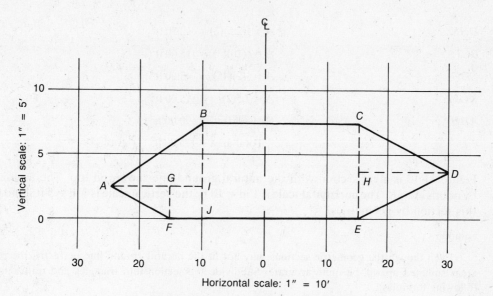

Fig. 11-8   Area bounded by *ABCDEF*.

CED:	$A = \frac{1}{2}(15)(7.5) = \;\;56.25\,\text{ft}^2$
FGIJ:	$A = 5(2.5) \quad\quad = \;\;12.50\,\text{ft}^2$
JECB:	$A = 25(7.5) \quad\; = 187.50\,\text{ft}^2$

Total area = 306.25 ft²   *Ans.*

11.5   Given: Area *ABCDE* (see Fig. 11-9) with a horizontal scale of 1 in = 10 ft and a vertical scale of 1 in = 2 ft. Find the area of *ABCDE*.

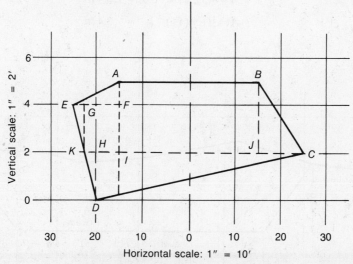

Fig. 11-9   Area bounded by *ABCDE*.

Solution

Divide *ABCDE* into triangles *EAF*, *EGK*, *DHK*, *BCJ*, and *CDH*, plus rectangles *KGFI* and *ABJI*, and add their areas. Use $A = \frac{1}{2}bh$ for the triangles and $A = bh$ for the rectangles.

EAF:	$A = \frac{1}{2}(10)(1) = \;\;5.00\,\text{ft}^2$
EGK:	$A = \frac{1}{2}(2.5)(2) = \;\;2.50\,\text{ft}^2$

DHK:	$A = \frac{1}{2}(2.5)(2) =$	2.50 ft^2
BCJ:	$A = \frac{1}{2}(10)(3) =$	15.00 ft^2
CDH:	$A = \frac{1}{2}(2)(45) =$	45.00 ft^2
KGFI:	$A = 7.5(2) =$	15.00 ft^2
ABJI:	$A = 30(3) =$	90.00 ft^2

Total area $= 175.00 \text{ ft}^2$ *Ans.*

11.6 Figure 11-10 is a cut section with the natural ground line shown on top. The area is enclosed by points *A* to *K*. The horizontal scale is 1 in = 10 ft; the vertical scale is 1 in = 5 ft. Find the area of this section by geometrics.

Solution

Even though the geometric sections may not fit the natural ground line perfectly, the result of the areas summed up will be quite accurate. Subdivide the section into triangles and trapezoids. Use the following formulas:

Area of a triangle: $A = \frac{1}{2}bh$

Area of a trapezoid: $A = \dfrac{h(\text{base } 1 + \text{base } 2)}{2}$

where A = area
  bases 1, 2 = the lengths defined by *KJ*, *BJ*, *EG*, as needed
     h = the perpendicular distance, or height, between parallel bases for a trapezoid and
       from base to vertex for a triangle.

AJK:	$A = \frac{1}{2}(5.7)(4.0)$	$= 11.4 \text{ ft}^2$
ABJ:	$A = \frac{1}{2}(5.2)(5.7 + 6.0) =$	30.4 ft^2
EFG:	$A = \frac{1}{2}(1.5)(4.5)$	$= 3.4 \text{ ft}^2$

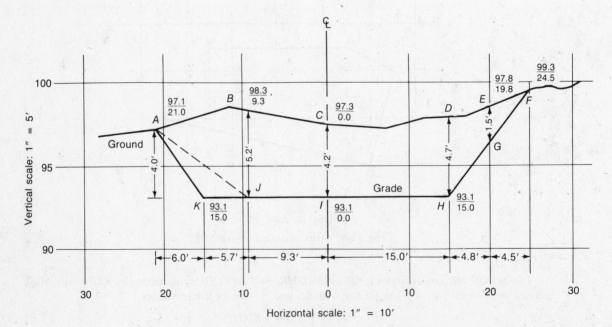

Fig. 11-10 Uneven cut area.

$BCIJ$: $A = \dfrac{9.3(5.2 + 4.2)}{2} = 43.7\ \text{ft}^2$

$CDHI$: $A = \dfrac{15.0(4.2 + 4.7)}{2} = 66.75\ \text{ft}^2$

$DEGH$: $A = \dfrac{4.8(4.7 + 1.5)}{2} = 14.9\ \text{ft}^2$

$$\text{Total area} = 170.6\ \text{ft}^2 \quad Ans.$$

In Probs. 11.7 to 11.10, use graph paper to set up the coordinates as given in the indicated figure.

11.7   Given: An area bounded by $ABCD$; see Fig. 11-5. The area has been found by counting the squares (Prob. 11.1), which gave an answer of 131 ft^2. By geometrics we found the area to be 131.9 ft^2 (Prob. 11.3). Find the area by the stripper method.

Solution

If you worked Example 11.3 you will have made a paper stripper. If not, make a stripper as explained in the example. Start the strippering of Fig. 11-5 at the first 5-square line. The location is 15 horizontally and 71.25 vertically. It is easiest to use the point of a dividers or compass point to set the point on your stripper and bring it forward to the next 5-square point.

First stripper figure	= 2.5
Second stripper figure	= 9.0
Third stripper figure	= 15.7
Fourth stripper figure	= 22.7
Fifth stripper figure	= 30.5
Sixth stripper figure	= 38.0
Seventh stripper figure	= 44.5
Eighth stripper figure	= 49.0
Ninth stripper figure	= 51.5
Tenth stripper figure	= 52.4 total

52.4 (stripper width of 5) = 262

1 in horizontally = 10 ft   1 in vertically = 5 ft

$10(5) = 50\ \text{ft}^2$ per 100 squares = 0.5 ft^2 per square

So   Area = 262(0.5) = 131.00 ft^2 Ans.

11.8   Given: Fig. 11-10. The area has an uneven natural ground line and is bounded by the points A to K. Find the area by the stripper method.

Solution

Although this appears to be a difficult problem because of the uneven natural ground, strippering is very simple and gives accurate results. Stripper the area starting with the first 5-square section at point K (15 horizontally, 93.1 vertically). The stripper reading totals 68. The stripper is 5 squares wide, so $68(5) = 340$. Horizontal scale = 10 ft; vertical scale = 5 ft. So $10(5) = 50\ \text{ft}^2$ per 100 squares = 0.5 ft^2 per square. So

$$\text{Strippered area} = 340(0.5) = 170\ \text{ft}^2 \quad Ans.$$

This answer is close to that found by geometrics, 170.6 ft^2 (see Prob. 11.6).

11.9 Given: An area bounded by *ABCDEF*; see Fig. 11-8. Horizontal scale is 1 in = 10 ft; vertical scale is 1 in = 5 ft. Find the area by the stripper method.

Solution

Strippering gives a total reading of 122. The stripper is 5 squares wide, so multiply: 122(5) = 610. Vertical scale = 5 ft; horizontal scale = 10 ft. So 10(5) = 50 ft² per 100 squares = 0.5 ft² per square. Thus

$$\text{Area} = 610(0.5) = 305 \text{ ft}^2 \qquad Ans.$$

11.10 Given: An area bounded by *ABCDKE*; see Fig. 11.9. Find the area by the stripper method.

Solution

Start the stripper at *D* (location 20 and 0). Total stripper reading is 174.4. The answer in Prob. 11.5 in which this area was calculated by geometrics was 175 ft². The answers are quite close, and the stripper method is very fast. It is for this reason that it is so universally used in road department offices for figuring areas.

11.11 Given: An area bounded by *ABCDEFG*; see Fig. 11-11. Vertical scale is 1 in = 2 ft; horizontal scale is 1 in = 10 ft. Find the area by the double-meridian-distance method.

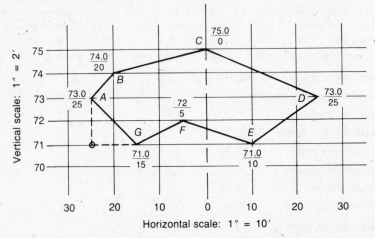

Course	Latitude	Departure	DMD	North Double Area (+)	South Double Area (−)
AB	+1.0	+5.0	5.0	5.0	
BC	+1.0	+20.0	30.0	30.0	
CD	−2.0	+25.0	75.0		150.0
DE	−2.0	−15.0	85.0		170.0
EF	+1.0	−15.0	55.0	55.0	
FG	−1.0	−10.0	30.0		30.0
GA	+2.0	−10.0	10.0	20.0	
				+110.0	−350.0

$$\text{Difference} = 350.0 - 110.0 = 240$$
$$\text{Area} = \frac{240}{2} = 120 \text{ ft}^2$$

Fig. 11-11 Double meridian distance method for calculating area.

Solution

Figure 11-11 shows all the computations. First from the figure compute the data for each point. Starting at point A, figure the latitude of course AB as $+1.0$ and the departure as $+5.0$. The first course has then a DMD of 5.0.

$$\text{North double area} = 5.0(1.0) = 5.0$$

Course BC has a latitude of $+1.0$ and a departure of $+20.0$. The DMD for this course is $5.0 + 5.0 + 20.0 = 30.0$.

$$\text{North double area} = 30.0(1.0) = 30.0$$

Compute all courses as shown in the chart in Fig. 11-11. As computed,

$$\text{Area} = 120\,\text{ft}^2 \qquad Ans.$$

11.12 Given: Area $ABCDEFG$; see Fig. 11-12. Compute the area by the double-meridian-distance method.

Solution

All computations are shown in the chart in Fig. 11-12.

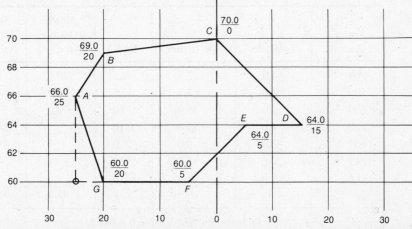

Course	Latitude	Departure	DMD	North Double Area (+)	South Double Area (−)
AB	$+3.0$	$+5.0$	5.0	15.0	
BC	$+1.0$	$+20.0$	30.0	30.0	
CD	-6.0	$+15.0$	65.0		390.0
DE	0.0	-10.0	70.0		
EF	-4.0	-10.0	50.0		200.0
FG	0.0	-15.0	25.0		
GA	$+6.0$	-5.0		30.0	
				75.0	590.0

Difference $= 590.0 - 75.0 = 515.0$

$$\text{Area} = \frac{515.0}{2} = 257.5\,\text{ft}^2$$

Fig. 11-12

11.13 Given: Area *ABCDEFG*; see Fig. 11-13. Compute the area by the double-meridian-distance method.

Solution

All computations are shown in the chart in Fig. 11-13.

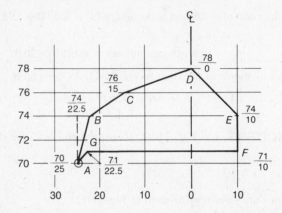

Course	Latitude	Departure	DMD	North Double Area (+)	South Double Area (−)
AB	+4.0	+2.5	2.5	10.0	
BC	+2.0	+7.5	12.5	25.0	
CD	+2.0	+15.0	35.0	70.0	
DE	−4.0	+10.0	60.0		240.0
EF	−3.0	0.0	70.0		210.0
FG	0.0	−32.5	37.5		
GA	−1.0	−2.5	2.5		2.5
				105.0	452.5

Difference = 452.5 − 105.0 = 347.5

$$\text{Area} = \frac{347.5}{2} = 173.75 \text{ ft}^2$$

Fig. 11-13

11.14 Given: Area *ABCDEF*; see Fig. 11-14. Compute the area by the double-meridian-distance method.

Solution

All computations are shown in the chart in Fig. 11-14.

11.15 Given: Fig. 11-11. Solve by the coordinate area method.

Solution

Set up a chart (see Table 11-3) with points, *x* and *y* coordinates, solid-line products, and dotted-line products. Carry out the multiplications for the solid- and dotted-line products. Sum each set of products. Divide the difference between the sums by 2. As shown in Table 11-3,

Area = 120 ft^2 *Ans.*

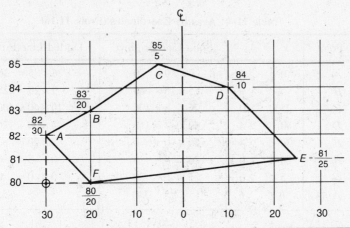

Course	Latitude	Departure	DMD	North Double Area (+)	South Double Area (−)
AB	+1.0	+10.0	10.0	10.0	
BC	+2.0	+15.0	35.0	70.0	
CD	−1.0	+15.0	65.0		65.0
DE	−3.0	+15.0	95.0		285.0
EF	−1.0	−45.0	65.0		65.0
FA	+2.0	−10.0	10.0	20.0	
				100.0	415.0

Difference = 415.0 − 100.0 = 315.0

$$\text{Area} = \frac{315}{2} = 157.5 \text{ ft}^2$$

Fig. 11-14

Table 11-3   Area by Coordinates (Prob. 11.15)

Point	x	y	Solid-Line Product	Dotted-Line Product
A	0	2.0	0 × 3 = 0	5 × 2 = 10
B	5	3.0	5 × 4 = 20	25 × 3 = 75
C	25	4.0	25 × 2 = 50	50 × 4 = 200
D	50	2.0	50 × 0 = 0	35 × 2 = 70
E	35	0	35 × 1 = 35	20 × 0 = 0
F	20	1.0	20 × 0 = 0	10 × 1 = 10
G	10	0	10 × 2 = 20	0 × 0 = 0
A	0	2.0	125	365

$$\text{Area} = \frac{365 - 125}{2} = \frac{240}{2} = 120 \text{ ft}^2$$

11.16 Given: Fig. 11-12. Find the area by the coordinate method.

Solution

Set up a chart (see Table 11-4) and perform the calculations as indicated in Prob. 11.15. As shown in the table,   Area = 257.5 ft²   *Ans.*

Table 11-4   Area by Coordinates (Prob. 11.16)

Point	x	y	Solid-Line Product	Dotted-Line Product
A	0	6	$0 \times 9 = 0$	$6 \times 5 = 30$
B	5	9	$5 \times 10 = 50$	$9 \times 25 = 225$
C	25	10	$25 \times 4 = 100$	$10 \times 40 = 400$
D	40	4	$40 \times 4 = 160$	$4 \times 30 = 120$
E	30	4	$30 \times 0 = 0$	$4 \times 20 = 80$
F	20	0	$20 \times 0 = 0$	$0 \times 5 = 0$
G	5	0	$5 \times 6 = 30$	$0 \times 0 = 0$
A	0	6	340	855

$$\text{Area} = \frac{855 - 340}{2} = \frac{515}{2} = 257.5 \text{ ft}^2$$

11.17   Given: Fig. 11-13. Find the area by the coordinate method.

Solution

Make a chart (see Table 11-5) and perform the calculations required. As shown in the table,

$$\text{Area} = 173.75 \text{ ft}^2 \quad \textit{Ans.}$$

Table 11-5   Area by Coordinates (Prob. 11.17)

Point	x	y	Solid-Line Product	Dotted-Line Product
A	0	0	$0 \times 4 = 0$	$0 \times 2.5 = 0$
B	2.5	4	$2.5 \times 6 = 15$	$4 \times 10 = 40$
C	10	6	$10 \times 8 = 80$	$6 \times 25 = 150$
D	25	8	$25 \times 4 = 100$	$8 \times 35 = 280$
E	35	4	$35 \times 1 = 35$	$4 \times 35 = 140$
F	35	1	$35 \times 1 = 35$	$1 \times 2.5 = 2.5$
G	2.5	1	$2.5 \times 0 = 0$	$1 \times 0 = 0$
A	0	0	265	612.5

$$\text{Area} = \frac{612.5 - 265.0}{2} = \frac{347.5}{2} = 173.75 \text{ ft}^2$$

11.18   Given: Fig. 11-14. Find the area by the coordinate method.

Solution

Set up a chart (see Table 11-6) and perform the calculations required. As shown in the table,

$$\text{Area} = 157.5 \text{ ft}^2 \quad \textit{Ans.}$$

11.19   Given: Two areas (sections of a road) 100 ft apart. The first end area is shown in Fig. 11-7. The second is not illustrated but the shape is the same as the first. We know the area of the second to be 138.9 ft². Find the volume of material contained between these two end areas in the 100-ft length.

Table 11-6 Area by Coordinates (Prob. 11.18)

Point	x	y	Solid-Line Product	Dotted-Line Product
A	0	2	$0 \times 3 = 0$	$2 \times 10 = 20$
B	10	3	$10 \times 5 = 50$	$3 \times 25 = 75$
C	25	5	$25 \times 4 = 100$	$5 \times 40 = 200$
D	40	4	$40 \times 1 = 40$	$4 \times 55 = 220$
E	55	1	$55 \times 0 = 0$	$1 \times 10 = 10$
F	10	0	$10 \times 2 = 20$	$0 \times 0 = 0$
A	0	2	210	525

$$\text{Area} = \frac{525 - 210}{2} = \frac{315}{2} = 157.5 \text{ ft}^2$$

Solution

The area of Fig. 11-7 was found to be approximately 109.4 ft² (see Prob. 11.2). Use the formula for cubic yards of material within a 100-ft length:

$$V = \frac{(A_1 + A_2)L}{54} = \frac{(109.4 + 138.9)100}{54} = 459.8 \text{ yd}^3 \qquad Ans.$$

11.20 Given: A section of roadway 100 ft long, illustrated by Fig. 11-10, and another section 100 ft farther up the road. The end area of the second section has been determined to be 223.5 ft². Find the volume of material contained between these two end areas in the 100-ft length.

Solution

The area of Fig. 11-10 has been found to be approximately 170 ft² (see Probs. 11.6 and 11.8). Use the formula for cubic yards of material in a 100-ft length:

$$V = \frac{(170 + 223.5)100}{54} = 728.7 \text{ yd}^3 \qquad Ans.$$

11.21 Given: Two roadway end sections. Both end areas have been worked out by the double-meridian-distance method; they are

$$A_1 = 657.6 \text{ ft}^2 \qquad A_2 = 743.2 \text{ ft}^2$$

Find the volume of material contained between these two end areas in the 100-ft length.

Solution

Both areas are known so apply the formula for volume (in cubic yards) within a 100-ft length:

$$V = \frac{(657.6 + 743.2)100}{54} = 2594 \text{ yd}^3 \qquad Ans.$$

11.22 Given: The end area shown in Fig. 11-7 and a second end area of 138.9 ft². Find the volume, using Table 11-2, and compare the answer with that found in Prob. 11.19.

Solution

$$\text{Area of Fig. 11-7} = 109.4 \text{ ft}^2$$
$$\text{Second end area} = 138.9 \text{ ft}^2$$
$$\text{Total} = 248.3 \text{ ft}^2$$

Look at Table 11-2. The figure 248 falls under the column headed 400.00 (volume). On the same row under the column headed 0.3 is 59.82. Add to find the volume:

$$400.00 + 59.82 = 459.8 \text{ yd}^3 \quad Ans.$$

This checks with the answer obtained in Prob. 11.19 by the volume formula (459.78 yd³).

11.23 Given: The end areas used in Prob. 11.20. Area of section in Fig. 11-10 = 170 ft², and area of a second section 100 ft distant = 223.5 ft². Using Table 11.2, find the volume between the two sections, and compare the answer with that found in Prob. 11.20.

Solution

Add: $$170 + 223.5 = 393.5 \text{ ft}^2$$

Find 393 ft² in the column headed 700 yd³. In the same row look under the column 0.5 to find an additional 28.70. Add:

$$700 + 28.70 = 728.70 \text{ yd}^3 \quad Ans.$$

This is the same answer obtained by the formula method in Prob. 11.20.

11.24 Given: Two stations 115 + 00 and 116 + 00. The coordinates for these stations are given in Fig. 11-15. The vertical scale is 1 in = 5 ft and the horizontal scale is 1 in = 10 ft. By strippering you find areas $A_1 = 170.0$ ft² and $A_2 = 101.25$ ft². Find the volume by the prismoidal formula.

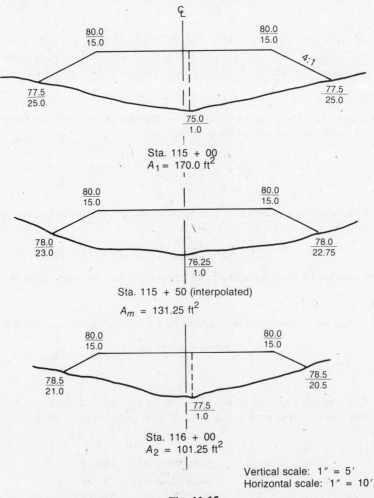

Fig. 11-15

Solution

A_m is obtained by first averaging the corresponding *dimensions* of A_1 and A_2, and the determining the area A_m—not by averaging the areas A_1 and A_2. Stripper A_m after averaging the coordinates as has been done in Fig. 11-15 to find $A_m = 131.25$ ft^2. Substitute into the prismoidal formula:

$$V = \frac{100}{6}[170.0 + 4(131.25) + 101.25] = 13\,270.8 \text{ ft}^3$$

Divide by 27 to put this answer into cubic yards:

$$\frac{13\,270}{27} = 491.51 \text{ yd}^3 \qquad Ans.$$

11.25 Given: Two stations $87 + 00$ and $88 + 00$; see Fig. 11-16. The coordinates are given for these stations. The vertical scale is 1 in = 5 ft and the horizontal scale is 1 in = 10 ft. Find the volume by the prismoidal formula.

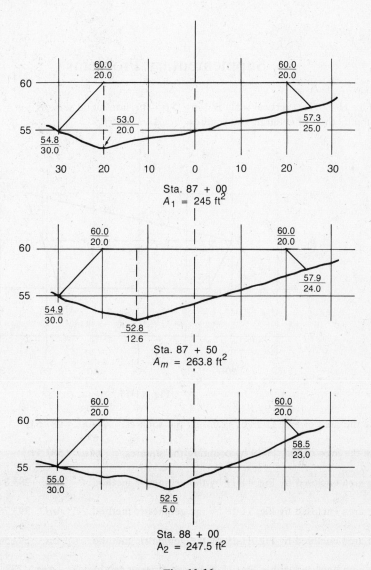

Fig. 11-16

Solution

Average the coordinates of station $87 + 00$ and $88 + 00$ in order to draw the midstation $87 + 50$:

$$\frac{25.0 + 23.0}{2} = 24.0 \qquad \frac{54.8 + 55.0}{2} = 54.9 \qquad \frac{53 + 52.5}{2} = 52.8 \qquad \frac{20 + 5.0}{2} = 12.5 \qquad \frac{57.3 + 58.5}{2} = 57.9$$

All the rest of the coordinates stay the same. Draw station $87 + 50$ between A_1 and A_2 (most easily done on graph paper). Stripper station $87 + 50$ to find the area of $263.8\ \text{ft}^2$. Figure 11-16 gives the areas of A_1 and A_2. You now have all the information you need to find the volume from the prismoidal formula. Substitute into the formula:

$$V = \frac{100}{6}[245 + 4(263.8) + 247.5] = 25\ 795\ \text{ft}^3$$

Divide this number by 27 (number of cubic feet in a cubic yard):

$$\frac{25\ 795}{27} = 955.37\ \text{yd}^3 \qquad Ans.$$

Supplementary Problems

11.26   Given: Fig. 11-17. The vertical scale is $1\ \text{in} = 5\ \text{ft}$. The horizontal scale is $1\ \text{in} = 10\ \text{ft}$. Find the enclosed area by the method of counting the squares.   *Ans.* $275\ \text{ft}^2$

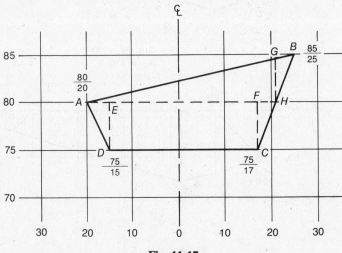

Fig. 11-17

11.27   Calculate the area of Fig. 11-18 by counting the squares.   *Ans.* $287.5\ \text{ft}^2$

11.28   Calculate the area of Fig. 11-19 by counting the squares.   *Ans.* $587.5\ \text{ft}^2$

11.29   Find the area enclosed by Fig. 11-17 by the geometric method.   *Ans.* $283.8\ \text{ft}^2$

11.30   Find the area enclosed by Fig. 11-18 by the geometric method.   *Ans.* $293.76\ \text{ft}^2$

11.31   Find the area enclosed by Fig. 11-19 by the geometric method.   *Ans.* $597.50\ \text{ft}^2$

11.32   Find the area enclosed within Fig. 11-17 by the stripper method.   *Ans.* $278.75\ \text{ft}^2$

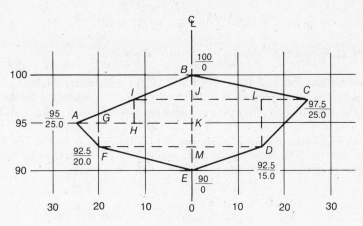

Fig. 11-18

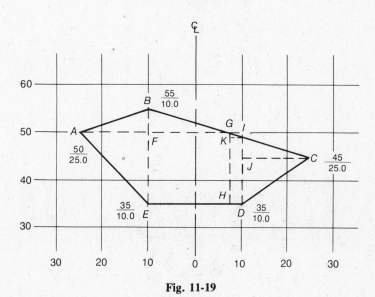

Fig. 11-19

11.33 Find the area enclosed within Fig. 11-19 by the stripper method. Notice the scales are both 1 in = 10 ft. *Ans.*   592.50 ft^2

11.34 Find the area enclosed within Fig. 11-17 by the double-meridian-distance method.   *Ans.*   285 ft^2

11.35 Find the area enclosed within Fig. 11-18 by the double-meridian-distance method.   *Ans.*   293.75 ft^2

11.36 Find the area enclosed within Fig. 11-19 by the double-meridian-distance method.   *Ans.*   600.0 ft^2

11.37 Given: Fig. 11-17 with the average end area figured by the DMD method as 285 ft^2; 100 ft away from this area on a roadway is another area which is computed to be 119.5 ft^2. Calculate the volume of material included between these two end areas.   *Ans.*   749.1 yd^3

11.38 Given: A roadway with an end section at station 96 + 00 with an area of 456.6 ft^2. The next station, 97 + 00, has an end-section area of 489.4 ft^2. Find the volume of material lying between these two sections by the average-end-area method.   *Ans.*   1752 yd^3

11.39 Given: Two road sections 100 ft apart. The first section has an end area of 341.8 ft^2. The second has an end area of 389.7 ft^2. Find the volume of material lying between these two sections by the average-end-area method.   *Ans.*   1355 yd^3

11.40   Work Prob. 11.38 by using Table 11-2.   *Ans.*   1751.84 yd^3

11.41   Work Prob. 11.39 by using Table 11-2.   *Ans.*   1,354.63 yd^3

11.42   Given: Two stations with end areas of 121 and 135 ft^2, respectively. Midway between them at the 50-ft point the area has been determined to be exactly 129.5 ft^2. Determine by the prismoidal formula the volume in the 100-ft stretch of highway between the two stations.   *Ans.*   477.8 yd^3

11.43   Given: Station end area of 234.6 ft^2 at station $56 + 00$. End area of station $57 + 00 = 257.2$ ft^2. It has been determined that the 50-ft point has an end area of 245.3 ft^2. Determine by the prismoidal formula the volume between stations $56 + 00$ and $57 + 00$.   *Ans.*   909.3 yd^3

Chapter 12

Horizontal Curves

12.1 INTRODUCTION

Curves may be simple, compound, reverse, or spiral. Compound and reverse curves are treated as a combination of two or more simple curves, whereas the spiral curve is based on a varying radius.

Curves of short radius (usually less than one tape length) can be established by holding one end of the tape at the center of the circle and swinging the tape in an arc, marking as many points as may be desired. As the radius and length of curve increases, the tape becomes impractical and the surveyor must use other methods. The common method is to measure angles and straight-line sight distances by which selected points, known as stations, may be located on the circumference of the arc.

12.2 TYPES OF HORIZONTAL CURVES

The four types of curves are described briefly as follows:

1. *Simple curve.* The simple curve is an arc of a circle (Fig. 12-1a). The radius of the circle determines the sharpness or flatness of the curve. The larger the radius, the flatter the curve. This type of curve is the most often used.

2. *Compound curve.* Frequently the terrain will necessitate the use of a compound curve. This curve normally consists of two simple curves joined together, but curving in the same direction; see Fig. 12-1b.

3. *Reverse curve.* A reverse curve consists of two simple curves joined together, but curving in opposite directions; see Fig. 12-1c. For safety reasons, this curve is seldom used in highway construction as it would tend to send an automobile off the road.

4. *Spiral curve.* The spiral is a curve which has a varying radius. It is used on railroads and some modern highways. Its purpose is to provide a transition from the tangent to a simple curve or between simple curves in a compound curve; see Fig. 12-1d. This curve will not be explained further as it is beyond the scope of this textbook.

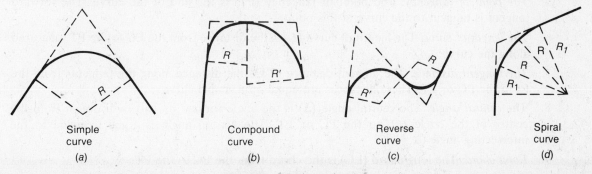

Simple curve	Compound curve	Reverse curve	Spiral curve
(a)	(b)	(c)	(d)

Fig. 12-1 Horizontal curves.

12.3 ELEMENTS OF A SIMPLE CURVE

Following are the main elements of a simple curve; see Fig. 12-2.

1. *Point of intersection.* The point of intersection (PI) is the point where the back and forward tangents intersect. It is one of the stations on the preliminary traverse.

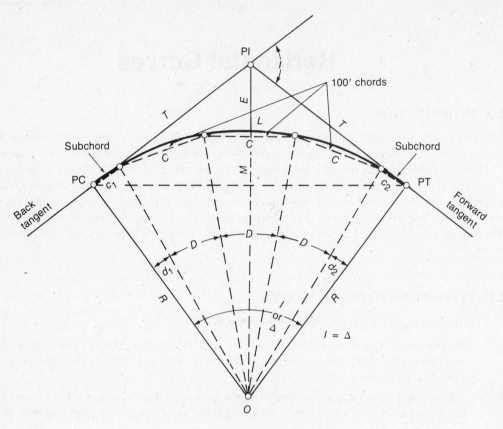

Fig. 12-2 Elements of a simple curve.

2. *The intersecting angle.* The intersecting angle (I) is the deflection angle at the PI. Its value is either computed from the preliminary traverse station angles or measured in the field.

3. *The radius.* The radius (R) is the radius of the circle of which the curve is an arc.

4. *The point of curvature.* The point of curvature (PC) is the point where the circular curve begins. The back tangent is tangent to the curve at this point.

5. *The point of tangency.* The point of tangency (PT) is the end of the curve. The forward tangent is tangent to the curve at this point.

6. *The length of curve.* The length of curve (L) is the distance from the PC to the PT measured along the curve.

7. *The tangent distance.* The tangent distance (T) is the distance along the tangents from the PI to the PC or PT. These distances are equal on a simple curve.

8. *The central angle.* The central angle (Δ) is the angle formed by two radii drawn from the center of the circle (O) to the PC or PT. The central angle is equal in value to the intersecting angle ($\Delta = I$).

9. *Long chord.* The long chord (LC) is the chord from the PC to the PT.

10. *External distance.* The external distance (E) is the distance from the PI to the midpoint of the curve. The external distance bisects the interior angle at the PI.

11. *Middle ordinate.* The middle ordinate (M) is the distance from the midpoint of the curve to the midpoint of the long chord. The extension of the middle ordinate bisects the central angle.

12. *Degree of curve.* The degree of curve (D) defines the "sharpness" or "flatness" of the curve. There are two common definitions for degree of curve (see Fig. 12-3), as follows:

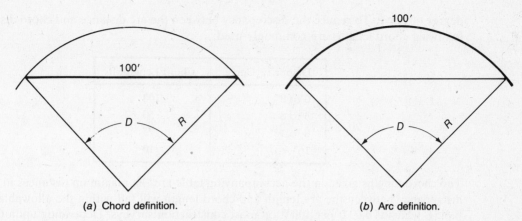

(a) Chord definition.   (b) Arc definition.

Fig. 12-3 Degree of curve.

(a) Chord definition. The chord definition states that the degree of curve is the angle formed by two radii drawn from the center of the circle to the ends of a chord 100 ft long; see Fig. 12-3a. The chord definition is used primarily for civilian railroad construction and is used by the military for both roads and railroads.

(b) Arc definition. The arc definition states that the degree of curve is the angle formed by two radii drawn from the center of the circle (point O in Fig. 12-2) to the ends of an arc 100 ft long. This definition is used primarily for highways; see Fig. 12-3b. Notice that the larger the degree of curve, the "sharper" the curve and the shorter the radius.

13. *Chords.* On a curve with a long radius it is impractical to stake the curve by locating the center of the circle and swinging the arc with a tape. Such curves are laid out by staking the ends of a series of chords; see Fig. 12-4. *Since the ends of the chords lie on the circumference of the curve, the arc is then defined in the field.* The length of the chords will vary with the

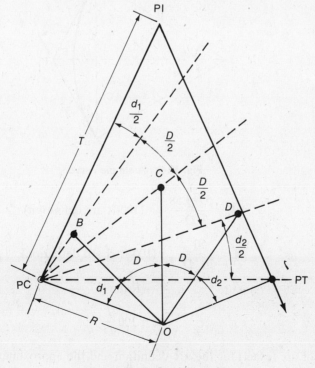

Fig. 12-4 Deflection angles.

degree of curve. To reduce the discrepancy between the arc distance and chord distance, the following chord lengths are commonly used:

Degree of Curve	Chord Length, ft
0 to 3°	100
3 to 8°	50
8 to 16°	25
16°	10

The chord lengths given in the accompanying table are the maximum distances in which the discrepancy between the arc length and chord length will fall within the allowable error for taping, which is 0.02 ft per 100 ft on most construction surveys. Depending upon the terrain and the needs of the project supervisor, the curve may be staked out with shorter or longer chords than the lengths recommended in the table.

14.   *Deflection angles.* The deflection angles are the angles between a tangent and the ends of chords, with the PC as vertex; see Fig. 12-4. They are used to locate the direction in which the chords are to be laid out. The total of the deflection angles is always equal to one-half the intersecting angle ($\frac{1}{2}I$). This total serves as a check on the computed deflection angles.

12.4   SIMPLE CURVE FORMULAS (ARC AND CHORD DEFINITIONS)

The following formulas are used in the computation of a simple curve. All the formulas apply to both the arc and chord definitions except those noted. See Fig. 12-5.

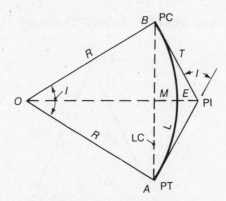

Fig. 12-5 A simple curve.

$$R = \frac{5729.58}{D} \quad \text{(100-ft arc definition)} \qquad (12\text{-}1)$$

$$R = \frac{50}{\sin\frac{1}{2}D} \quad \text{(100-ft chord definition)} \qquad (12\text{-}2)$$

$$T = R \tan\frac{1}{2}I \qquad (12\text{-}3)$$

$$L = 100\,\frac{I}{D} \qquad (12\text{-}4)$$

where L = length of arc (exact) for the arc definition and the approximate distance along the chord for the chord definition.

$$PC = PI - T \qquad (12\text{-}5)$$

$$PT = PC + L \qquad (12\text{-}6)$$

$$E = \frac{R}{\cos\frac{1}{2}I} - R \qquad (12\text{-}7)$$

$$E = R \text{ exsec}\tfrac{1}{2}I \qquad \text{where} \qquad \text{exsec}\,\tfrac{1}{2}I = \sec\tfrac{1}{2}I - 1 \qquad (12\text{-}8)$$

$$M = R - (R\cos\tfrac{1}{2}I) \qquad (12\text{-}9)$$

$$M = R \text{ vers}\tfrac{1}{2}I \qquad \text{where} \qquad \text{vers}\,\tfrac{1}{2}I = 1 - \cos\tfrac{1}{2}I \qquad (12\text{-}10)$$

Deflection Angles

$$d = \frac{D}{2}\left(\frac{C}{100}\right) \qquad (12\text{-}11a)$$

$$d = 0.3CD \qquad (12\text{-}11b)$$

where d = deflection angle, minutes
$\quad\quad\quad C$ = chord length, ft
$\quad\quad\quad D$ = degree of curvature

Equations (12-11a) and (12-11b) are exact for the 100-ft arc definition and approximate for the chord definition.

$$\sin d = \frac{C}{2R} \qquad (12\text{-}11c)$$

where C = chord length, ft
$\quad\quad\quad R$ = radius

$$d = 1719\,\frac{a}{R} \qquad (12\text{-}11d)$$

where d = deflection angle, minutes
$\quad\quad\quad a$ = length of arc, ft
$\quad\quad\quad R$ = radius

Equation (12-11c) is exact for the chord definition.

$$D = \frac{5729.58}{R} \qquad \text{(arc definition)} \qquad (12\text{-}12)$$

$$\sin\tfrac{1}{2}D = \frac{50}{R} \qquad \text{(chord definition)} \qquad (12\text{-}13)$$

$$LC = 2R \sin\tfrac{1}{2}I \qquad (12\text{-}14)$$

12.5 SOLUTION OF A SIMPLE CURVE

To solve a simple curve, three elements must be known: the point of intersection PI, the intersecting angle I, and the degree of curvature. Normally the degree of curve is given in the project specifications or computed using one of the elements which has been limited by the terrain. The PI and I are normally determined on the preliminary traverse for the road or other project, but may also be determined by triangulation when the PI is inaccessible.

EXAMPLE 12.1 Assume that the following is known about a curve: PI = 18 + 00, $I = 75°$, and $D = 15°$. (a) Solve the curve, using the arc definition. (b) Solve the curve, using the chord definition.

Solution: Draw a sketch for reference; see Fig. 12-5. Since a hand calculator does not round off, results will be slightly more accurate if you use a hand calculator instead of a table to find the mathematical functions.

(*a*)

$$R = \frac{5729.58}{D} = \frac{5729.58}{15} = 381.97 \text{ ft}$$

$$T = R \tan \tfrac{1}{2}I = 381.97(0.76733) = 293.10 \text{ ft}$$

$$PC = PI - T$$

$$
\begin{aligned}
PI &= \quad 18 + 00.00 \\
-T &\quad - (2 + 93.10) \\
\hline
PC &= \quad 15 + 06.90
\end{aligned}
$$

$$L = 100\,\frac{I}{D} = 100\left(\frac{75}{15}\right) = 500 \text{ ft}$$

$$PT = PC + L$$

$$
\begin{aligned}
PC &= \quad 15 + 06.90 \\
L &= \quad + \ 5 + 00.00 \\
\hline
PT &= \quad 20 + 06.90
\end{aligned}
$$

$$E = R \operatorname{exsec} \tfrac{1}{2}I = 381.97(0.26047) = 99.49 \text{ ft}$$

$$M = R \operatorname{vers} \tfrac{1}{2}I = 381.97(0.20665) = 78.93 \text{ ft}$$

$$LC = 2R \sin \tfrac{1}{2}I = 2(381.97)(0.60876) = 465.06 \text{ ft}$$

(*b*)

$$R = \frac{50}{\sin \tfrac{1}{2}D} = \frac{50}{0.1305262} = 383.065 \text{ ft}$$

$$T = R \tan \tfrac{1}{2}I = 383.065(0.76733) = 293.94 \text{ ft}$$

$$PC = PI - T$$

$$
\begin{aligned}
PI &= \quad 18 + 00.00 \\
-T &\quad - (2 + 93.94) \\
\hline
PC &= \quad 15 + 06.06
\end{aligned}
$$

$$L = 100\,\frac{I}{D} = 100\left(\frac{75}{15}\right) = 500 \text{ ft}$$

$$PT = PC + L$$

$$
\begin{aligned}
PC &= \quad 15 + 06.06 \\
+L &= \quad + \ 5 + 00.00 \\
\hline
PT &= \quad 20 + 06.06
\end{aligned}
$$

$$E = R \operatorname{exsec} \tfrac{1}{2}I = 383.05(0.26047) = 99.78 \text{ ft}$$

$$M = R \operatorname{vers} \tfrac{1}{2}I = 383.05(0.20665) = 79.16 \text{ ft}$$

$$LC = 2R \sin \tfrac{1}{2}I = 2(383.065)(0.60876) = 466.39 \text{ ft}$$

12.6   STAKING OUT A SIMPLE HORIZONTAL CURVE

Surveyors are often required to stake out horizontal curves. This may be done in several ways. The procedure given here is often used. If this procedure is understood, the other systems can be followed.

A simple horizontal curve is composed of a single arc. Usually the curve must be laid out so as to join two straight lines called *tangents* which are marked on the ground by PTs (points on tangent); see Fig. 12-6*a*. The first, or *back*, tangent has usually been marked off in stations. These tangents are run to intersection; thus location of the PI (point of intersection) is established.

The plus of the PI and the angle *x* are measured; see Fig. 12-6*b*. With these values and any given value of *R* (the radius desired for the curve), the data required for staking out the curve can be computed.

The transit is set up at the PC (point of curve) at a distance *T* (tangent distance) from the PI and aimed at the PI with the vernier set at zero. Points on the curve, usually at 50-ft intervals, are then staked out by measuring the computed chord from each previous point and taking line from the transit set at the proper deflection angle, as shown in Fig. 12-7. The point where the curve ends at the forward tangent is the PT (point of tangency).

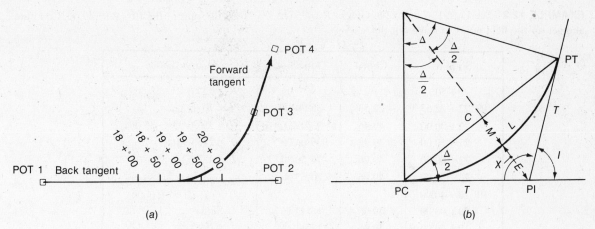

Fig. 12-6 Procedure for staking out a curve.

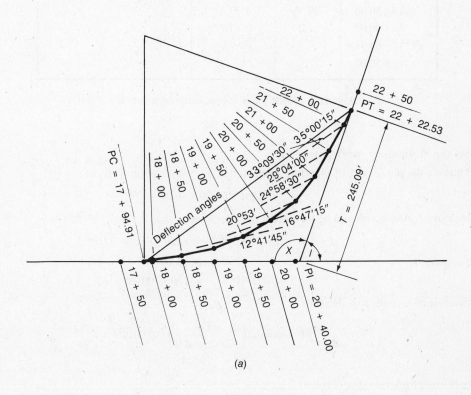

(a)

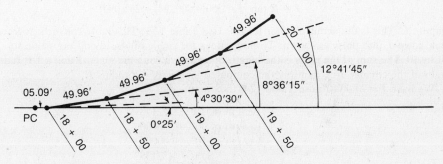

(b) Enlarged start of curve

Fig. 12-7 Method of staking out a curve of 350-ft radius.

EXAMPLE 12.2   See Figs. 12-7 and 12-8. Given: $R = 350$ ft; $I = 70°00'10''$; plus of the PI = 20 + 40.00. Compute and set up the field notes to stake out this curve.

Station	Chord	Deflection	Curve Data
+ 50			$R = 350$ ft
PT 22 + 22.53	22.53	35°00'15''	$I = 70°00'10''$
22 + 00.00	49.96	33°09'30''	$\frac{1}{2}I = 35°00'05'' = 35.001°$
21 + 50.00	49.96	29°04'00''	$T = 245.09$ ft
21 + 00.00	49.96	24°58'30''	$L = 427.62$ ft
20 + 50.00	49.96	20°53'00''	$I = 70° = $ 1.22173
PI 20 + 40.00			10'' = 0.00005
20 + 00.00	49.96	16°47'15''	Sum = 1.22178
19 + 50.00	49.96	12°41'45''	$\times R = \times$ 350
19 + 00.00	49.96	8°36'15''	$L = $ 427.62300 ft
18 + 50.00	49.96	4°30'30''	
18 + 00.00	5.09	0°25'00''	
PC 17 + 94.91		0°	
17 + 50.00			

Fig. 12-8   Field notes for staking out a curve.

Solution:

1. Set the PI. Intersect the two tangents.

2. Measure the plus of the PI. Measure from station 20 + 00.00 to get 40 ft, so

    $$\text{Plus PI} = 20 + 40.00$$

3. Measure I. Angle x should be measured and computed.

    $$I = 180° - x = 179°59'60'' - 109°59'50'' = 70°00'10''$$

4. Compute T. $\qquad\qquad\qquad\qquad T = R \tan \frac{1}{2}I$

    $$I = 70°00'10'' \qquad \frac{1}{2}I = 35°00'05''$$

    Convert 05 seconds to a decimal:

    $$(5 \text{ seconds})\left(\frac{1°}{3600 \text{ seconds}}\right) = 0.0014°$$

    So $\qquad\qquad\qquad\qquad\qquad \frac{1}{2}I = 35.0014°$

    Substitute back into the formula:

    $$T = R \tan \frac{1}{2}I = 350(0.700244) = 245.09 \text{ ft}$$

5. Compute L. This is the actual circular length. Use Table 12-1. This table converts degrees to radians. Break down I into parts as shown. To find values for single digits, divide the value for 10 times the digit by 10. The sum of the parts is the length of a curve having the given I and a 1-ft radius. To get L (length of the circular arc), multiply the computed length by the radius R.
    We have $I = 70°00'10''$; use Table 12-1 to compute L.

    $$\text{Length of } 70° = 1.22173$$
    $$\text{Length of } 00' = 0.00000$$
    $$\text{Length of } 10'' = \underline{0.00005}$$
    $$\text{Sum} = 1.22178$$

    $$L = 1.22178R = 1.22178(350) = 427.62 \text{ ft}$$

As a check, use Eqs. (12-12) and (12-4):

$$D = \frac{5729.58}{R} = \frac{5729.58}{350} = 16.3702°$$

$$L = 100\frac{I}{D} = 100\left(\frac{70.0028}{16.3702}\right) = 427.62 \text{ ft}$$

Table 12-1   Length of Curve for a 1-ft Radius

I, °	Length	I, minutes	Length	I, seconds	Length
10	0.17453	10	0.00291	10	0.00005
20	0.34907	20	0.00582	20	0.00010
30	0.52360	30	0.00873	30	0.00015
40	0.69813	40	0.01164	40	0.00019
50	0.87266	50	0.01454	50	0.00024
60	1.04720	60	0.01745	60	0.00029
70	1.22173	70	0.02036	70	0.00034
80	1.39626	80	0.02327	80	0.00039
90	1.57080	90	0.02618	90	0.00044
100	1.74533	100	0.02910	100	0.00048

Example Computation

Given: $I = 37°35'49''$ $R = 500$ ft

$$30° = 0.52360$$
$$7° = 0.12217$$
$$30' = 0.00873$$
$$5' = 0.00145$$
$$40'' = 0.00019$$
$$9'' = \underline{0.00004}$$
$$\text{Sum} = 0.65618$$

$$L = \text{sum} \times R = 0.65618(500) = 328.09 \text{ ft}$$

6. Compute the pluses. The stationing is usually continuous from the back tangent, around the curve, and on the forward tangent; see Fig. 12-7.

$$
\begin{array}{ll}
\text{PI} = & 20 + 40.00 \\
-T & -(2 + 45.09) \\
\text{PC} = & 17 + 94.91 \\
+L & +\ \ 4 + 27.62 \\
\text{PT} = & 22 + 22.53
\end{array}
$$

7. Compute the deflection angles. The first deflection angle subtends an arc from the PC to the first 50-ft point on the curve. First compute the arc length.

$$
\begin{array}{ll}
\text{First 50-ft point} = & 18 + 00.00 \\
-\text{ the plus of the PC} & -(17 + 94.91) \\
\text{Arc length} = & 05.09 \text{ ft}
\end{array}
$$

Now use Eq. (12-11d):

$$D = 1719\frac{a}{R} = 1719\left(\frac{5.09}{350}\right) = 24.99 \text{ minutes} = 0°25'00''$$

Note where this angle is recorded in the field notes (Fig. 12-8).

The second deflection angle equals the first deflection angle plus the angle which subtends 50 ft.

$$d = 1719\frac{a}{R} = 1719\left(\frac{50}{350}\right) = 245.57 \text{ minutes}$$

Convert minutes to degrees: There are 60 minutes in 1°. The nearest multiple of 60 under 245.57 is 4(60) = 240. Thus:

$$245.57' - 240.00' = 5.57'$$

This now gives 4°5.57'. Convert 0.57 minutes into seconds:

$$(0.57 \text{ minute})(60 \text{ seconds per minute}) = 34.2 \text{ seconds}$$

Thus: $245.57' = 4°5'34.2''$

In field notes the seconds are set at the nearest multiple of 15 seconds. Thus this would be recorded in the field book as 4°5'30''.

The value of 4°05.57' is added for each 50-ft point until the point just previous to the PT, 22 + 22.53, is reached. The arc to the PT is 22.53 ft. The deflection angle for this arc is computed.

$$d = 1719\frac{a}{R} = 1719\left(\frac{22.53}{350}\right) = 110.65 \text{ minutes}$$
$$110.65 \text{ minutes} = 1°50.65' = 1°50'39''$$

With these values the computation of deflection angles can be made (see the accompanying table). The deflection angle computed for the PT should equal $\frac{1}{2}I$, or 35°00'05''. The table shows this angle as 35°00'15''. This small error is due to rounding off. A large error indicates a mistake in computation.

8. Compute the chord lengths. Since the length of each chord is slightly less than the length of the arc it subtends, the chord lengths must be computed. Use Table 12-2. Look under the R column and find 350 (the radius). Go across the row and stop under the column headed 20 (close to 22.53 ft). Read 0.003. Subtract the correction to find the length of the chord:

$$22.53 - 0.003 = 22.527 \approx 22.53$$

Repeat this process for each arc. Values are recorded in the accompanying table.

Arc, ft	Correction, ft	Chord, ft
22.53	−0.003	22.53
50.00	−0.042	49.96
05.09	−0.000	05.09

9. Set the PT. Measure $T = 245.09$ ft from the PI and set PT on line with the forward tangent.

10. Set the PC. Measure 94.91 ft from station 17 and line in.

11. Set the stations. Set up at the PC. Set the vernier at zero and aim at the PI. Turn left the deflection angle for 18 + 00.00 (0°25'00'') and set 18 + 00 on line at the chord distance from the PC (5.09). Set the vernier at the next deflection angle (4°30'30'') and set 18 + 50.00 on line and at the chord distance (49.96 ft) from 18 + 00.00. Continue in this way, setting off each successive deflection angle and measuring out the required chord length from the previous point until the last point is set previous to the PT (22 + 22.53).

12. Measure to the PT. Assume that from 22 + 00.00 to the PT measured 22.40 ft instead of the chord length of 22.53 ft, giving an error of 0.13 ft. Turn the transit to the deflection angle for the PT (35°00'05''). Measure the perpendicular distance from the line of sight to the PT. Assume this is 0.152 ft. Compute the total error. Use the following formula:

$$\text{Total error} = \sqrt{(\text{chord length error}) + (\text{line-of-sight error})^2}$$
$$\text{Error} = \sqrt{(0.13)^2 + (0.152)^2} = 0.2000 \text{ ft}$$

Computation of Deflection Angles for Example 12.2

Station	Deflection	Deflection to 15"
PC 17 + 94.91	00°	0°
18 + 00.00	0°24.99' + 4°05.57'	0°25'00"
18 + 50.00	4°30.56' + 4°30.57'	4°30'30"
19 + 00.00	8°36.13' + 4°05.57'	8°36'15"
19 + 50.00	12°41.70' + 4°05.57'	12°41'45"
20 + 00.00	16°47.27' + 4°05.57'	16°47'15"
20 + 50.00	20°52.84' + 4°05.57'	20°53'00"
21 + 00.00	24°58.41' + 4°05.57'	24°58'30"
21 + 50.00	28°63.98' + 4°05.57'	29°04'00"
22 + 00.00	33°09.55' + 1°50.65'	33°09'30"
PT 22 + 22.53	35°00.20'	35°00'15"

13. Compute the error ratio. Use the following formula:

$$\text{Error ratio} = \frac{\text{total error of survey}}{\text{total length of survey}}$$

Total length of the survey $= 2T + L = 2(245.09) + 427.62 = 917.80$ ft

Error ratio:

$$\frac{0.2000}{917.80} = \frac{1}{x}$$

$$x = 4589$$

Thus the error ratio is 1:4589. The maximum ratio that should be accepted is 1:3000 so this error is acceptable.

12.7 THE COMPOUND CURVE

Often two curves of different radii are joined together, as in Fig. 12-9. Point P is called the *point of compound curve* (PCC). GH is a common tangent. The subscript 1 refers to the curve of the smaller radius.

Table 12-2   Corrections to Subarcs for Subchords, Given *R* (Subtract)

	Lengths of Arcs in Feet									
R	5	10	15	20	25	30	35	40	45	50
100	0.001	0.004	0.014	0.033	0.065	0.112	0.178	0.266	0.379	0.519
110	0.000	0.003	0.012	0.028	0.054	0.093	0.147	0.220	0.313	0.429
120	0.000	0.003	0.010	0.023	0.045	0.078	0.124	0.185	0.263	0.361
130	0.000	0.002	0.008	0.020	0.038	0.066	0.106	0.158	0.224	0.308
140	0.000	0.002	0.007	0.017	0.033	0.057	0.091	0.136	0.193	0.265
150	0.000	0.002	0.006	0.015	0.029	0.050	0.079	0.119	0.168	0.231
160	0.000	0.002	0.005	0.013	0.025	0.044	0.070	0.104	0.148	0.203
170	0.000	0.001	0.005	0.011	0.023	0.039	0.062	0.092	0.131	0.180
180	0.000	0.001	0.004	0.010	0.020	0.035	0.055	0.082	0.117	0.161
190	0.000	0.001	0.004	0.009	0.018	0.031	0.049	0.074	0.105	0.144
200	0.000	0.001	0.004	0.008	0.016	0.028	0.045	0.067	0.095	0.130
225	0.000	0.001	0.003	0.007	0.013	0.022	0.035	0.053	0.075	0.103
250	0.000	0.001	0.002	0.005	0.010	0.018	0.029	0.043	0.061	0.083
275	0.000	0.001	0.002	0.004	0.009	0.015	0.024	0.035	0.050	0.069
300	0.000	0.000	0.002	0.004	0.007	0.013	0.020	0.030	0.042	0.058
325	0.000	0.000	0.001	0.003	0.006	0.011	0.017	0.025	0.036	0.049
350	0.000	0.000	0.001	0.003	0.005	0.009	0.015	0.022	0.031	0.042
375	0.000	0.000	0.001	0.002	0.005	0.008	0.013	0.019	0.027	0.037
400	0.000	0.000	0.001	0.002	0.004	0.007	0.011	0.017	0.024	0.033
425	0.000	0.000	0.001	0.002	0.004	0.006	0.010	0.015	0.021	0.029
450	0.000	0.000	0.001	0.002	0.003	0.006	0.009	0.013	0.018	0.026
475	0.000	0.000	0.001	0.001	0.003	0.005	0.008	0.012	0.017	0.023
500	0.000	0.000	0.001	0.001	0.003	0.004	0.007	0.011	0.015	0.021
550	0.000	0.000	0.000	0.001	0.002	0.004	0.006	0.009	0.012	0.017
600	0.000	0.000	0.000	0.001	0.002	0.003	0.005	0.007	0.011	0.014
650	0.000	0.000	0.000	0.001	0.002	0.003	0.004	0.006	0.009	0.012
700	0.000	0.000	0.000	0.001	0.001	0.002	0.004	0.005	0.008	0.011
750	0.000	0.000	0.000	0.001	0.001	0.002	0.003	0.005	0.007	0.009
800	0.000	0.000	0.000	0.000	0.001	0.002	0.003	0.004	0.006	0.008
850	0.000	0.000	0.000	0.000	0.001	0.002	0.002	0.004	0.005	0.007
900	0.000	0.000	0.000	0.000	0.001	0.001	0.002	0.003	0.005	0.006
950	0.000	0.000	0.000	0.000	0.001	0.001	0.002	0.003	0.004	0.006
1000	0.000	0.000	0.000	0.000	0.001	0.001	0.002	0.003	0.004	0.005
1100	0.000	0.000	0.000	0.000	0.001	0.001	0.001	0.002	0.003	0.004
1200	0.000	0.000	0.000	0.000	0.000	0.001	0.001	0.002	0.003	0.004
1300	0.000	0.000	0.000	0.000	0.000	0.001	0.001	0.002	0.002	0.003
1400	0.000	0.000	0.000	0.000	0.000	0.001	0.001	0.001	0.002	0.003
1500	0.000	0.000	0.000	0.000	0.000	0.001	0.001	0.001	0.002	0.002
1600	0.000	0.000	0.000	0.000	0.000	0.000	0.001	0.001	0.001	0.002
1700	0.000	0.000	0.000	0.000	0.000	0.000	0.001	0.001	0.001	0.002
1800	0.000	0.000	0.000	0.000	0.000	0.000	0.001	0.001	0.001	0.002
1900	0.000	0.000	0.000	0.000	0.000	0.000	0.000	0.001	0.001	0.001
2000	0.000	0.000	0.000	0.000	0.000	0.000	0.000	0.001	0.001	0.001
2100	0.000	0.000	0.000	0.000	0.000	0.000	0.000	0.001	0.001	0.001
2200	0.000	0.000	0.000	0.000	0.000	0.000	0.000	0.001	0.001	0.001
2300	0.000	0.000	0.000	0.000	0.000	0.000	0.000	0.001	0.001	0.001
2400	0.000	0.000	0.000	0.000	0.000	0.000	0.000	0.000	0.001	0.001
2500	0.000	0.000	0.000	0.000	0.000	0.000	0.000	0.000	0.001	0.001
2600	0.000	0.000	0.000	0.000	0.000	0.000	0.000	0.000	0.001	0.001
2700	0.000	0.000	0.000	0.000	0.000	0.000	0.000	0.000	0.001	0.001
2800	0.000	0.000	0.000	0.000	0.000	0.000	0.000	0.000	0.000	0.001
2900	0.000	0.000	0.000	0.000	0.000	0.000	0.000	0.000	0.000	0.001
3000	0.000	0.000	0.000	0.000	0.000	0.000	0.000	0.000	0.000	0.001
3100	0.000	0.000	0.000	0.000	0.000	0.000	0.000	0.000	0.000	0.001
3200	0.000	0.000	0.000	0.000	0.000	0.000	0.000	0.000	0.000	0.001
3300	0.000	0.000	0.000	0.000	0.000	0.000	0.000	0.000	0.000	0.000
3400	0.000	0.000	0.000	0.000	0.000	0.000	0.000	0.000	0.000	0.000
3500	0.000	0.000	0.000	0.000	0.000	0.000	0.000	0.000	0.000	0.000
3600	0.000	0.000	0.000	0.000	0.000	0.000	0.000	0.000	0.000	0.000
3700	0.000	0.000	0.000	0.000	0.000	0.000	0.000	0.000	0.000	0.000
3800	0.000	0.000	0.000	0.000	0.000	0.000	0.000	0.000	0.000	0.000
3900	0.000	0.000	0.000	0.000	0.000	0.000	0.000	0.000	0.000	0.000
4000	0.000	0.000	0.000	0.000	0.000	0.000	0.000	0.000	0.000	0.000
5000	0.000	0.000	0.000	0.000	0.000	0.000	0.000	0.000	0.000	0.000

Table 12-2   (*cont.*)

	Lengths of Arcs in Feet									
R	55	60	65	70	75	80	85	90	95	100
100	0.691	0.896	1.138	1.420	1.745	2.116	2.536	3.007	3.532	4.115
110	0.571	0.741	0.942	1.175	1.444	1.752	2.099	2.489	2.925	3.408
120	0.480	0.623	0.792	0.988	1.215	1.473	1.766	2.095	2.462	2.869
130	0.409	0.531	0.675	0.843	1.036	1.256	1.506	1.787	2.100	2.447
140	0.353	0.458	0.582	0.727	0.894	1.084	1.300	1.542	1.812	2.112
150	0.308	0.399	0.507	0.633	0.779	0.945	1.133	1.344	1.580	1.842
160	0.270	0.351	0.446	0.557	0.685	0.831	0.996	1.182	1.389	1.620
170	0.240	0.311	0.395	0.494	0.607	0.736	0.883	1.047	1.231	1.436
180	0.214	0.277	0.353	0.440	0.541	0.657	0.788	0.935	1.099	1.281
190	0.192	0.249	0.317	0.395	0.486	0.590	0.707	0.839	0.987	1.150
200	0.173	0.225	0.286	0.357	0.439	0.532	0.638	0.757	0.891	1.038
225	0.137	0.178	0.226	0.282	0.347	0.421	0.505	0.599	0.704	0.821
250	0.111	0.144	0.183	0.228	0.281	0.341	0.409	0.485	0.571	0.665
275	0.092	0.119	0.151	0.189	0.232	0.282	0.338	0.401	0.472	0.550
300	0.077	0.100	0.127	0.159	0.195	0.235	0.284	0.337	0.396	0.462
325	0.066	0.085	0.108	0.135	0.166	0.202	0.242	0.287	0.338	0.394
350	0.056	0.073	0.093	0.117	0.143	0.174	0.209	0.248	0.291	0.340
375	0.049	0.064	0.081	0.102	0.125	0.152	0.182	0.216	0.254	0.294
400	0.043	0.056	0.072	0.089	0.110	0.133	0.160	0.190	0.223	0.260
425	0.038	0.050	0.063	0.079	0.097	0.118	0.142	0.168	0.198	0.231
450	0.034	0.044	0.057	0.071	0.087	0.105	0.126	0.150	0.176	0.206
475	0.031	0.040	0.051	0.063	0.078	0.095	0.113	0.135	0.158	0.185
500	0.028	0.036	0.046	0.057	0.070	0.085	0.102	0.121	0.143	0.167
550	0.023	0.030	0.038	0.047	0.058	0.071	0.085	0.100	0.118	0.138
600	0.019	0.025	0.032	0.040	0.049	0.059	0.071	0.085	0.099	0.116
650	0.016	0.021	0.027	0.034	0.042	0.050	0.061	0.072	0.085	0.099
700	0.014	0.018	0.023	0.029	0.036	0.044	0.052	0.062	0.073	0.085
750	0.012	0.016	0.020	0.025	0.031	0.038	0.046	0.054	0.063	0.074
800	0.011	0.014	0.018	0.022	0.027	0.033	0.040	0.048	0.056	0.065
850	0.010	0.012	0.016	0.020	0.024	0.030	0.035	0.042	0.049	0.057
900	0.009	0.011	0.014	0.018	0.022	0.026	0.032	0.037	0.044	0.052
950	0.008	0.010	0.013	0.016	0.020	0.024	0.028	0.034	0.040	0.046
1000	0.007	0.009	0.011	0.014	0.018	0.021	0.026	0.030	0.036	0.042
1100	0.006	0.007	0.009	0.012	0.015	0.018	0.021	0.025	0.030	0.035
1200	0.005	0.006	0.008	0.010	0.012	0.015	0.018	0.021	0.025	0.029
1300	0.004	0.005	0.007	0.008	0.010	0.013	0.015	0.018	0.021	0.025
1400	0.004	0.005	0.006	0.007	0.009	0.011	0.013	0.015	0.018	0.021
1500	0.003	0.004	0.005	0.006	0.008	0.010	0.011	0.013	0.016	0.018
1600	0.003	0.004	0.004	0.006	0.007	0.008	0.010	0.012	0.014	0.016
1700	0.002	0.003	0.004	0.005	0.006	0.007	0.009	0.011	0.012	0.014
1800	0.002	0.003	0.004	0.004	0.005	0.007	0.008	0.009	0.011	0.013
1900	0.002	0.002	0.003	0.004	0.005	0.006	0.007	0.008	0.010	0.012
2000	0.002	0.002	0.003	0.004	0.004	0.005	0.006	0.008	0.009	0.011
2100	0.002	0.002	0.003	0.003	0.004	0.005	0.006	0.007	0.008	0.009
2200	0.001	0.002	0.002	0.003	0.004	0.004	0.004	0.005	0.006	0.007	0.009
2300	0.001	0.002	0.002	0.003	0.003	0.004	0.005	0.006	0.007	0.008
2400	0.001	0.002	0.002	0.002	0.003	0.004	0.004	0.005	0.006	0.007
2500	0.001	0.001	0.002	0.002	0.003	0.003	0.004	0.005	0.006	0.007
2600	0.001	0.001	0.002	0.002	0.003	0.003	0.004	0.004	0.005	0.006
2700	0.001	0.001	0.002	0.002	0.002	0.003	0.003	0.004	0.005	0.006
2800	0.001	0.001	0.001	0.002	0.002	0.003	0.003	0.004	0.005	0.005
2900	0.001	0.001	0.001	0.002	0.002	0.003	0.003	0.004	0.004	0.005
3000	0.001	0.001	0.001	0.002	0.002	0.002	0.003	0.003	0.004	0.005
3100	0.001	0.001	0.001	0.001	0.002	0.002	0.003	0.003	0.004	0.004
3200	0.001	0.001	0.001	0.001	0.002	0.002	0.003	0.003	0.003	0.004
3300	0.001	0.001	0.001	0.001	0.002	0.002	0.002	0.003	0.003	0.004
3400	0.001	0.001	0.001	0.001	0.002	0.002	0.002	0.003	0.003	0.004
3500	0.001	0.001	0.001	0.001	0.001	0.002	0.002	0.003	0.003	0.003
3600	0.001	0.001	0.001	0.001	0.001	0.002	0.002	0.002	0.003	0.003
3700	0.001	0.001	0.001	0.001	0.001	0.002	0.002	0.002	0.003	0.003
3800	0.000	0.001	0.001	0.001	0.001	0.001	0.002	0.002	0.002	0.003
3900	0.000	0.001	0.001	0.001	0.001	0.001	0.002	0.002	0.002	0.003
4000	0.000	0.001	0.001	0.001	0.001	0.001	0.002	0.002	0.002	0.003
5000	0.000	0.000	0.000	0.001	0.001	0.001	0.001	0.001	0.001	0.002

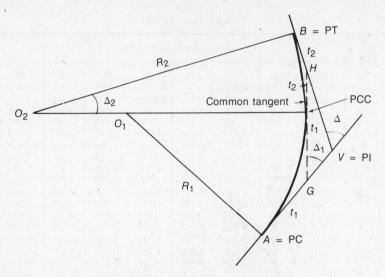

Fig. 12-9 A compound curve.

The angle is measured: R_1 and R_2, and/or Δ_1 and Δ_2, are given. To find the curve data for the two curves, the following variables are computed. From the figure:

$$\Delta_1 = \Delta - \Delta_2 \quad \text{or} \quad \Delta_2 = \Delta - \Delta_1 \tag{12-15}$$

$$t_1 = R_1 \tan \frac{\Delta_1}{2} \qquad t_2 = R_2 \tan \frac{\Delta_2}{2} \tag{12-16}$$

$$GH = t_1 + t_2 \tag{12-17}$$

$$VG = (\sin \Delta_2)\left(\frac{GH}{\sin \Delta}\right) \tag{12-18}$$

$$VH = (\sin \Delta_1)\left(\frac{GH}{\sin \Delta}\right) \tag{12-19}$$

$$T_1 = VA = VG + t_1 \tag{12-20}$$

$$T_2 = VB = VH + t_2 \tag{12-21}$$

Staking the Curves

To stake out the curves, the deflection angles and the chords are computed for the two curves separately. When P is reached, the transit is oriented to the second curve by aiming it so that the vernier reads zero when pointed along the imaginary common tangent GH. To accomplish this, aim at any point on the first curve with the telescope reversed and the vernier set to the right (if the first curve is a left curve), at the deflection angle of the PCC on the first curve minus the deflection angle of the point sighted. Once you are oriented in this way, the deflection angles computed for the second curve can be used.

EXAMPLE 12.3 We have a compound curve for which the following are known: Plus of $PI = 22 + 87.93$; $\Delta = 88°32'54''$; $R_1 = 400$ ft; $\Delta_1 = 42°18'34''$; $R_2 = 600$ ft. Refer to Fig. 12-9. Compute the pluses of PC, PCC, and PT, and the length T_2.

Solution:

1. Angle Δ is measured.
2. Find Δ_2.

$$\Delta_2 = \Delta - \Delta_1 = 88°32'54'' - 42°18'34'' = 46°14'20''$$

Note: For finding trigonometric functions put the angles in decimal degrees.

$$(20 \text{ seconds})\left(\frac{1°}{3600 \text{ seconds}}\right) = 0.0056°$$

$$(14 \text{ minutes})\left(\frac{1°}{60 \text{ minutes}}\right) = 0.2333°$$

So $\Delta_2 = 46°14'20'' = 46.2389°$

3. Find t_1.

$$\Delta_1 = 42°18'34'' \qquad \frac{\Delta_1}{2} = 21°09'17'' = 21.1547°$$

$$\tan 21.1547° = 0.3869651$$

Substitute these values into Eq. (*12-16*).

$$t_1 = R_1 \tan \frac{\Delta_1}{2} = 400(0.3869651) = 154.79 \text{ ft}$$

Find t_2.

$$\Delta_2 = 46°14'20'' \qquad \frac{\Delta_2}{2} = 23°07'10''$$

Put into decimal degrees and find the tangent.

$$23°07'10'' = 23.1195° \qquad \tan 23.1195° = 0.4269385$$

$$t_2 = R_2 \tan \frac{\Delta_2}{2} = 600(0.4269385) = 256.16 \text{ ft}$$

4. Find *GH*.

$$GH = t_1 + t_2 = 154.79 + 256.16 = 410.95 \text{ ft}$$

5. Find *VG*.

$$\Delta_2 = 46°14'20'' = 46.2389° \qquad \sin 46.2389° = 0.72223$$
$$\Delta = 88°32'54'' = 88.5483° \qquad \sin 88.5483° = 0.999679$$

Substitute into Eq. (*12-18*).

$$VG = (\sin \Delta_2)\left(\frac{GH}{\sin \Delta}\right) = (0.72223)\left(\frac{410.95}{0.999679}\right) = (0.72223)(411.08196) = 296.90 \text{ ft}$$

6. Find *VH*.

$$\Delta_1 = 42°18'34'' = 42.3094° \qquad \sin 42.3094° = 0.6731338$$

Substitute into Eq. (*12-19*).

$$VH = (\sin \Delta_1)\left(\frac{GH}{\sin \Delta}\right) = (0.6731338)(411.08196) = 276.71 \text{ ft}$$

7. Find T_1. Substitute into Eq. (*12-20*).

$$T_1 = VG + t_1 = 296.90 + 154.79 = 451.69 \text{ ft}$$

8. Find T_2. Substitute into Eq. (*12-21*).

$$T_2 = VH + t_2 = 276.71 + 256.16 = 532.87 \text{ ft}$$

9. Find L_1 and L_2 (the lengths of the circular arcs of the two curves). Use Table 12-1 to find the lengths of the angle increments.

$$\Delta_1 = 42°18'34''$$
$$40° = 0.69813$$
$$2° = 0.03491$$
$$10' = 0.00291$$
$$8' = 0.00233$$
$$30'' = 0.00015$$
$$4'' = \underline{0.00002}$$
$$\text{Sum} = 0.73845$$

$$L_1 = 0.73845R, = 0.73845(400) = 295.38 \text{ ft}$$
$$\Delta_2 = 46°14'20''$$
$$40° = 0.69813$$
$$6° = 0.10472$$
$$10' = 0.00291$$
$$4' = 0.00116$$
$$20'' = \underline{0.00010}$$
$$\text{Sum} = 0.80702$$

$$L_2 = 0.80702R_2 = 0.80702(600) = 484.21 \text{ ft}$$

10. Example summary:

Plus PI =	2287.93
Less T_1	− 451.69
Plus PC =	1836.24
Add L_1	+ 295.38
Plus PCC =	2231.62
Add L_2	+ 484.21
Plus PT =	2715.83

$$T_2 = 532.87 \text{ ft}$$

Solved Problems

In solving these problems, draw a sketch and state all formulas required.

12.1 Given: PI = station 104 + 50.00; $I = 8°00'00''$ rt (rt means the arc of the curve is turning to the right); $D = 2°00'00''$. Draw simple arcs for the centerline of survey and centerline of construction. Solve the centerline of survey for R, T, PC, L, PT, E, M, and LC by the arc definition.

Solution

Draw the curve; see Fig. 12-10a. Insert values known into the formulas for the simple curve. Use Eq. (*12-1*) for the radius:

$$R = \frac{5729.58}{D} = \frac{5729.58}{2} = 2864.79 \text{ ft}$$

Use Eq. (*12-3*) to find the tangent distance T:

$$T = R \tan \tfrac{1}{2}I = 2864.79(0.069927) = 200.32617 \approx 200.33 \text{ ft}$$

Find the point of the curve (PC), using Eq. (*12-5*).

$$PC = PI - T$$

PI station =	104 + 50.00
− T	− (2 + 00.33)
PC station =	102 + 49.67

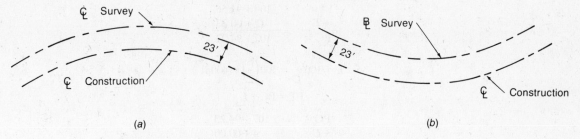

Fig. 12-10   Sketch of centerlines of survey and construction.

Find the length L of the curve, using Eq. (12-4).

$$L = 100 \frac{I}{D} = 100\left(\frac{8}{2}\right) = 400 \text{ ft}$$

Find the point of tangent PT of the curve, using Eq. (12-6).

$$PT = PC + L$$

PC station =	$102 + 49.67$
$+ L$	$+ \quad 4 + 00.00$
PT station =	$106 + 49.67$

Find the external distance E, using Eq. (12-8).

$$E = R \sec \tfrac{1}{2}I = 2864.79(0.00244) = 6.99 \text{ ft}$$

Find the middle ordinate.

$$M = R \text{ vers} \tfrac{1}{2}I = 2864.79(0.00244) = 6.98 \text{ ft}$$

Find the long chord, using Eq. (12-14).

$$LC = 2R \sin \tfrac{1}{2}I = 2(2864.79)(0.069756) = 399.68 \text{ ft}$$

12.2   Given: Same data as given in Prob. 12.1. The centerline of construction is 23 ft right of the centerline of survey; see Fig. 12-10a. Solve the centerline of construction for tangent (T), radius (R) and long chord (LC).

Solution

Calculate the new radius R.

$$\text{Radius from Prob. 12.1} = 2864.79 \text{ ft}$$

Since the centerline of construction is 23 ft right, subtract to find the new radius:

$$\text{Radius of centerline construction} = 2864.79 - 23 = 2841.79 \text{ ft}$$
$$T = R \tan \tfrac{1}{2}I = 2841.79(0.069927) = 198.72 \text{ ft}$$
$$LC = 2R \sin \tfrac{1}{2}I = 2(2841.79)(0.069756) = 396.46 \text{ ft}$$

12.3   Given: The following data on the baseline of survey—PI station $= 109 + 94.56$; $I = 8°00'00''$ lt (left); $D = 2°00'00''$. Find R, T, PC, L, PT, and LC by the arc definition.

Solution

$$R = \frac{5729.58}{D} = \frac{5729.58}{2} = 2864.79 \text{ ft}$$
$$T = R \tan \tfrac{1}{2}I = 2864.79(0.06993) = 200.33 \text{ ft}$$
$$PC = PI - T$$

$$PI = \qquad 109 + 94.56$$
$$-T = \underline{\quad - \ (2 + 00.33)}$$
$$PC = \qquad 107 + 94.23$$

$$L = 100\,\frac{I}{D} = 100\left(\frac{8}{2}\right) = 400 \text{ ft}$$

$$PT = PC + L$$

$$PC = \qquad 107 + 94.23$$
$$+L \qquad \underline{+ \quad 4 + 00.00}$$
$$PT = \qquad 111 + 94.23$$

$$LC = 2R \sin \tfrac{1}{2}I = 2(2864.79)(0.0697565) = 399.68 \text{ ft}$$

12.4 Given: The same data as given in Prob. 12.3 except the centerline of construction is 23 ft right of the baseline of survey, and the new degree of curve D is not given for the centerline of construction. Find R, T, PC, D, L, PT and LC by the arc definition.

Solution

Find the new radius R of the centerline of construction.

$$\begin{aligned}
R \text{ for centerline survey} = \qquad & 2864.79 \text{ ft}\\
\text{Add 23 ft rt for centerline construction} \quad & \underline{+ \quad 23.00 \text{ ft}}\\
R \text{ for centerline of construction} = \qquad & 2887.79 \text{ ft}
\end{aligned}$$

$$T = R \tan \tfrac{1}{2}I = 2887.79(0.069927) = 201.93 \text{ ft}$$

$$PC = PI - T$$

$$PI = \qquad 109 + 94.56$$
$$-T \qquad \underline{- \ (2 + 01.94)}$$
$$PC = \qquad 107 + 92.62$$

$$D = \frac{5729.58}{R} = \frac{5729.58}{2887.79} = 1.98407°$$

$$L = 100\,\frac{I}{D} = 100\left(\frac{8}{1.98407}\right) = 403.21 \text{ ft}$$

$$PT = PC + L$$

$$PC = \qquad 107 + 92.62$$
$$+L \qquad \underline{+ \quad 4 + 03.21}$$
$$PT = \qquad 111 + 95.83$$

$$LC = 2R \sin \tfrac{1}{2}I = 2(2887.79)(0.069756) = 402.88 \text{ ft}$$

12.5 Given: The information listed for a horizontal curve—PI station $= 100 + 00.00$; $I = 23°08'08''$ lt (left); $D = 2°00'00''$. Find R, T, PC, L, PT, M, and LC.

Solution

$$R = \frac{5729.58}{D} = \frac{5729.58}{2} = 2864.79 \text{ ft}$$

$$I = 23°08'08'' \quad \text{so} \quad \tfrac{1}{2}I = 11°34'04'' = 11.5678° \quad \text{and} \quad \tan 11.5678° = 0.20468$$
$$T = R \tan \tfrac{1}{2}I = 2864.79(0.20468) = 586.376 = 586.38 \text{ ft}$$

$$PC = PI - T$$

$$PI = \qquad 100 + 00.00$$
$$-T \qquad \underline{- \ (5 + 86.38)}$$
$$PC = \qquad 94 + 13.62$$

$$I = 23°08'08'' = 23.1356°$$

$$L = 100 \frac{I}{D} = 100\left(\frac{23.1356}{2}\right) = 1156.78 \text{ ft}$$

$$PT = PC + L$$

$$
\begin{array}{ll}
PC = & 94 + 13.62 \\
+L & + \ 11 + 56.78 \\
\hline
PT = & 105 + 70.40
\end{array}
$$

Find the ordinate:

$$M = R \text{ vers} \tfrac{1}{2}I = 2864.79(0.02032) = 58.21 \text{ ft}$$

$$LC = 2R \sin \tfrac{1}{2}I = 2(2864.79)(0.20053) = 1148.94 \text{ ft}$$

The following series of problems are to be solved by the chord method.

12.6   Given: PI station $= 104 + 50.00$; $I = 8°00'00''$; $D = 2°00'00''$; see Fig. 12-10a. Determine the centerline of survey by the chord definition. Find R, T, PC, L, E, M, and LC.

Solution

Insert known values into the chord definition formulas for the simple curve.

$$R = \frac{50}{\sin \tfrac{1}{2}D} = \frac{50}{0.01745245} = 2864.94 \text{ ft}$$

$$T = R \tan \tfrac{1}{2}I = 2864.94(0.069927) = 200.34 \text{ ft}$$

$$PC = PI - T$$

$$
\begin{array}{ll}
PI = & 104 + 50.00 \\
-T & - \ (2 + 00.35) \\
\hline
PC = & 102 + 49.65
\end{array}
$$

$$L = 100 \frac{I}{D} = 100\left(\frac{8}{2}\right) = 400 \text{ ft}$$

$$PT = PC + L$$

$$
\begin{array}{ll}
PC = & 102 + 49.65 \\
+L & + \ \ 4 + 00.00 \\
\hline
PT = & 106 + 49.65
\end{array}
$$

$$E = R \text{ exsec} \tfrac{1}{2}I = 2864.94(0.00240) = 6.99 \text{ ft}$$

$$M = R \text{ vers} \tfrac{1}{2}I = 2864.94(0.00243) = 6.96 \text{ ft}$$

$$LC = 2R \sin \tfrac{1}{2}I = 2(2864.94)(0.06976) = 399.72 \text{ ft}$$

12.7   Given: Data of Prob. 12.6; note $I = 8°00'00''$. The centerline of construction is located 23 ft right of the centerline of survey; see Fig. 12-10a. Solve for the centerline of construction. Find R, T, and LC.

Solution

$$
\begin{array}{lr}
\text{Radius from Prob. 12.6} = & 2864.94 \text{ ft} \\
\text{Subtract CL of construction 23 ft rt} & - \ \ \ 23.00 \\
\hline
\text{Radius CL construction} = & 2841.94 \text{ ft}
\end{array}
$$

$$T = R \tan \tfrac{1}{2}I = 2841.94(0.069927) = 198.73 \text{ ft}$$

$$LC = 2R \sin \tfrac{1}{2}I = 2(2841.94)(0.06976) = 396.51 \text{ ft}$$

12.8   A centerline of construction is offset a distance 23 ft to the right of the baseline of the survey (this problem is very common in road design). Given: Centerline of survey data—PI station = $109 + 94.56$; $I = 8°00'00''$ lt; $D = 2°00'00''$. Solve the baseline of survey by the chord definition. Find R, T, PC, PT, and LC.

Solution

Draw a sketch of the arcs; see Fig. 12-10b.

$$R = \frac{50}{\sin \frac{1}{2}I} = \frac{50}{0.01745} = 2865.33 \text{ ft}$$

$$T = R \tan \tfrac{1}{2}I = 2865.33(0.06993) = 200.33 \text{ ft}$$

$$PC = PI - T$$

$$
\begin{array}{ll}
PI = & 109 + 94.56 \\
-T & - \;\;(2 + 00.34) \\
\hline
PC = & 107 + 94.22
\end{array}
$$

$$L = 100 \frac{I}{D} = 100\left(\frac{8}{2}\right) = 400 \text{ ft}$$

$$PT = PC + L$$

$$
\begin{array}{ll}
PC = & 107 + 94.22 \\
+L & + \;\;\;\; 4 + 00.00 \\
\hline
PT = & 111 + 94.22
\end{array}
$$

$$LC = 2R \sin \tfrac{1}{2}I = 2(2865.33)(0.06976) = 399.77 \text{ ft}$$

12.9   Given: The same data as given in Prob. 12.8 except the centerline of construction is 23 ft right of the baseline of survey. The degree of curve D is not given, and since the radius R will have changed, D must be refigured. Find R, D, T, PC, L, PT, and LC.

Solution

Draw a sketch of the arcs; see Fig. 12-10b.

$$
\begin{array}{ll}
R \text{ for BL of survey, Prob. 12.8} = & 2865.33 \text{ ft} \\
\text{Add distance to right of CL} & + \;\;\;\; 23.00 \\
\hline
R \text{ for CL of construction} = & 2888.33 \text{ ft}
\end{array}
$$

$$D = \frac{5729.58}{R} = \frac{5729.58}{2888.33} = 1.983699°$$

$$T = R \tan \tfrac{1}{2}I = 2888.33(0.0699268) = 201.97 \text{ ft}$$

$$PC = PI - T$$

$$
\begin{array}{ll}
PI = & 109 + 94.56 \\
-T & - \;\;(2 + 01.97) \\
\hline
PC = & 107 + 92.59
\end{array}
$$

$$L = 100 \frac{I}{D} = 100\left(\frac{8}{1.983699}\right) = 403.29 \text{ ft}$$

$$PT = PC + L$$

$$
\begin{array}{ll}
PC = & 107 + 92.59 \\
+L & + \;\;\;\; 4 + 03.29 \\
\hline
PT = & 111 + 95.88
\end{array}
$$

$$LC = 2R \sin \tfrac{1}{2}I = 2(2888.33)(0.069757) = 402.96 \text{ ft}$$

12.10   Given: PI station = $100 + 00.00$; $I = 23°08'08''$ lt; $D = 2°00'00''$. Find R, T, PC, L, PT, M, and LC by the chord method.

Solution

$$R = \frac{50}{\sin \frac{1}{2}D} = \frac{50}{0.0174524} = 2864.93 \text{ ft}$$

To find $\tan \frac{1}{2}I$, first convert $\frac{1}{2}I$ to decimal degrees.

$$I = 23°08'08'' \qquad \frac{1}{2}I = \frac{23°08'08''}{2} = 11°34'04'' = 11.5678°$$

$$T = R \tan \tfrac{1}{2}I = 2864.93(0.20468) = 586.40 \text{ ft}$$

$$PC = PI - T$$

$$
\begin{array}{lr}
PI = & 100 + 00.00 \\
-T & -\ (5 + 86.40) \\
\hline
PC = & 94 + 13.60
\end{array}
$$

$$L = 100\,\frac{I}{D} = 100\left(\frac{23°08'08''}{2}\right) = 100(11.5678°) = 1156.78 \text{ ft}$$

$$PT = PC + L$$

$$
\begin{array}{lr}
PT = & 94 + 13.60 \\
+L & +\ 11 + 56.78 \\
\hline
PT = & 105 + 70.38
\end{array}
$$

Find the middle ordinate:

$$M = R \text{ vers } \tfrac{1}{2}I = 2864.93(0.02032) = 58.22 \text{ ft}$$
$$LC = 2R \sin \tfrac{1}{2}I = 2(2864.93)(0.20053) = 1148.98 \text{ ft}$$

12.11 Given: The following data on a simple curve. Compute and set up the field notes. $R = 300$ ft; $I = 76°46'36''$; PI = 7 + 00.00.

Solution

1. Set the PI.

2. Measure the plus of the PI. Measure 40.00 ft from station 7 + 00.00: PI = 7 + 40.00.

3. Measure I. Angle x should be measured, usually two direct and reverse sights, at least 1 DR, and computed.

$$I = 180° - x = 179°59'60'' - 103°13'24'' = 76°46'36''$$

4. Compute T. $\qquad\qquad\qquad\qquad T = R \tan \tfrac{1}{2}I$

Convert $\frac{1}{2}I$ into decimal degrees for the tangent function in the hand calculator.

$$I = 76°46'36'' \qquad \tfrac{1}{2}I = 38°23'18'' = 38.39°$$

Substituting into $T = R \tan \tfrac{1}{2}I$ gives

$$T = 300(0.792306) = 237.69 \text{ ft}$$

5. Compute L. Use Table 12-1. Separate I into incremental lengths as shown:

$$
\begin{array}{rl}
70° = & 1.22173 \\
6° = & 0.10472 \\
40' = & 0.01164 \\
6' = & 0.00174 \\
30'' = & 0.00015 \\
6'' = & \underline{0.00003} \\
\text{Sum} = & 1.34001
\end{array}
$$

$$L = (\text{length of curve for 1-ft radius})R = 1.34001(300) = 402.00 \text{ ft}$$

6. Compute the pluses.
$$
\begin{array}{lr}
PI = & 7 + 40.00 \\
-T & -(2 + 37.68) \\
\hline
PC = & 5 + 02.32 \\
+L & +\ 4 + 02.00 \\
\hline
PT = & 9 + 04.32
\end{array}
$$

7. Compute the deflection angles.

$$\begin{array}{lr}\text{First 50-ft point} = & 5 + 50 \\ -\text{ plus of the PC} & - (5 + 02.32) \\ \text{First arc length} = & 47.68 \text{ ft}\end{array}$$

Use formula:
$$d = 1719\frac{a}{R} = 1719\left(\frac{47.68}{300}\right) = 273.186 \text{ minutes}$$

$$273.19 \text{ minutes} = 4°33.19' = 4°33'11''$$

Compute for 50-ft chords.

$$d = 1719\frac{a}{R} = 1719\left(\frac{50}{300}\right) = 286.50 \text{ minutes} = 4°46.50'$$

Compute for the final arc.

$$\begin{array}{lr}\text{PT station} = & 9 + 04.32 \\ -\text{previous station} & - (9 + 00.00) \\ \text{Last arc length} = & 04.32 \text{ ft}\end{array}$$

$$d = 1719\frac{a}{R} = 1719\left(\frac{04.32}{300}\right) = 24.70 \text{ minutes}$$

Computation of Deflection Angles for Prob. 12.11

Station	Deflection	Deflection to 15″
PC 5 + 02.32	0°	0°
5 + 50.00	4°33.19' + 4°46.50'	4°33'15″
6 + 00.00	9°19.69' + 4°46.50'	9°19'45″
6 + 50.00	14°06.19' + 4°46.50'	14°06'15″
7 + 00.00	18°52.69' + 4°46.50'	18°52'45″
7 + 50.00	23°39.19' + 4°46.50'	23°39'15″
8 + 00.00	28°25.69' + 4°46.50'	28°25'45″
8 + 50.00	33°12.19' + 4°46.50'	33°12'15″
9 + 00.00	37°58.69' + 24.70'	37°58'45″
	38°23.39'	38°23'30″

Thus: Total deflection angle = 38°23'30″

As a check, $\frac{1}{2}I = 38°23'18''$. The small difference between these figures is due to rounding. See Table 12-3 for the field book notes.

Table 12-3   Field Book Notes for Prob. 12.11

Station	Chord	Deflection	Curve Data
+ 50			$R = 300$ ft
PT 09 + 04.31	04.31	38°23′30″	$I = 76°46′36″$
09 + 00.00	49.94	37°58′45″	$\frac{1}{2}I = 38°23′18″ = 38°23.30′$
08 + 50.00	49.94	33°12′15″	$T = 237.60$ ft
08 + 00.00	49.94	28°25′45″	$I = 70° = \quad 1.22173$
07 + 50.00	49.94	23°39′15″	$6° = \quad 0.10472$
PI 07 + 40.00			$40′ = \quad 0.01164$
07 + 00.00	49.94	18°52′45″	$6′ = \quad 0.00174$
06 + 50.00	49.94	14°06′15″	$30″ = \quad 0.00015$
06 + 00.00	49.94	09°19′45″	$6″ = \quad 0.00003$
05 + 50.00	47.63	04°33′15″	Sum = $\quad 1.34001$
PC 05 + 02.31		0°	$\times R \quad \times \quad 300$
5 + 00.00			$L \quad = \quad 402.0030$ ft

8. Compute the chord lengths. Use Table 12-2.

Arc, ft	Correction, ft	Chord, ft
47.68	−0.050	47.63
50.00	−0.058	49.94
4.31	−0.000	4.31

12.12 Given: $D = 15°$; PI = 18 + 00; $I = 75°$. Stake out this simple curve and make field notes.

Solution

1. Set the PI.
2. Measure the plus of the PI. Measure from station 18; it is +00.00 ft, so the plus of the PI is 0.
3. Measure I. Angle x should be measured and I computed.

$$I = 180° − x = 180° − 105° = 75°$$

4. Compute T. First find the radius (arc definition).

$$R = \frac{5729.58}{D} = \frac{5729.58}{15} = 381.97 \text{ ft}$$

$$\tfrac{1}{2}I = \tfrac{1}{2}(75°) = 37.5°$$

$$T = R \tan \tfrac{1}{2}I = R \tan 37.5° = 381.97(0.767327) = 293.10 \text{ ft}$$

5. Compute L. Use Table 12-1. Separate $I = 75°$ into increments as shown.

$$70° = 1.22173$$
$$5° = 0.08727$$
$$\text{Sum} = 1.30900$$

$$L = 1.30900R = 1.30900(381.97) = 499.9987 = 500 \text{ ft}$$

6. Compute the pluses.

$$
\begin{array}{ll}
\text{PI} = & 18 + 00.00 \\
-T & - (2 + 93.10) \\
\text{PC} = & 15 + 06.90 \\
+L & + \; 5 + 00.00 \\
\text{PT} = & 20 + 06.90
\end{array}
$$

7. Compute the deflection angles.

$$
\begin{aligned}
\text{First 50-ft point} &= \qquad 15 + 50.00 \\
- \text{ plus of the PC} &\quad - (15 + 06.90) \\
\text{First arc length} &= \qquad 43.10 \text{ ft}
\end{aligned}
$$

Use formula: $\quad d = 1719\dfrac{a}{R} = 1719\left(\dfrac{43.10}{381.97}\right) = 193.9651 \text{ minutes} = 3°13.97'$

For 50-ft arcs: $\quad d = 1719\left(\dfrac{50}{381.97}\right) = 225.0176 \text{ minutes} = 3°45.02'$

Last arc:

$$
\begin{aligned}
\text{PT station} &= \qquad 20 + 06.90 \\
- \text{ previous station} &\quad - (20 + 00.00) \\
\text{Last arc length} &= \qquad 06.90 \text{ ft}
\end{aligned}
$$

$$d = 1719\left(\dfrac{06.90}{381.97}\right) = 31.05 \text{ minutes} = 0°31.05'$$

Computation of Deflection Angles for Prob. 12.12

Station	Deflection	Deflection to 15″
PC 15 + 00.00	0°0′	
15 + 06.90	0°	0°
15 + 50.00	3°13.97′ + 3°45.02′	3°14′00″
16 + 00.00	6°58.99′ + 3°45.02′	6°59′00″
16 + 50.00	10°44.01′ + 3°45.02′	10°44′00″
17 + 00.00	14°29.03′ + 3°45.02′	14°29′00″
17 + 50.00	18°14.05′ + 3°45.02′	18°14′00″
18 + 00.00	21°59.07′ + 3°45.02′	21°59′00″
18 + 50.00	25°44.09′ + 3°45.02′	25°44′00″
19 + 00.00	29°29.11′ + 3°45.02′	29°29′00″
19 + 50.00	33°14.13′ + 3°45.02′	33°14′00″
20 + 00.00	36°59.15′ + 0°31.05′	36°59′00″
PT 20 + 06.90	37°30.05′	37°30′15″

Note that the deflection angle for each station is a cumulative total. See Table 12-4 for the field book notes.

Table 12-4   Field Book Notes for Prob. 12.12

Station	Chord	Deflection	Curve Data
20 + 50			$R = 381.97$ ft
PT 20 + 06.90	06.90	37°30′15″	$I = 75°$
20 + 00.00	49.96	36°59′00″	$\frac{1}{2}I = 37.5° = 37°30′$
19 + 50.00	49.96	33°14′00″	$T = 293.10$ ft
19 + 00.00	49.96	29°29′00″	$L = 500$ ft
18 + 50.00	49.96	25°44′00″	$I = 75°$
PI 18 + 00.00	49.96	21°59′00″	70° = 1.22173
17 + 50.00	49.96	18°14′00″	5° = 0.08727
17 + 00.00	49.96	14°29′00″	Sum = 1.30900
16 + 50.00	49.96	10°44′00″	$\times R$ $\times$ 381.97
16 + 00.00	49.96	06°59′00″	L = 499.9987 ft
15 + 50.00	43.08	03°14′00″	$L = 500$ ft (rounded)
PC 15 + 06.90		0°	
15 + 00.00			

8. Compute the chord lengths.

Arc, ft	Correction, ft	Chord, ft
43.10	−0.024	43.08
50.00	−0.037	49.96
06.90	−0.000	06.90

9. Set the PT. Measure $T = 293.10$ ft from the PI and set the PT on line with the forward tangent.

10. Set the PC. Measure 06.90 ft from station 15 + 00 and line in.

11. Set the stations. Set up at the PC. Set the vernier at zero and aim at the PI. Turn left the deflection angle for 15 + 50 (3°14′00″) and set 15 + 50 on line at the chord distance from the PC (43.08 ft). Set the vernier at the next deflection angle (6°59′00″) and set 16 + 00 on line at the chord distance of 49.96 ft from station 15 + 50. Continue in this way, setting off each successive deflection angle for 15 + 50 (3°14′00″) and set 15 + 50 on line at the chord distance from the PC last point is set previous to the PT (20 + 06.90).

12.13 Compute the necessary data and set up the field notes for the following horizontal curve. Use 50-ft stations. Given: Station PI = 54 + 32.2; $I = 24°20′$; $D = 4°00′$.

Solution

1. Set the PI. Intersect the two tangents.

2. Measure the plus of the PI. From station 54 + 00 measure 32.2 ft; thus

$$\text{Plus PI} = 54 + 32.2$$

3. Measure I. Angle x should be measured and I computed.

$$I = 180° - x = 179°60′ - 155°40′ = 24°20′$$

4. Compute T. First find R. $R = \dfrac{5729.58}{D} = \dfrac{5729.58}{4} = 1432.40$ ft

$$\tfrac{1}{2}I = \tfrac{1}{2}(24°20′) = 12°10′ = 12.167°$$
$$T = R \tan \tfrac{1}{2}I = R \tan 12.167° = 1430.40(0.21560) = 308.83 \text{ ft}$$

5. Compute L. By formula,

$$L = 100 \frac{I}{D} = 100\left(\frac{24.333}{4}\right) = 608.32 \text{ ft}$$

6. Compute the pluses.

$$
\begin{array}{ll}
PI = & 54 + 32.20 \\
-T & -(3+08.83) \\
\hline
PC = & 51 + 23.37 \\
+L & +\ 6+08.32 \\
\hline
PT = & 57 + 31.69
\end{array}
$$

7. Compute the deflection angles.

$$
\begin{array}{ll}
\text{First 50-ft point} = & 51 + 50.00 \\
-\text{ plus of the PC} & -(51+23.37) \\
\hline
\text{First arc length} = & 26.63 \text{ ft}
\end{array}
$$

Use formula: $d = 1719 \dfrac{a}{R} = 1719\left(\dfrac{26.63}{1432.40}\right) = 31.96 \text{ minutes} = 0°32'00''$

50-ft arcs: $d = 1719\left(\dfrac{50}{1432.40}\right) = 60.0044 \text{ minutes} = 1°00'$

Last arc:

$$
\begin{array}{ll}
\text{PT station} = & 57 + 31.69 \\
-\text{ previous station} & -(57+00.00) \\
\hline
\text{Last arc length} = & 31.69 \text{ ft}
\end{array}
$$

$$d = 1719\left(\frac{31.69}{1432.40}\right) = 38.03 \text{ minutes} = 0°38'05''$$

Note: The deflection angle at PT = 12°10′, which is the value of $\frac{1}{2}I$. This checks the angles. See Table 12-5 for the field book notes.

Table 12-5   Field Book Notes for Prob. 12.13

Station	Chord	Deflection	Curve Data
PT 57 + 31.62	31.62	12°10′	R = 1432.40 ft
57 + 00.00	50.00	11°32′	I = 24°20′
56 + 50.00	50.00	10°32′	$\frac{1}{2}I$ = 12°10′ = 12.17°
56 + 00.00	50.00	9°32′	T = 308.83 ft
55 + 50.00	50.00	8°32′	L = 608.25 ft
55 + 00.00	50.00	7°32′	I = 24°20′
54 + 50.00	50.00	6°32′	
PI 54 + 00.00	50.00	5°32′	20° = 0.34907
53 + 50.00	50.00	4°32′	4° = 0.06981
53 + 00.00	50.00	3°32′	20′ = 0.00582
52 + 50.00	50.00	2°32′	Sum = 0.42470
52 + 00.00	50.00	1°32′	$\times R$ $\times$ 1432.39
51 + 50.00	26.63	0°32′	L = 608.33 ft
PC 51 + 23.37		0°00′	
51 + 00.00			

Computation of Deflection Angles for Prob. 12.13

Station	Deflection	Deflection to 15″
PC 51 + 23.37	0°	0°
51 + 50	0°32′ + 1°00′	0°32′
52 + 00	1°32′ + 1°00′	1°32′
52 + 50	2°32′ + 1°00′	2°32′
53 + 00	3°32′ + 1°00′	3°32′
53 + 50	4°32′ + 1°00′	4°32′
54 + 00	5°32′ + 1°00′	5°32′
54 + 50	6°32′ + 1°00′	6°32′
55 + 00	7°32′ + 1°00′	7°32′
55 + 50	8°32′ + 1°00′	8°32′
56 + 00	9°32′ + 1°00′	9°32′
56 + 50	10°32′ + 1°00′	10°32′
57 + 00	11°32′ + 0°38′	11°32′
PT 57 + 31.62	12°10′	12°10′

8. Compute the chord lengths.

Arc, ft	Correction, ft	Chord, ft
26.63	−0.000	26.63
50.00	−0.003	50.00
31.69	−0.001	31.69

Note: The radius is so long that the corrections are small enough to be negligible.

9. Set the PT. Measure $T = 308.83$ ft from the PI and set on line with the forward tangent.

10. Set the PV − PC. Measure 23.37 ft from station 51 + 00.00 and line in.

11. Set the stations as in the previous problems.

Computation of Deflection Angles for Prob. 12.13

Station	Deflection	Deflection to 15″
PC 37 + 58.48	0°	0°
38 + 00.00	0°49′49″ + 2°00′00″	0°49′45″
39 + 00.00	2°49′49″ + 2°	2°49′45″
40 + 00.00	4°49′49″ + 2°	4°49′45″
41 + 00.00	6°49′49″ + 2°	6°49′45″
42 + 00.00	8°49′49″ + 2°	8°49′45″
43 + 00.00	10°49′49″ + 0°49′20″	10°49′45″
PT 43 + 40.98	11°39′09″	11°39′00″

12.14 Given: $I = 23°18′$; $D = 4°00′$; station PI = $40 + 53.87$. Calculate the necessary data, using the chord definition, for this curve, and set up the field book notes.

Solution

1. Set the PI.

2. Measure the plus of the PI. From station 40 measure 53.87 ft. Thus, Plus PI = $40 + 53.87$.

3. Measure I. Angle x should be measured and I computed.

$$I = 180° - x = 179°60′ - 156°42′ = 23°18′$$

4. Compute T. First find R (chord definition):

$$R = \frac{50}{\sin \frac{1}{2}D} = \frac{50}{\sin 2} = \frac{50}{0.0348995} = 1432.69 \text{ ft}$$

$$I = 23°18′ \quad \text{so} \quad \tfrac{1}{2}I = 11°39′ = 11.65°$$

$$T = R \tan \tfrac{1}{2}I = R \tan 11.65° = 1432.69(0.20618) = 295.39 \text{ ft}$$

5. Compute L. Use Table 12-1. Break down $I = 23°18′$ into incremental lengths.

$$
\begin{aligned}
20° &= 0.34907 \\
3° &= 0.05236 \\
10′ &= 0.00291 \\
8′ &= \underline{0.00233} \\
\text{Sum} &= 0.40667
\end{aligned}
$$

$$L = 0.40667R = 0.40667(1432.69) = 582.62 \text{ ft}$$

6. Compute the pluses.

$$
\begin{array}{lr}
\text{PI} = & 40 + 53.87 \\
-T & -(2 + 95.39) \\
\text{PC} = & 37 + 58.48 \\
+L & +\ 5 + 82.50 \\
\text{PT} = & 43 + 40.98
\end{array}
$$

7. Compute the deflection angles.

$$\begin{array}{ll} \text{First 100-ft point} = & 38 + 00.00 \\ - \text{ plus of PC} & - (37 + 58.48) \\ \hline \text{First arc length} = & 41.52 \text{ ft} \end{array}$$

$$d = 1719 \frac{a}{R} = 1719\left(\frac{41.52}{1432.69}\right) = 49.81 \text{ minutes} = 49'49''$$

100-ft arcs:

$$d = 1719\left(\frac{100}{1432.69}\right) = 119.9841' = 2°00'$$

Last arc:

$$\begin{array}{ll} \text{PT station} = & 43 + 40.98 \\ - \text{ previous station} & - (43 + 00.00) \\ \hline \text{Last arc length} = & 40.98 \text{ ft} \end{array}$$

$$d = 1719\left(\frac{40.98}{1432.69}\right) = 49.17 \text{ minutes} = 49°10''$$

Note: Total deflection angle of 11°39'00'' checks with the $\frac{1}{2}I$ angle. See Table 12-6 for the field book notes.

Table 12-6 · Field Book Notes for Prob. 12.14

Station	Chord	Deflection	Curve Data
44 + 00.00			$R = 1432.69$ ft
PT 43 + 40.98	40.99	11°39'00''	$I = 23°18'$
43 + 00.00	99.98	10°49'45''	$\frac{1}{2}I = 11°39' = 11.65°$
42 + 00.00	99.98	8°49'45''	$T = 295.39$ ft
41 + 00.00	99.98	6°49'45''	$L = 582.50$ ft
PI 40 + 53.87			$I = 23°18'$
40 + 00.00	99.98	4°49'45''	20° = 0.34907
39 + 00.00	99.98	2°49'45''	3° = 0.05236
38 + 00.00	41.53	0°49'45''	10' = 0.00291
PC 37 + 58.48			8' = 0.00233
37 + 00.00		0°	Sum = 0.40667
			$\times R$ $\times 1432.69$
			L = 582.63 ft

8. Compute the chord lengths.

Nominal Chord, ft	Correction, ft	Chord, ft
41.52	+0.007	41.53
100.00		100.00
40.98	+0.007	40.99

9. Set the PI. Measure $T = 295.39$ ft from the PI and set the PT on line with the forward tangent.

10. Set the PC. Measure 58.48 ft from station 37 and line in.

11. Set the station as in the previous problems.

12.15 Given these facts about a compound curve: Plus of PI = 16 + 30.12; $\Delta = 96°42'40''$; the first radius $R_1 = 500$ ft; $\Delta_1 = 62°22'20''$; $R_2 = 800$ ft. Compute the following: Pluses of the PC, PCC, and PT, and length T_2.

Solution

1. Measure angle Δ.

2. Find Δ_2. $\Delta_2 = \Delta - \Delta_1 = 96°42'40'' - 62°22'20'' = 34°20'20'' = 34.3389°$

3. Find t_1 and t_2.

   $\Delta_1 = 62°22'20''$ so $\tfrac{1}{2}\Delta_1 = 31°11'10'' = 31.1861°$ and $\tan 31.1861° = 0.60529$

   $$t_1 = R_1 \tan \tfrac{1}{2}\Delta_1 = 500(0.60529) = 302.65 \text{ ft}$$

   $\Delta_2 = 34°20'20''$ so $\tfrac{1}{2}\Delta_2 = 17°10'10'' = 17.1695°$ and $\tan 17.1695° = 0.308968$

   $$t_2 = R_2 \tan \tfrac{1}{2}\Delta_2 = 800(0.308968) = 247.17 \text{ ft}$$

4. Find GH. $GH = t_1 + t_2 = 302.65 + 247.17 = 549.82 \text{ ft}$

5. Find VG.

   $\Delta_2 = 34°20'20'' = 34.3389°$ and $\sin 34.3389° = 0.564087$

   $\Delta = 96°42'40'' = 96.7111°$ and $\sin 96.7111° = 0.993148$

   $$VG = (\sin \Delta_2)\left(\frac{GH}{\sin \Delta}\right) = (0.564087)\left(\frac{549.82}{0.993148}\right) = 0.564087(553.61) = 312.28 \text{ ft}$$

6. Find VH.

   $\Delta_1 = 62°22'20'' = 62.3723°$ and $\sin 62.3723° = 0.885980$

   $$VH = (\sin \Delta_1)\left(\frac{GH}{\sin \Delta}\right) = (0.885980)\left(\frac{549.82}{0.993148}\right) = 0.885980(553.61) = 490.49 \text{ ft}$$

7. Find T_1. $T_1 = VG + t_1 = 312.28 + 302.65 = 614.93 \text{ ft}$

8. Find T_2. $T_2 = VH + t_2 = 490.49 + 247.17 = 737.66 \text{ ft}$

9. Find L_1 and L_2. Use Table 12-1.

   $$\Delta_2 = 62°22'20''$$

   $$
   \begin{aligned}
   60° &= 1.04720 \\
   2° &= 0.03491 \\
   20' &= 0.00582 \\
   2' &= 0.00058 \\
   20'' &= \underline{0.00010} \\
   \text{Sum} &= 1.08861
   \end{aligned}
   $$

   $$L_1 = 1.08861 R_1 = 1.08861(500) = 544.31 \text{ ft}$$

   $$\Delta_2 = 34°20'20''$$

   $$
   \begin{aligned}
   30° &= 0.52360 \\
   4° &= 0.06981 \\
   20' &= 0.00582 \\
   20'' &= \underline{0.00010} \\
   \text{Sum} &= 0.59933
   \end{aligned}
   $$

   $$L_2 = 0.59933 R_2 = 0.59933(800) = 479.46 \text{ ft}$$

10. Summary of the problem:

Plus PI =	1630.12
Less T_1	− 614.93
Plus PC =	1015.19
Add L_1	+ 544.31
Plus PCC =	1559.50
Add L_2	+ 479.46
Plus PT =	2038.96

    $$T_2 = 737.66 \text{ ft}$$

12.16 Given for a compound curve: Plus of PI = 24 + 16.10; $\Delta = 92°44'42''$; $R_1 = 400$ ft; $\Delta_1 = 60°24'20''$; $R_2 = 600$ ft. Compute the pluses of PC, PCC, and PT, and length T_2.

Solution

1. Measure angle Δ.

2. Find Δ_2. $\Delta_2 = \Delta - \Delta_1 = 92°44'42'' - 60°24'20'' = 32°20'22'' = 32.3394°$

3. Find t_1 and t_2.

    $\Delta_1 = 60°24'20''$ so $\tfrac{1}{2}\Delta_1 = 30°12'20'' = 30.2028°$ and $\tan 30.2028° = 0.582079$

    $t_1 = R_1 \tan \tfrac{1}{2}\Delta_1 = 400(0.582079) = 232.83$ ft

    $\Delta_2 = 32°20'22''$ so $\tfrac{1}{2}\Delta_2 = 16°10'11'' = 16.1698°$ and $\tan 16.1698° = 0.289955$

    $t_2 = R_2 \tan \tfrac{1}{2}\Delta_2 = 600(0.289955) = 173.97$ ft

4. Find GH. $GH = t_1 + t_2 = 232.83 + 173.97 = 406.80$ ft

5. Find VG. $\Delta_2 = 32°20'22'' = 32.3394°$ $\sin 32.3394 = 0.534934$

    $\Delta = 92°44'42'' = 92.745°$ $\sin 92.745° = 0.998853$

$$VG = (\sin \Delta_2)\left(\frac{GH}{\sin \Delta}\right) = (0.534934)\left(\frac{406.80}{0.998853}\right) = 0.534934(406.80) = 217.86 \text{ ft}$$

6. Find VH.

    $\Delta_1 = 60°24'20'' = 60.4056°$ $\sin 60.4056° = 0.869543$

$$VH = (\sin \Delta_1)\left(\frac{GH}{\sin \Delta}\right) = (0.869543)\left(\frac{406.80}{0.998853}\right) = 0.869543(406.80) = 354.14 \text{ ft}$$

7. Find T_1. $T_1 = VG + t_1 = 217.86 + 232.83 = 450.69$ ft

8. Find T_2. $T_2 = VH + t_2 = 354.14 + 173.97 = 528.11$ ft

9. Find L_1 and L_2. $\Delta_1 = 60°24'20''$

$$
\begin{aligned}
60° &= 1.04720 \\
20' &= 0.00582 \\
4' &= 0.00116 \\
20'' &= \underline{0.00010} \\
\text{Sum} &= 1.05428
\end{aligned}
$$

$$L_1 = 1.05428 R_1 = 1.05428(400) = 421.71 \text{ ft}$$
$$\Delta_2 = 32°20'22''$$

$$
\begin{aligned}
30° &= 0.52360 \\
2° &= 0.03491 \\
20' &= 0.00582 \\
20'' &= 0.00010 \\
2'' &= \underline{0.00001} \\
\text{Sum} &= 0.56444
\end{aligned}
$$

$$L_2 = 0.56444 R_2 = 0.56444(600) = 338.66 \text{ ft}$$

10. Summary of the problem:

$$
\begin{array}{lr}
\text{Plus PI} = & 2416.10 \\
\text{Less } T_1 & -\ \ 450.69 \\
\hline
\text{Plus PC} = & 1965.41 \\
\text{Add } L_1 & +\ \ 421.71 \\
\hline
\text{Plus PCC} = & 2387.12 \\
\text{Add } L_2 & +\ \ 338.66 \\
\hline
\text{Plus PT} = & 2725.78 \\
\end{array}
$$

$$T_2 = 528.11 \text{ ft}$$

Supplementary Problems

12.17 Given the following data for a simple curve: Station PI = 106 + 50.12; $I = 6°00'00''$ rt; $D = 2°00'00''$. Find R, T, PT, and LC by the arc definition.

Ans. $R = 2864.79$ ft; $T = 150.14$ ft; PT = 107 + 99.98; LC = 299.86 ft

12.18 Given the following data for a simple curve: Station PI = 110 + 80.25; $I = 4°00'00''$ lt; $D = 3°00'00''$. Find R, T, PC, PT, and LC by the arc definition.

Ans. $R = 1909.86$ ft; $T = 66.69$ ft; PC = 110 + 13.56; PT = 111 + 46.89; LC = 133.33 ft

12.19 Given the following data for a simple curve: Station PI = 100 + 00.00; $I = 27°10'12''$ lt; $D = 2°00'00''$. Find T, PC, L, and M by the arc definition.

Ans. $T = 692.27$ ft; PC = 93 + 07.73; $L = 1358.50$ ft; $M = 80.15$ ft

12.20 Given the following data for a simple curve: Station PI = 105 + 56.18; $I = 8°00'00''$ rt; $D = 4°00'00''$. Find PC, L, PT, E, and LC by the chord definition.

Ans. PC = 104 + 56.00; $L = 200$ ft; PT = 106 + 56.00; $E = 3.50$ ft; LC = 199.88 ft

12.21 Given the following data for a simple curve: Station PI = 109 + 94.56; $I = 8°00'00''$ lt; $D = 2°00'00''$. Find R, PC, L, and LC by the chord definition.

Ans. $R = 2864.93$ ft; PC = 107 + 94.22; $L = 400$ ft; LC = 399.70 ft

12.22 Given the following data for a simple curve: Station PI = 100 + 00.00; $I = 23°08'08''$ lt; $D = 2°00'00''$. Find T, PC, M, and LC by the chord definition.

Ans. $T = 586.41$ ft; PC = 94 + 13.59; $M = 58.19$ ft; LC = 1148.99 ft

12.23 Set up the field notes for staking out the following curve, and solve the curve by the arc definition. Given: $R = 400$ ft; $I = 66°18'24''$; PI = 68 + 25.32.

Ans. $T = 261.29$ ft; $L = 462.91$ ft; first deflection angle = 2°34.58′; deflection angle for 50-ft arc = 3°34.88′; deflection angles at the following stations: 66 + 50 = 6°09′26″; 68 + 00 = 16°54′00″; at the PT = 33°09′12″

12.24 Set up field notes for staking out the following curve. Given: $R = 500$ ft; $I = 58°08'40''$; PI = 78 + 17.25.

Ans. $T = 277.98$; $L = 507.41$ ft; PC = 75 + 39.27; deflection angle for 50-ft arc = 2°51.90′; deflection angle at stations: 76 + 00 = 3°28′45″; 77 + 50 = 12°04′30″; 80 + 00 = 26°23′51″

12.25 Set up field notes for staking out the following curve, and solve the curve by the arc definition. Given: $R = 600$ ft; $I = 42°34'28''$; PI = 08 + 37.42.

Ans. $T = 233.78$ ft; $L = 445.84$ ft; PC = 6 + 03.64; PT = 10 + 49.48; deflection angle for a 50-ft arc = 2°23.24′

12.26 The following facts are known about a compound curve: PI = 32 + 87.93; $\Delta = 68°32'54''$; $R_1 = 350$ ft; $\Delta_1 = 40°18'34''$; $R_2 = 500$ ft. Compute the pluses of the PC, PCC, and PT, and length T_2.

Ans. $\Delta_2 = 28°14'20''$; $t_1 = 128.46$ ft; $GM = 254.23$ ft; $VH = 176.71$ ft; $T_2 = 302.48$ ft. See Fig. 12-11.

12.27 The following facts are known about a compound curve: PI = 30 + 26.20; $\Delta = 80°44'42''$; $R_1 = 400$ ft; $\Delta_1 = 40°24'20''$; $R_2 = 500$ ft. Compute the pluses of the PC, PCC, and PT, and length T_2.

Ans. $t_1 = 147.19$ ft; $t_2 = 183.66$ ft; $GH = 330.85$ ft; $VG = 216.99$ ft; $VH = 217.28$ ft; $T_2 = 400.94$ ft. See Fig. 12-12.

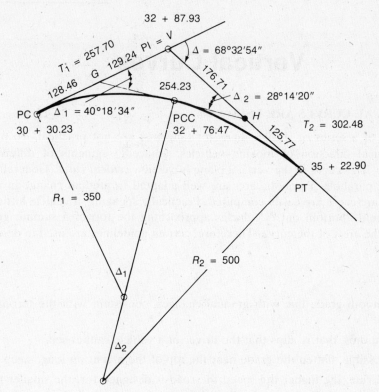

Fig. 12-11

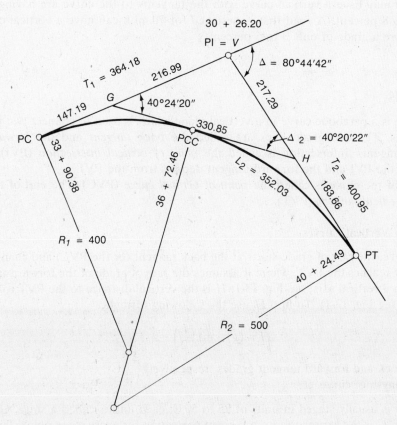

Fig. 12-12

Chapter 13

Vertical Curves

13.1 WHY VERTICAL CURVES ARE USED

Roads made up of a series of straight lines (or tangents) are not practical. To prevent abrupt changes in the vertical direction of moving vehicles, adjacent segments of differing grade are connected by a curve. This curve in the vertical plane is called a *vertical curve*. Generally, the vertical curve is the arc of a parabola. Parabolic arcs are well adapted to gradual change in direction, and elevations along the arc curve are easily computed. Practically speaking, a vehicle hitting the bottom of sag alignment would "bottom out." Vehicles approaching the top of a summit gradient would become airborne at the crest of the curve. Therefore, certain guidelines are used in designing vertical curves.

General Guidelines

1. Provide a smooth grade line with gradual changes, consistent with the terrain and type of highway.
2. Avoid hidden dips, that is, dips that the driver of a vehicle cannot see.
3. Whenever possible, flatten the grade near the top of the ascent on long, steep grades.
4. As a general rule, the higher the speed the road is designed for, the smaller the percent of grade that is allowed. For example, a road designed for a maximum speed of 30 miles per hour (mi/h) may have a vertical curve with the tangents to the curve arc having a grade as high as 6 to 8 percent. A road that is designed for 70 mi/h can have a vertical curve whose tangents have a grade of only 3 to 5 percent.

13.2 DEFINITIONS

A vertical curve is a parabolic curve in a vertical plane which is used to connect two numerically different grade lines. The two grade lines are called the *back tangent* and the *forward tangent*, respectively. The tangents' intersection is called the *point of vertical intersection* (PVI). The back tangent approaches the PVI and the forward tangent departs from the PVI.

The beginning of the curve is called the *point of vertical curve* (PVC). The end of the curve is called the *point of vertical tangent* (PVT).

Elements Governing Vertical Curves

In a vertical curve, the rate of grade starts at the back tangent (or the PVC) and changes at some constant rate until it reaches the PVT, where it assumes the rate of grade of the forward tangent. It is a true parabola with a vertical axis; see Fig. 13-1. H is the vertical distance to the PVT from the back tangent prolonged (see Fig. 13-1). To find H use the following formula:

$$H = (g_2 - g_1)\left(\frac{L}{2}\right) \tag{13-1}$$

where g_1, g_2 = back and forward tangent grades, respectively
             L = length of curve, ft

The length of curve is usually staged in units of 25 to 50 ft; each unit is called a *stage*. Grade stakes are set at each stage. C is the expression for a tangent correction for every stage point. To find C, use the following formula:

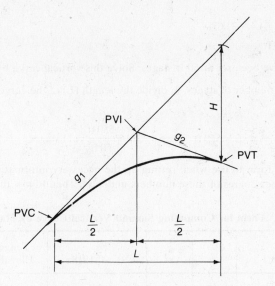

Fig. 13-1 Elements of a vertical curve.

$$C = \frac{H}{n^2} x^2 \qquad (13\text{-}2)$$

where H = vertical distance of PVT from back tangent prolonged [Eq. (13-1)]
           n = number of stages
           x = distance from the PVC (measured in stages) to the stage point for which C is being determined

To investigate drainage conditions, cover over pipes and storm drains, clearance beneath overhead structures, etc., it is sometimes necessary to determine the elevation and location of the lowest point on a "sag" curve or the highest point on a "summit" curve. The formula is the same for both computations:

$$a = g_1 \left(\frac{L}{g_1 - g_2} \right) \qquad (13\text{-}3)$$

where a = distance from PVC, ft
           L = curve length, ft
  g_1, g_2 = back and forward tangent grades, respectively, expressed as decimals

Once a has been determined, the tangent correction C may be found by the following formula:

$$C = \left(\frac{a}{L} \right)^2 H \qquad (13\text{-}4)$$

13.3 COMPUTING VERTICAL CURVES: THE STAGE METHOD

There are several methods of computing vertical curves. The stage method will be described first. The stage method of vertical curve computation is best completed with a constant-memory hand calculator. When computing a vertical curve, refer to the information that is available on the plan and profile sheet on a set of road plans. All the necessary information usually is included on the bottom graph of the sheet.

EXAMPLE 13.1 The following information about a vertical curve appears in the road plans:

PVI station = 11 + 02.43 at elevation 43.32 ft
Back tangent grade (g_1) = +5.00%

Forward tangent grade $(g_2) = -3.00\%$

Length of curve $L = 550$ ft

Assume that grade stakes have been set at 50-ft stages. Solve this vertical curve by the stage method.

Solution: To determine the number of stages, n, divide the length (L) of the curve (from the plans) by the stage distance:

$$L = 550 \text{ ft} \qquad \text{so} \qquad n = \frac{550 \text{ ft}}{50 \text{ ft}} = 11 \text{ stages}$$

Table 13-1 shows the proper form to use when setting up the necessary information. Place the numbers of the stage points under x. Place the squares of these numbers under x^2. Then follow the steps listed below.

Table 13-1   Form for Computing Summit Vertical Curve by Stages (Example 13.1)

Station	x	x^2	$C = -0.1818x^2$	Tangent Elevation	Curve Elevation
PVC 8 + 27.43	0	0	0.0	29.57	29.57
8 + 77.43	1	1	-0.18	32.07	31.89
9 + 27.43	2	4	-0.73	34.57	33.84
9 + 77.43	3	9	-1.64	37.07	35.43
10 + 27.43	4	16	-2.91	39.57	36.66
10 + 77.43	5	25	-4.55	42.07	37.52
11 + 27.43	6	36	-6.54	44.57	38.03
11 + 77.43	7	49	-8.91	47.07	38.16
12 + 27.43	8	64	-11.64	49.57	37.93
12 + 77.43	9	81	-14.73	52.07	37.34
13 + 27.43	10	100	-18.18	54.57	36.39
PVT 13 + 77.43	11	121	-22.00	57.07	35.07

1. Compute the station and elevation of the PVC and the PVT. Note that the horizontal distances to each from the PVI are equal to $L/2$ ($L = 550$ ft, so $L/2 = 275$ ft). The values used are given in the plans.

PVI station =	$11 + 02.43$	PVI station =	$11 + 02.43$
$-L/2$	$-(02 + 75.00)$	$+L/2$	$+02 + 75.00$
PVC station =	$8 + 27.43$	PVT station =	$13 + 77.43$
PVI elevation =	43.32	PVI elevation =	43.32
$-5\%(L/2)$	-13.75	$-3\%(L/2)$	-8.25
PVC elevation =	29.57	PVT elevation =	35.07

Note: Always draw a sketch of the curve (see Fig. 13-2). The back tangent is always extended beyond the last stage in the horizontal direction so that you can compute H, which is the vertical distance to the PVT from the back tangent prolonged.

2. Compute H. Use Eq. (*13-1*).

$$H = (g_2 - g_1)\left(\frac{L}{2}\right) = (-0.03 - 0.05)(275) = -22$$

3. Compute C. Use Eq. (*13-2*).

$$C = \frac{H}{n^2}x^2 = \left(\frac{-22}{121}\right)x^2 = -0.1818x^2$$

4. Compute the values in the C column (see Table 13-1) by multiplying each number in the x^2 column by -0.1818. Note that C for $PVT = H$.

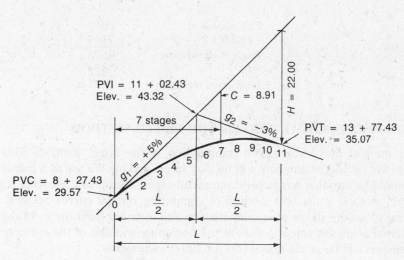

Fig. 13-2 Vertical curve on a summit.

5. Compute the elevations on the back tangent at the stage points, starting with the elevation of the PVC and adding $5\% \times 50 = 2.5$ ft successively. See Table 13-1.

6. Compute the curve elevations by applying the tangent correction C to the tangent elevations; see Table 13-1. For example:

$$
\begin{array}{lr}
\text{Tangent elevation at station } 8 + 77.43 = & 32.07 \\
-\text{correction} & -\ 0.18 \\
\hline
\text{Curve elevation} = & 31.89
\end{array}
$$

7. Compute the elevation at the required stage. Examine Fig. 13-2 and Table 13-1. Notice that Fig. 13-2 shows a vertical line indicating the position of the seventh stage. (This could be any location; stage 7 has been picked arbitrarily.) Calculate the curve elevation at stage 7:

$$
\begin{array}{lr}
\text{Tangent elevation at stage } 7 = & 47.07 \\
-\text{correction } C \text{ stage } 7 & -\ 8.91 \\
\hline
\text{Curve elevation at stage } 7 = & 38.16
\end{array}
$$

With a table such as Table 13-1 the elevation of any point on a vertical summit curve may be found.

8. To back-check the computations made, note the following procedures: Compute the PVI elevation using the forward tangent prolonged; see Fig. 13-2 and Table 13-1. Given: PVT = 13 + 77.43 at curve elevation 35.07; $g_2 = -3\%$. Go backward from the PVT; use the following formula:

$$
\text{Rise} = g_2\left(\frac{L}{2}\right) = 0.03(275) = 8.25
$$

Thus the rise from PVT to PVI is 8.25 ft. So PVT elevation of $35.07 + 8.25$ rise $= 43.32$ ft. This checks the PVI elevation.

9. Compute the high point of the curve. Use Eq. (*13-3*).

$$
a = g_1\left(\frac{L}{g_1 - g_2}\right) = 0.05\left[\frac{550}{0.05 - (-0.03)}\right] = 343.75 \text{ ft}
$$

Find the tangent correction C, and apply it to the tangent elevation that occurs at 343.75 ft from the PVC.

$$
C = \left(\frac{a}{L}\right)^2 H = \left(\frac{343.75}{550}\right)^2 (-22) = -8.59 \text{ ft}
$$

10. Find the station and elevation of the high point.

$$
\begin{array}{lr}
\text{PVC station} = & 8 + 27.43 \\
+ a & +\ 3 + 43.75 \\
\hline
\text{High point station} = & 11 + 71.18
\end{array}
$$

PVC elevation	29.57
+ 5%(343.75)	+ 17.19
Tangent elevation =	46.76
− C	− 8.59
High point elevation =	38.17 ft

13.4 COMPUTING VERTICAL CURVES: THE DIRECT METHOD

Another method of vertical curve computation is the direct method. This method has one disadvantage; the calculated numbers will be very large, making the use of a calculator essential. The calculator should be one that has storage and recall capabilities plus a squaring feature.

The direct method is the best method of computing vertical curves because in field work it is more practical to set the stakes at regular stations and measure from there. In the direct method an irregular interval of measurement occurs at the beginning and end of the curve only. All the rest of the measurements will be at the normal 50- or 100-ft increments.

The elevations found by the stage method must be staked out at the stage points, which often are not located at the regular stations. For example, in Example 13.1, PVC = 8 + 27.43, 8 + 77.43, 9 + 77.43, etc., which are not the regular stations. If the plus of the PVI and the length L had been chosen so that the PVC fell at a 50-ft point (as is often the case), the regular stations could be used. If this is not the case (as in Example 13.1) and if it is absolutely essential to set the grade stakes at regular stations, the direct method should be used.

We will apply the direct method to the same curve on which we used the stage method.

EXAMPLE 13.2 The following values are found on the profile section (graph portion) of a plan and profile sheet. (*Note*: They are the same values given in Example 13.1 for the stage method of computing a vertical summit curve.)

PVI station = 11 + 02.43 at elevation of 43.32 ft

Back tangent grade (g_1) = +5%

Forward tangent grade (g_2) = −3%

Length of curve L = 550 ft

Find PVC and PVT stations and elevations, H, C, elevations on the back tangent at station points, curve elevations and high point.

Solution: Draw a sketch of the curve (see Fig. 13-3) and make a computation table (see Table 13-2).

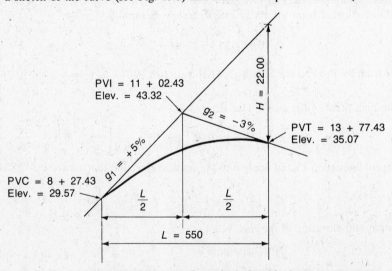

Fig. 13-3 Summit vertical curve for calculation by the direct method.

Table 13-2　Form for Computing Curve Elevations by the Direct Method (Example 13.2)

Station	a	a^2	$C = -0.00007273a^2$	Tangent Elevation	Curve Elevation
PVC　8 + 27.43	0	0	0.0	29.57	29.57
8 + 50	22.6	510	−0.04	30.70	30.66
9 + 00	72.6	5 270	−0.38	33.20	32.82
9 + 50	122.6	15 031	−1.09	35.70	34.61
10 + 00	172.6	29 791	−2.17	38.20	36.03
10 + 50	222.6	49 551	−3.60	40.70	37.10
11 + 00	272.6	74 311	−5.40	43.20	37.80
11 + 50	322.6	104 071	−7.57	45.70	38.13
12 + 00	372.6	138 831	−10.10	48.20	38.10
12 + 50	422.6	178 591	−12.99	50.70	37.71
13 + 00	472.6	223 351	−16.24	53.20	36.96
13 + 50	522.6	273 111	−19.86	55.70	35.84
PVT　13 + 77.43	550.0	302 500	−22.00	57.07	35.07

1. Compute the station and elevation of the PVC and PVT. Note that the horizontal distances to each from the PVI are both equal to $L/2$.

PVI =	11 + 02.43	PVI station =	11 + 02.43
$-L/2$	$-$ (2 + 75.00)	$+L/2$	$+$ 2 + 75.00
PVC station =	8 + 27.43	PVT station =	13 + 77.43
PVI elevation =	43.32	PVI elevation =	43.32
$-5\%(L/2)$	$-$ 13.75	$-3\%(L/2)$	$-$ 8.25
PVC elevation =	29.57	PVT elevation =	35.07

The values will be worked out in feet; see Table 13-2. The values under a are the distances from the PVC to the nearest 0.1 ft.

2. Compute H (the vertical distance to the PVT from the back tangent prolonged).

$$H = (g_2 - g_1)\left(\frac{L}{2}\right) = (-0.03 - 0.05)(275) = -22$$

3. Compute C to four significant figures.

$$C = \frac{H}{L^2}a^2 = -0.00007273a^2$$

4. Compute the values in the C column by multiplying each value for a^2 by -0.00007273.

5. Compute the elevations on the back tangent at the regular station points. Remember $g_1 =$ back tangent $= 5\%$. This makes the first elevation $22.6(0.05) = 1.13$ ft and the last elevation $27.4(0.05) = 1.37$ ft. All other elevations $= 50(0.05) = 2.5$ ft. Put the tangent elevations into the computation table; see Table 13-2.

6. Compute the curve elevations by applying the tangent corrections C to the tangent elevations.

7. Find the high point.

$$a = g_1\left(\frac{L}{g_1 - g_2}\right) = 0.05\left(\frac{550}{0.05 + 0.03}\right) = 343.75\text{ ft}$$

Find the tangent correction C, and apply it to the tangent elevation that occurs at 343.75 ft from the PVC.

$$C = \left(\frac{a}{L}\right)^2 H = \left(\frac{343.75}{550}\right)^2(-22) = -8.59\text{ ft}$$

8. Find station and elevation of the high point.

$$
\begin{array}{lr}
\text{PVC station} = & 8 + 27.43 \\
+a & + \ 3 + 43.75 \\
\hline
\text{High point station} = & 11 + 71.18 \\
\\
\text{PVC elevation} = & 29.57 \\
+5\% (343.75) & + 17.19 \\
\hline
\text{Tangent elevation} = & 46.76 \\
-C & - \ 8.59 \\
\hline
\text{High point elevation} = & 38.17
\end{array}
$$

9. Check on the computations. Compute the PVI elevation using the forward tangent prolonged; see Fig. 13-3. Given: PVT = 13 + 77.43 at elevation 35.07; $g_2 = -3\%$. Go backward from the PVT; use formula:

$$\text{Rise} = g_2 \left(\frac{L}{2}\right) = 0.03(275) = 8.25\text{-ft rise from PVT to PVI}$$

So PVT elevation of 35.07 + 8.25 = 43.32 ft; this checks the PVI elevation.

13.5   VERTICAL CURVE IN SAG

On highways and railways, in order that there may be no abrupt change in the vertical direction of the moving vehicles at the bottom of a curve, the principle of the parabola is again employed. Design of crest and sag vertical curves is a function of the grades of the intersecting tangents. Other considerations are stopping or passing sight distance, drainage of the roadway, and headlight beam distance. A sag vertical curve may be solved by either the stage method or the direct method.

The Stage Method

EXAMPLE 13.3   Given on plans:

PVI station = 21 + 25.00 at elevation 82.79 ft

Back tangent grade $(g_1) = -4\%$

Forward tangent grade $(g_2) = +3\%$

Length of curve $L = 450$ ft

Assume that grade stakes are set every 50 ft; each 50-ft distance is a stage. Make the necessary computations to design a sag vertical curve by the stage method.

Solution: Sketch the curve; see Fig. 13-4. Compute the number of stages n:

$$n = \frac{L}{\text{stage length}} = \frac{450}{50} = 9 \text{ stages}$$

1. Compute the station and elevation of the PVC and PVT.

$$
\begin{array}{lr\qquad lr}
\text{PVI station} = & 21 + 25.00 & \text{PVI station} = & 21 + 25.00 \\
-L/2 & - (2 + 25.00) & +L/2 & + \ 2 + 25.00 \\
\hline
\text{PVC station} = & 19 + 00.00 & \text{PVT station} = & 23 + 50.00 \\
\\
\text{PVI elevation} = & 82.79 & \text{PVI elevation} = & 82.79 \\
+4\% (L/2) & + \ 9.00 & +3\% (L/2) & + \ 6.75 \\
\hline
\text{PVC elevation} = & 91.79 & \text{PVT elevation} = & 89.54
\end{array}
$$

2. Compute H. $\quad H = (g_2 - g_1)\left(\dfrac{L}{2}\right) = [0.03 - (-0.04)]\left(\dfrac{450}{2}\right) = 15.75$

3. Compute C. $\quad C = \dfrac{H}{n^2}x^2 = \dfrac{15.75}{81}x^2 = 0.1944x^2$

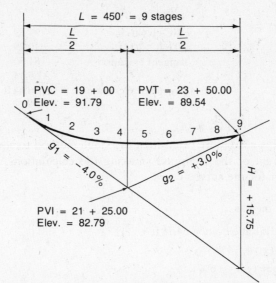

Fig. 13-4 Vertical curve in sag for calculation by the stage method.

4. Compute the values in C column by multiplying each x^2 by 0.1944. See Table 13-3.

5. Compute the elevations on the back tangent at the stage points, starting with the elevation at the PVC and subtracting 4%(50 ft) = 2.99 ft successively.

6. Compute the curve elevations by applying the tangent corrections C to the tangent elevations. Fill the computation table; see Table 13-3.

7. Find the low point on the curve.

$$a = g_1\left(\frac{L}{g_1 - g_2}\right) = (-0.04)\left(\frac{450}{-0.04 - 0.03}\right) = 257.14 \text{ ft}$$

$$C = \left(\frac{a}{L}\right)^2 H = \left(\frac{257.14}{450}\right)^2 (15.75) = 5.14 \text{ ft}$$

8. Find station and elevation of the low point.

PVC station =	$19 + 00.00$
$+a =$	$+ 2 + 57.14$
Low point station =	$21 + 57.14$

Table 13-3 Form for Computing Vertical Curve in Sag by the Stage Method
(Example 13.3)

Station	x	x^2	$C = 0.1944x^2$	Tangent Elevation	Curve Elevation
PVC $19 + 00$	0	0	0.0	91.79	91.79
$19 + 50$	1	1	0.19	89.78	89.97
$20 + 00$	2	4	0.78	87.78	88.56
$20 + 50$	3	9	1.75	85.78	87.53
$21 + 00$	4	16	3.11	83.78	86.89
$21 + 50$	5	25	4.86	81.78	86.64
$22 + 00$	6	36	7.00	79.78	86.78
$22 + 50$	7	49	9.53	77.78	87.31
$23 + 00$	8	64	12.44	75.78	88.22
PVT $23 + 50$	9	81	15.75	73.78	89.54

PVC elevation =	91.79
$-0.04(257.14)$	-10.29
Tangent elevation =	81.50
$+C$	$+\ 5.14$
Low point elevation =	86.64 ft

The Direct Method

The direct method is used more often than the stage method to compute a vertical curve in a sag because, with the advent of the computer, making the calculations is easy, and measuring of the chords is much simpler for the survey crew.

EXAMPLE 13.4   Given on plans:

PVI station $= 32 + 11.61$ at elevation 64.18 ft

Back tangent grade $(g_1) = -4\%$

Forward tangent grade $(g_2) = +6\%$

Length of curve $L = 600$ ft

Note: This is a theoretical problem. The grade percent would seldom be this high over so short a curve length; the resulting roller coaster effect would be poor highway design.

Find PVC and PVT stations and elevations; compute H, C, elevation on the back tangent at regular station points, curve elevations, and low point. Sketch the curve and set up the computation table.

Solution: Sketch the curve; see Fig. 13-5.

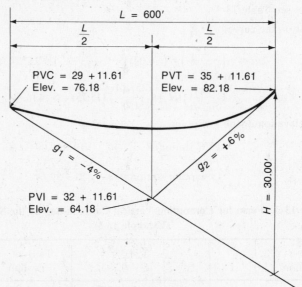

Fig. 13-5   Vertical sag curve.

1.   Compute the station and elevation of the PVC and PVT.

PVI station =	$32 + 11.61$		PVI station =	$32 + 11.61$
$-L/2$	$-(3 + 00.00)$		$+L/2$	$+\ 3 + 00.00$
PVC station =	$29 + 11.61$		PVT station =	$35 + 11.61$

PVI elevation =	64.18		PVI elevation =	64.18
$+4\%(L/2)$	$+12.00$		$+6\%(L/2)$	$+18.00$
PVC elevation =	76.18		PVT elevation =	82.18

2.   Compute H. $H = (g_2 - g_1)\left(\dfrac{L}{2}\right) = [0.06 - (-0.04)](300) = 30.00$ ft

3. Compute C.

$$C = \frac{H}{L^2} a^2 = \frac{30}{360\,000} a^2 = 0.00008333a^2$$

4. Compute the values in the C column by multiplying each value for a^2 by 0.00008333. See Table 13-4.

5. Compute the elevations on the back tangent at the regular station points. First subtract $4\%(a) = 0.04(38.4) = 1.54$ ft from PVC elevation: $76.18 - 1.54 = 76.64$. Then subtract $0.04(50) = 2.00$ ft successively until the last station, which is $(11.6\,\text{ft})(0.04) = 0.46$ ft. Put all figures into Table 13-4.

6. Compute the curve elevations by applying the tangent corrections C to the tangent elevations. Fill the computation table; see Table 13-4.

Table 13-4 Form for Computing Curve Elevations for Example 13.4

Station	a	a^2	$C = 0.00008333a^2$	Tangent Elevation	Curve Elevation
PVC 29 + 11.61	0	0	0.0	76.18	76.18
29 + 50	38.4	1 500	0.12	74.64	74.76
30 + 00	88.4	7 800	0.65	72.64	73.29
30 + 50	138.4	19 200	1.60	70.64	72.24
31 + 00	188.4	35 500	2.96	68.64	71.60
31 + 50	238.4	56 800	4.74	66.64	71.38
32 + 00	288.4	83 200	6.93	64.64	71.57
32 + 50	338.4	114 500	9.54	62.64	72.81
33 + 00	388.4	150 900	12.57	60.64	73.21
33 + 50	438.4	192 200	16.01	58.64	74.65
34 + 00	488.4	238 500	19.87	56.64	76.51
34 + 50	538.4	289 900	24.16	54.64	78.80
35 + 00	588.4	346 200	28.84	52.64	81.48
PVT 35 + 11.61	600	360 000	29.99	52.18	82.18

7. Calculate the lowest point.

$$a = g_1\left(\frac{L}{g_1 - g_2}\right) = (-0.04)\left(\frac{600}{-0.04 - 0.06}\right) = 240.00 \text{ ft}$$

$$C = \left(\frac{a}{L}\right)^2 H = \left(\frac{240}{600}\right)^2 (30) = 4.80$$

8. Find station and elevation of the low point.

PVC station =	29 + 11.61
$+ a$	+ 2 + 40.00
Low point station =	31 + 51.61
PVC elevation =	76.18
$-(-0.04)(240)$	− 9.60
Tangent elevation =	66.58
$+ C$	+ 4.80
Low point elevation =	71.38 ft

Solved Problems

13.1   The profile section of a road plan sheet provided the following information about a vertical curve:

> PVI station = 17 + 60.03 with an elevation of 703.21 ft
>
> Back tangent grade $g_1 = +4\%$
>
> Forward tangent grade $g_2 = -2\%$
>
> Length of the summit vertical curve $L = 600$ ft

Assume that the stages will be every 50 ft. Solve this curve by the stage method. Sketch the curve and make a computation form to be filled in after finding the following: Station and elevations of PVC and PVT, H, C, elevations on the back tangent at the stage points, and curve elevations. Find the high point of the curve.

Solution

$$\text{Number of stages } n = \frac{\text{length of curve } L}{\text{stage length}} = \frac{600}{50} = 12$$

Draw a sketch of the curve; see Fig. 13-6. Set up a computation form similar to the one shown in Example 13.1; see Table 13-5. Repeat all the steps shown in the example.

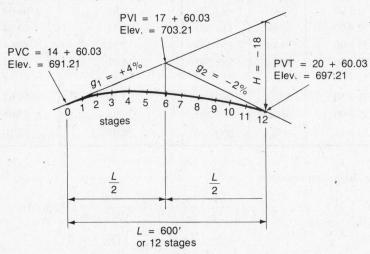

Fig. 13-6   Vertical curve on a summit.

1. Compute the station and elevation of the PVC and the PVT. Note the horizontal distances to each from the PVI are both equal to $L/2$.

PVI station =	17 + 60.03	PVI station =	17 + 60.03
$-L/2$	$-(3 + 00.00)$	$+L/2$	$+\ 3 + 00.00$
PVC station =	14 + 60.03	PVT station =	20 + 60.03
PVI elevation =	703.21	PVI elevation =	703.21
$-4\%(L/2)$	$-\ 12.00$	$-2\%(L/2)$	$-\ 6.00$
PVC elevation =	691.21	PVT elevation =	697.21

2. Compute H. $H = (g_2 - g_1)\left(\dfrac{L}{2}\right) = (-0.02 - 0.04)\left(\dfrac{600}{2}\right) = (-0.06)(300) = -18\,\text{ft}$

3. Compute C. $C = \dfrac{H}{n^2}x^2 = \dfrac{-18}{12^2}x^2 = -0.125x^2$

Put C into the computation form (Table 13-5).

Table 13-5 Computation Form for Prob. 13.1

Station	x	x^2	$C = -0.125x^2$	Tangent Elevation	Curve Elevation
PVC 14 + 60.03	0	0	0.0	691.21	691.21
15 + 10.03	1	1	−0.125	693.21	693.09
15 + 60.03	2	4	−0.500	695.21	694.71
16 + 10.03	3	9	−1.125	697.21	696.09
16 + 60.03	4	16	−2.000	699.21	697.21
17 + 10.03	5	25	−3.125	701.21	698.09
PVI 17 + 60.03	6	36	−4.500	703.21	698.71
18 + 10.03	7	49	−6.125	705.21	699.09
18 + 60.03	8	64	−8.000	707.21	699.21
19 + 10.03	9	81	−10.125	709.21	699.09
19 + 60.03	10	100	−12.500	711.21	698.71
20 + 10.03	11	121	−15.125	713.21	698.09
PVT 20 + 60.03	12	144	−18.000	715.21	697.21

4. Compute the values in the C column by multiplying each x^2 value by -0.125 ft. Note that C for PVT = H. Fill in all the C's in the computation form.

5. Compute the elevations on the back tangent at the stage points, starting with the elevation of the PVC and adding $4\% \times 50$ ft = 2.00 ft successively.

6. Compute the curve elevations by applying the tangent corrections C to the tangent elevations. Check the PVI by using the forward tangent prolonged back to the PVI.

$$
\begin{array}{lr}
\text{PVT elevation} = & 697.21 \\
+0.02(300) & +\;\;\;6.00 \\
\hline
\text{PVI elevation} = & 703.21
\end{array}
$$

This checks with the given PVI elevation.

7. Compute the high point of the curve (a = distance in feet from PVC).

$$a = g_1 \left(\frac{L}{g_1 - g_2} \right) = 0.04 \left(\frac{600}{0.04 + 0.02} \right) = 400 \text{ ft}$$

Find the tangent correction C and apply it to the tangent elevation that occurs at 400 ft from the PVC.

$$C = \left(\frac{a}{L} \right)^2 H = \left(\frac{400}{600} \right)^2 (-18) = -8.00 \text{ ft}$$

$$
\begin{array}{lr}
\text{PVC station} = & 14 + 60.03 \\
+a & +\;\;4 + 00.00 \\
\hline
\text{High point station} = & 18 + 60.03
\end{array}
$$

$$
\begin{array}{lr}
\text{PVC elevation} = & 691.21 \\
+4\%(400) & +\;\;16.00 \\
\hline
\text{Tangent elevation} = & 707.21 \\
-C & -\;\;\;8.00 \\
\hline
\text{High point elevation} = & 699.21 \text{ ft}
\end{array}
$$

13.2 On the basis of information listed below, find the station and elevation of the PVC and PVT. Sketch the curve and set up and fill out the computation form after finding the following: H, C, elevations on the back tangent at the stage points, and curve elevations. Find the high point. Assume that the stages will be every 50 ft. Given:

Summit curve with length $L = 600$ ft

PVI station $= 36 + 00$ with an elevation of 55.29 ft

Back tangent grade $g_1 = +1.5\%$

Forward tangent grade $g_2 = -2.75\%$

Solution

1. Calculate PVC and PVT.

PVI station $=$	$36 + 00$	PVI station $=$	$36 + 00$
$-L/2$	$-(3 + 00)$	$+L/2$	$+3 + 00$
PVC station $=$	$33 + 00$	PVT station $=$	$39 + 00$
PVI elevation $=$	55.29	PVI elevation $=$	55.29
$-0.015(300)$	-4.50	$-0.0275(300)$	-8.25
PVC elevation $=$	50.79	PVT elevation $=$	47.04

Draw a sketch (see Fig. 13-7) and set up a computation table (see Table 13-6).

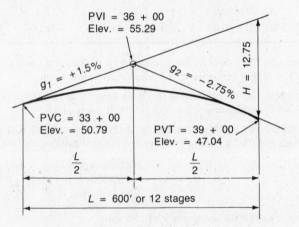

Fig. 13-7   Vertical curve on a summit.

Table 13-6   Computation Form for Prob. 13.2

Station	x	x^2	$C = -0.0885x^2$	Tangent Elevation	Curve Elevation
PVC $33 + 00$	0	0	0.0	50.79	50.79
$33 + 50$	1	1	-0.09	51.54	51.45
$34 + 00$	2	4	-0.35	52.29	51.94
$34 + 50$	3	9	-0.80	53.04	52.24
$35 + 00$	4	16	-1.42	53.79	52.37
$35 + 50$	5	25	-2.21	54.54	52.33
$36 + 00$	6	36	-3.19	55.29	52.10
$36 + 50$	7	49	-4.34	56.04	51.70
$37 + 00$	8	64	-5.66	56.79	51.13
$37 + 50$	9	81	-7.17	57.54	50.37
$38 + 00$	10	100	-8.85	58.29	49.44
$38 + 50$	11	121	-10.71	59.04	48.33
PVT $39 + 00$	12	144	-12.74	59.79	47.05

2. Compute H.

$$H = (g_2 - g_1)\left(\frac{L}{2}\right) = (-0.0275 - 0.015)\left(\frac{600}{2}\right) = -12.75 \text{ ft}$$

3. Compute C. (*Note*: Number of stages $n = 600/50 = 12$.)

$$C = \frac{H}{n^2}x^2 = \left(\frac{-12.75}{144}\right)x^2 = -0.0885x^2$$

4. Compute C column values and enter into Table 13-6. Multiply x^2 by -0.0885.
5. Compute tangent elevation. Start at PVC elevation and add $1.5\% \times 50$ ft $= 0.75$ ft successively.
6. Compute curve elevations by applying tangent corrections to tangent elevations.
7. Check the PVT elevation. A method of finding out if the offsets from the curve are figured correctly is to use the forward tangent prolonged and the H dimension.

PVC elevation =	50.79
+1.5(600)	+ 9.00
Elevation at 600 ft tangent prolonged =	50.79
$-H$	− 12.75
	47.04

This checks with the PVT elevation of 47.04. These features can be easily seen from the sketch (Fig. 13-7).

8. Compute the high point.

$$a = g_1\left(\frac{L}{g_1 - g_2}\right) = 0.015\left(\frac{600}{0.015 + 0.0275}\right) = 211.76 \text{ ft}$$

$$C = \left(\frac{a}{L}\right)^2 H = \left(\frac{211.76}{600}\right)^2(-12.75) = -1.59 \text{ ft}$$

PVC station =	33 + 00
$+a$	+ 2 + 11.76
High point station =	35 + 11.76

PVC elevation =	50.79
+1.5%(211.76)	+ 3.18
Tangent elevation	53.97
$-C$	− 1.59
High point elevation =	52.38 ft

13.3 This problem is adapted from an actual construction project in which a road was built by a state transportation department. Figure 13-8 shows a small section of a plan and profile sheet from the job. Several things are apparent. First, the PI ($=$ PVI) is located at station 349. The PI elevation is given as 202.87. The back tangent grade shows as $+3\%$. The length of the vertical curve shows at the bottom of the profile view as 1800 ft. Assume each stage is 100 ft; you have all the information required to work this problem. Find all the elements required to design a vertical curve of 1800 ft with the tangent grades and PVI given in Fig. 13-8. Determine the necessary quantities and log into a computation form.

Solution

$$\text{Number of stages } n = \frac{\text{length of curve}}{\text{stage length}} = \frac{1800 \text{ ft}}{100 \text{ ft}} = 18$$

1. Compute the station and elevation of the PVC and PVT.

PVI station =	349 + 00	PVI station =	349 + 00
$-L/2$	− (9 + 00)	$+L/2$	+ 9 + 00
PVC station =	340 + 00	PVT station =	358 + 00

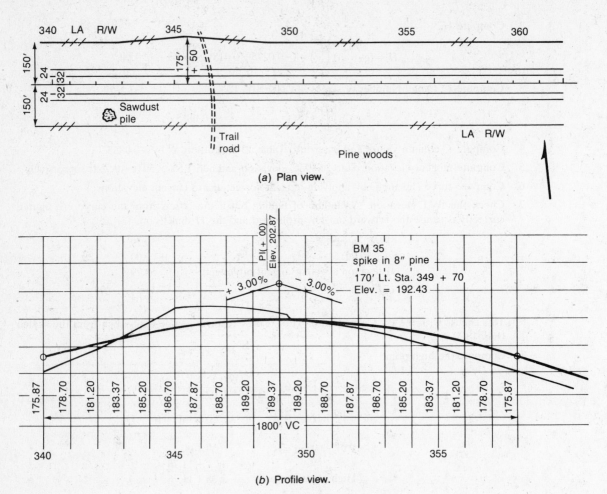

(a) Plan view.

(b) Profile view.

Fig. 13-8   Views of a summit curve.

PVI elevation =	202.87	PVI elevation =	202.87
− 3% (L/2)	− 27.00	− 3% (L/2)	− 27.00
PVC elevation =	175.87	PVT elevation =	175.87

Note that the PVT elevation of 175.87 does check the profile view in Fig. 13-8. The PVC and PVT elevations are both 175.87 ft and the stationing is also the same. The actual PVC and PVT are not generally called out on the plan and profile views in a set of road plans. Draw a sketch (see Fig. 13-9) and set up a computation table (see Table 13-7) even though we have a profile view of the curve. The sketch should be shown at the beginning of the problem with the computation of PVC and PVT stations and elevations. Notice in this problem that the front and back slopes are equal and the PVC and PVT are at the same elevation.

2. Compute H. $H = (g_2 - g_1)\left(\dfrac{L}{2}\right) = (-0.03 - 0.03)\left(\dfrac{1800}{2}\right) = -54$ ft

3. Compute C. $C = \dfrac{H}{n^2} x^2 = \dfrac{-54}{18^2} x^2 = -0.1667 x^2$

4. Compute the values in the C column of Table 13-7 by multiplying each x^2 value by −0.1667.

5. Compute the elevations on the back tangent at the stage points starting with the elevation at the PVC and adding (3%)(100 ft) = 3 ft successively. Add to computation form.

6. Compute the curve elevations by applying the tangent correction C to the tangent elevations. Add to computation form.

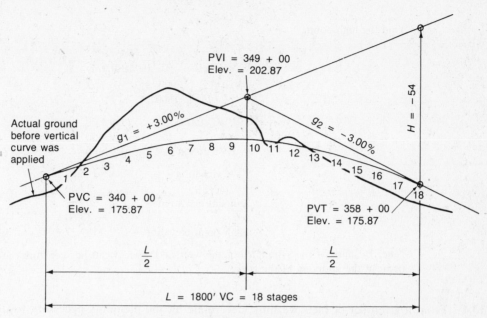

Fig. 13-9 Summit vertical curve with natural ground superimposed.

Table 13-7   Computation Form for Prob. 13.3

Station	x	x^2	$C = -0.1667x^2$	Tangent Elevation	Curve Elevation
PVC 340 + 00	0	0	0.0	175.87	175.87
341 + 00	1	1	−0.17	178.87	178.70
342 + 00	2	4	−0.67	181.87	181.20
343 + 00	3	9	−1.50	184.87	183.37
344 + 00	4	16	−2.67	187.87	185.20
345 + 00	5	25	−4.17	190.87	186.70
346 + 00	6	36	−6.00	193.87	187.87
347 + 00	7	49	−8.17	196.87	188.70
348 + 00	8	64	−10.67	199.87	189.20
349 + 00	9	81	−13.50	202.87	189.37
350 + 00	10	100	−16.67	205.87	189.20
351 + 00	11	121	−20.17	208.87	188.70
352 + 00	12	144	−24.00	211.87	187.87
353 + 00	13	169	−28.17	214.87	186.70
354 + 00	14	196	−32.67	217.87	185.20
355 + 00	15	225	−37.51	220.87	183.37
356 + 00	16	256	−42.68	223.87	181.20
357 + 00	17	289	−48.18	226.87	178.70
PVT 358 + 00	18	324	−54.01	229.87	175.87

7. Check the PVT elevation. From Fig. 13-9 it is apparent that $L/2$, or 900 ft, times the −3% grade will show the PVT elevation.

$$
\begin{array}{lr}
\text{PVI elevation} = & 202.87 \\
+(-0.03)(900) & -\ 27.00 \\
\hline
\text{PVT elevation} = & 175.87
\end{array}
$$

This checks with the PVT point elevation of 175.87.

8. Find the high point.

$$a = g_1 \left(\frac{L}{g_1 - g_2}\right) = 0.03 \left(\frac{1800}{0.03 + 0.03}\right) = 900 \text{ ft}$$

$$C = \left(\frac{a}{L}\right)^2 H = \left(\frac{900}{1800}\right)^2 (-54) = -13.5 \text{ ft}$$

PVC station =	$340 + 00$
$+a$	$+\quad 9 + 00$
PVI station =	$349 + 00$

PVC elevation =	175.87
$+0.03(900)$	$+\quad 27.00$
Tangent elevation =	202.87
$-C$	$-\quad 13.50$
High point elevation =	189.37

Note: Because of the symmetry of this vertical curve, it can be seen that the high point must occur at the midpoint of the curve.

13.4 From a given set of facts we must determine all the elements necessary to design a vertical curve by the direct method. Draw a sketch of the curve and set up the computation form. Given:

PVI = 472 + 25.00 at elevation 610.20 ft

Back tangent grade $g_1 = +2.82\%$

Forward tangent grade $g_2 = -3.50\%$

Length of curve $L = 800$ ft

Solution

Draw a sketch of the curve; see Fig. 13-10.

1. Compute the station and elevation of the PVC and PVT.

PVI station =	$472 + 25.00$	PVI station =	$472 + 25.00$
$-L/2$	$-\quad (4 + 00.00)$	$+ L/2$	$4 + 00.00$
PVC station =	$468 + 25.00$	PVT station =	$476 + 25.00$

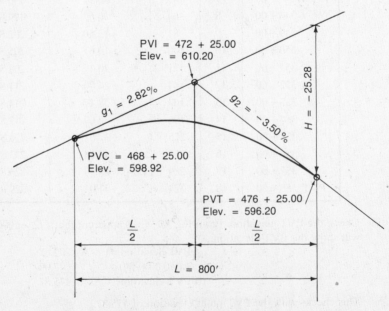

Fig. 13-10 Summit curve solved by the direct method.

	PVI elevation =	610.20	PVI elevation =	610.20
	$-2.8\%(L/2)$	$-$ 11.28	$-3.5\%(L/2)$	$-$ 14.00
	PVC elevation =	598.92	PVT elevation =	596.20

2. Compute H.

$$H = (g_2 - g_1)\left(\frac{L}{2}\right) = (-0.035 - 0.0282)\left(\frac{800}{2}\right) = -25.28 \text{ ft}$$

3. Compute C.

$$C = \frac{H}{L^2}\, a^2 = \frac{-25.28}{640\,000}\, a^2 = -0.00003950 a^2$$

4. Compute the elevations on the back tangent at the regular station points. First elevation at station $468 + 50$ will be $2.8\% \times 25$ ft $= 0.705$ ft. The remaining elevations, up to station $476 + 00$, go up by 50-ft intervals, or

$$0.0282(50) = 1.41 \text{ ft per station}$$

Put the computations into Table 13-8.

Table 13-8   Form for Calculating Elevations on a Summit Curve by the Direct Method (Prob. 13.4)

Station	a	a^2	$C = -0.00003950a^2$	Tangent Elevation	Curve Elevation
PVC $468 + 25.00$	0.0	0	0.0	598.92	598.92
$468 + 50.00$	25.0	625	-0.02	599.63	599.61
$469 + 00.00$	75.0	5 625	-0.22	601.04	600.82
$469 + 50.00$	125.0	15 625	-0.62	602.45	601.83
$470 + 00.00$	175.0	30 625	-1.21	603.86	602.65
$470 + 50.00$	225.0	50 625	-2.00	605.27	603.27
$471 + 00.00$	275.0	75 625	-2.99	606.68	603.69
$471 + 50.00$	325.0	105 625	-4.17	608.05	603.88
$472 + 00.00$	375.0	140 625	-5.55	609.50	603.95
$472 + 50.00$	425.0	180 625	-7.13	610.91	603.78
$473 + 00.00$	475.0	225 625	-8.91	612.32	603.41
$473 + 50.00$	525.0	275 625	-10.89	613.73	602.84
$474 + 00.00$	575.0	330 625	-13.06	615.14	602.08
$474 + 50.00$	625.0	390 625	-15.43	616.55	601.12
$475 + 00.00$	675.0	455 625	-18.00	617.96	599.96
$475 + 50.00$	725.0	525 625	-20.76	619.37	598.61
$476 + 00.00$	775.0	600 625	-23.72	620.78	597.06
PVT $476 + 25.00$	800.0	640 000	-25.28	621.49	596.20

5. Compute the curve elevations by applying the tangent correction C to the tangent elevations. To check, compute the PVI elevation using the forward tangent prolonged.

	PVT elevation =	596.20
	$+(L/2)(0.035)$	$+$ 14.00
		610.20

This checks the given PVI elevation of 610.20.

6. Find the high point.

$$a = g_1\left(\frac{L}{g_1 - g_2}\right) = 0.0282\left[\frac{800}{0.0282 - (-0.035)}\right] = 356.96 \text{ ft}$$

Find the tangent correction C, and apply it to the tangent elevation that occurs at 356.96 ft from the PVC.

$$C = \left(\frac{a}{L}\right)^2 H = \left(\frac{356.96}{800}\right)^2 (-25.28) = -5.03 \text{ ft}$$

PVC station =	468 + 25.00
+ a	+ 3 + 56.96
High point station =	471 + 81.96

PVC elevation =	598.92
+2.82% (356.96)	+ 10.07
Tangent elevation =	608.99
− C	− 5.03
High point elevation =	603.96

13.5 Solve Prob. 13.1 by the direct method. Data given on the profile section of the road plan sheet:

PVI station = 17 + 60.03 at elevation of 703.21 ft

Back tangent grade $g_1 = +4\%$

Forward tangent grade $g_2 = -2\%$

Length of curve $L = 600$ ft

Solution

Sketch the vertical curve; see Fig. 13-11.

1. Compute the station and elevation of the PVC and PVT.

PVI station =	17 + 60.03	PVI station =	17 + 60.03
− L/2	− (3 + 00.00)	+ L/2	+ 3 + 00.00
PVC station =	14 + 60.03	PVT station =	20 + 60.03

PVI elevation =	703.21	PVI elevation =	703.21
−4% (L/2)	− 12.00	−4% (L/2)	− 6.00
PVC elevation =	691.21	PVT elevation =	697.21

2. Compute H. $H = (g_2 - g_1)\left(\frac{L}{2}\right) = (-0.02 - 0.04)\left(\frac{600}{2}\right) = (-0.06)(300) = -18 \text{ ft}$

3. Compute C. $C = \frac{H}{L^2} a^2 = \frac{-18}{360\,000} a^2 = -0.00005000 a^2$

Note the number of significant figures to which the answer is carried.

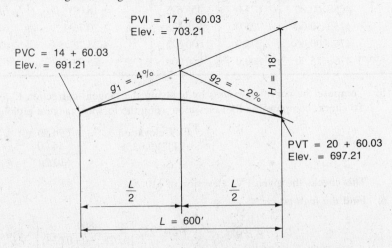

Fig. 13-11 Summit curve solved by the direct method.

4. Compute C values by multiplying each a^2 value by -0.00005000.

5. Compute the elevations on the back tangent at the regular station points ($0.04a$ for each station).

6. Compute the curve elevations by applying the tangent correction C to the tangent elevation. Complete the computation table; see Table 13-9.

Table 13-9   Form for Calculating Curve Elevations by the Direct Method (Prob. 13.5)

Station	a	a^2	$C = -0.00005a^2$	Tangent Elevation	Curve Elevation
PVC 14 + 60.03	0.0	0	0.0	691.21	691.21
15 + 00.00	40.0	1 600	−0.08	692.81	692.73
15 + 50.00	90.0	8 100	−0.41	694.81	694.40
16 + 00.00	140.0	19 600	−0.98	696.81	695.83
16 + 50.00	190.0	36 100	−1.81	698.81	697.00
17 + 00.00	240.0	57 600	−2.88	700.81	697.93
17 + 50.00	290.0	84 100	−4.21	702.81	698.60
18 + 00.00	340.0	115 600	−5.78	704.81	699.03
18 + 50.00	390.0	152 100	−7.61	706.81	699.20
19 + 00.00	440.0	193 600	−9.68	708.81	699.13
19 + 50.00	490.0	240 100	−12.01	710.81	698.80
20 + 00.00	540.0	291 600	−14.58	712.81	698.23
20 + 50.00	590.0	348 100	−17.41	714.81	697.40
PVT 20 + 60.03	600.0	360 000	−18.00	715.21	697.21

7. Find the high point.

$$a = g_1 \left(\frac{L}{g_1 - g_2}\right) = 0.04 \left[\frac{600}{0.04 - (-0.02)}\right] = 400 \text{ ft}$$

$$C = \left(\frac{a}{L}\right)^2 H = \left(\frac{400}{600}\right)^2 (-18) = -8.00$$

PVC station =	14 + 60.03
+ a	+ 4 + 00.00
High point station =	18 + 60.03
PVC elevation =	691.21
+4%(400)	+ 16.00
Tangent elevation =	707.21
− C	− 8.00
High point elevation =	699.21 ft

13.6   Solve the following vertical curve by the direct method. Given:

PVC = 34 + 25.00 at elevation of 30.15

PVI = 36 + 25.00 at elevation of 31.60

PVT = 38 + 25.00 at elevation of 29.75

Back tangent grade $g_1 = +0.727\%$

Forward tangent grade $g_2 = -0.923\%$

Curve length $L = 400$ ft

Find H, C, elevations on the back tangent at station points, curve elevations. Draw a sketch of the curve and make a complete computation table.

Solution

Since the PVC and PVT are given along with the PVI, the first step of finding the PVC and PVI is eliminated. The sketch of the curve is easily drawn (only the H value is missing); see Fig. 13-12. The first step is thus to compute H.

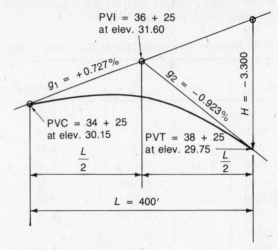

Fig. 13-12 Summit curve solved by the direct method.

1. Compute H.

$$H = (g_2 - g_1)\left(\frac{L}{2}\right) = (-0.00923 - 0.00727)\left(\frac{400}{2}\right) = -3.300 \text{ ft}$$

2. Compute C to four significant figures.

$$C = \frac{H}{L^2} a^2 = \frac{-3.300}{160\,000} a^2 = -0.00002063a^2$$

3. Compute values for C column by multiplying each a^2 value by -0.00002063.

4. Compute the elevations on the back tangent at the regular station points. The first step to station $34 + 50$ is only 25 ft, so $25(0.00727) = 0.18$ ft. The following steps are each 50 ft, so add $50(0.00727) = 0.36$ ft successively.

5. Compute curve elevations by applying the tangent corrections C to the tangent elevations. Fill in the computation table; see Table 13-10.

Table 13-10   Calculation Form for Prob. 13.6

Station	a	a^2	$C = -0.00002063a^2$	Tangent Elevation	Curve Elevation
PVC $34 + 25.00$	0.0	0	0.0	30.15	30.15
$34 + 50$	25.0	625	-0.01	30.33	30.32
$35 + 00$	75.0	5 625	-0.12	30.69	30.57
$35 + 50$	125.0	15 625	-0.32	31.05	30.73
$36 + 00$	175.0	30 625	-0.63	31.42	30.79
$36 + 50$	225.0	50 625	-1.04	31.78	30.74
$37 + 00$	275.0	75 625	-1.56	32.14	30.58
$37 + 50$	325.0	105 625	-2.18	32.51	30.33
$38 + 00$	375.0	140 625	-2.90	32.87	29.97
PVT $38 + 25$	400.0	160 000	-3.30	33.05	29.75

13.7　Given on plan view of project 54001-3402 (see Fig. 13-13):

> PVI station $= 332 + 00$ at elevation 151.87
>
> Back tangent grade $g_1 = +1.4\%$
>
> Forward tangent grade $g_2 = +3.00\%$
>
> Length of sag curve $L = 800$ ft

Use 100-ft stages ($800/100 = 8$ stages). Without drawing a sketch, work all elements of the profile view of Fig. 13-13 on a sag curve with PVI station of $332 + 00$. Compare your results with those shown in the figure. Set up and fill out a computation form.

Solution

1.　Compute the station and elevation of the PVC and PVT.

PVI station =	$332 + 00$		PVI station =	$332 + 00$
$+L/2$	$+\quad 4 + 00$		$-L/2$	$-\quad (4 + 00)$
PVT station =	$336 + 00$		PVC station =	$328 + 00$
PVI elevation =	157.87		PVI elevation =	151.87
$+0.03(400)$	$+\quad 12.00$		$-0.0142(400)$	$-\quad 5.68$
PVT elevation =	163.87		PVC elevation =	146.19

2.　Compute H.

$$H = (g_2 - g_1)\left(\frac{L}{2}\right) = (0.03 - 0.142)\left(\frac{L}{2}\right) = 0.0158(400) = 6.32 \text{ ft}$$

3.　Compute C.

$$C = \frac{H}{n^2} x^2 = \frac{6.32}{64} x^2 = 0.09875 x^2$$

4.　Compute the values in the C column by multiplying each x^2 value by 0.09875. (See Table 13-11.)

5.　Compute the elevations on the back tangent at the stage points, starting with the elevation at the PVC and adding $(1.42\%)(100 \text{ ft}) = 1.42$ ft successively.

6.　Compute the curve elevations by applying the tangent corrections C to the tangent elevations. Note that your calculations check with the curve elevations shown on project 54001-3402 (Fig. 13-13) between stations $328 + 00$ and $336 + 00$.

Since there was a very clear profile view of this project, we eliminated the sketch. Unless the profile view is very clear the sketch should be drawn.

13.8　A sag curve is to be designed on project 54001-3402 (see Prob. 13.7, Fig. 13-13). Again work without a sketch. Try to visualize the curve from the profile view. The only thing missing is the extended back tangent and H. From Fig. 13-13 pick out the following:

> PVI station $= 316 + 50$ at elevation 129.86
>
> Back tangent grade $g_1 = -1.08\%$
>
> Forward tangent grade $g_2 = +1.42\%$
>
> Length of sag curve $L = 800$ ft

Use 50-ft increments for stages. Use all the figures available in the profile view to work out the components necessary to design this sag curve.

Solution

$$\text{Number of stages } n = \frac{\text{length of curve}}{\text{length of stage}} = \frac{800 \text{ ft}}{50 \text{ ft}} = 16$$

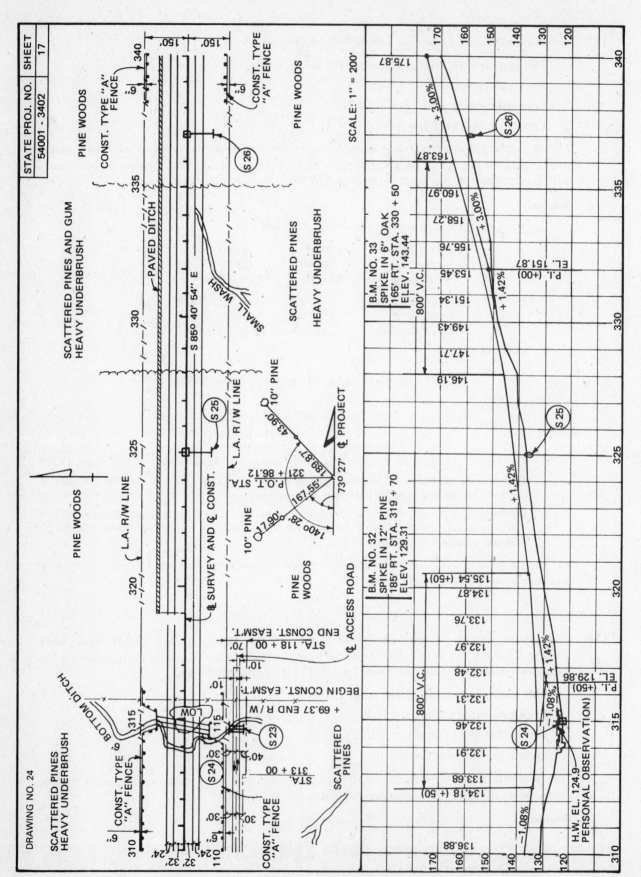

Fig. 13-13 Plan and profile sheet of project 54001-3402.

Table 13-11 Computation Form for Vertical Curve on Project 54001-3402 (Prob. 13.7)

Station	x	x^2	$C = 0.09875x^2$	Tangent Elevation	Curve Elevation
PVC 328 + 00	0	0	0.0	146.19	146.19
329 + 00	1	1	0.09875	147.61	147.71
330 + 00	2	4	0.395	149.03	149.43
331 + 00	3	9	0.88875	150.45	151.34
332 + 00	4	16	1.5800	151.87	153.45
333 + 00	5	25	2.4687	153.29	155.76
334 + 00	6	36	3.555	154.71	158.27
335 + 00	7	49	4.8387	156.13	160.97
PVT 336 + 00	8	64	6.32	157.55	163.87

1. Compute the station and elevation of the PVC and PVT.

PVI station =	316 + 50	PVI station =	316 + 50
− L/2	− (4 + 00)	+ L/2	+ 4 + 00
PVC station =	312 + 50	PVT station =	320 + 50
PVI elevation =	129.86	PVI elevation =	129.86
+0.0108(L/2)	+ 4.32	+0.0142(L/2)	+ 5.68
PVC elevation =	134.18	PVT elevation =	135.54

Note: The state job sheet shows an error in the PVC elevation. It reads 134.08. It should read 134.18. No one is perfect—not even professional engineers.

2. Compute H.

$$H = (g_2 - g_1)\left(\frac{L}{2}\right) = (0.0142 + 0.0108)\left(\frac{800}{2}\right) = 0.025(400) = 10.00 \text{ ft}$$

3. Compute C.

$$C = \frac{H}{n^2}x^2 = \frac{10}{256}x^2 = 90.03906x^2$$

4. Compute values of C by multiplying each x^2 value by 0.03906.

5. Compute the elevation on the back tangent at the stage points, starting with the elevation at the PVC and subtracting 1.08% × 50 ft = 0.54 ft successively.

6. Compute the curve elevations by applying the tangent correction C to the tangent elevations. Fill out the computation form; see Table 13-12.

7. Compute the lowest point.

$$a = g_1\left(\frac{L}{g_1 - g_2}\right) = -0.0108\left(\frac{800}{-0.0108 - 0.0142}\right) = 345.60 \text{ ft}$$

$$C = \left(\frac{a}{L}\right)^2 H = \left(\frac{345.60}{800}\right)^2(10) = 1.86 \text{ ft}$$

PVC station =	312 + 50
+a	+ 3 + 45
Low point station =	315 + 95
PVC elevation =	134.18
+(−0.0108)(345.6)	− 3.73
	130.41
+C	+ 1.86
Low point elevation =	132.27

Table 13-12   Computation Form for Elevations on Project 54001-3402 (Prob. 13.8)

Station	x	x^2	$C = 0.03906x^2$	Tangent Elevation	Curve Elevation
PVC 312 + 50	0	0	0.0	134.18	134.18
313 + 00	1	1	0.03906	133.64	133.68
313 + 50	2	4	0.15624	133.10	133.26
314 + 00	3	9	0.3515	132.56	132.91
314 + 50	4	16	0.6250	132.02	132.65
315 + 00	5	25	0.9765	131.48	132.46
315 + 50	6	36	1.4062	130.94	132.35
316 + 00	7	49	1.9139	130.40	132.31
316 + 50	8	64	2.4999	129.86	132.36
317 + 00	9	81	3.1639	129.32	132.48
317 + 50	10	100	3.906	128.78	132.69
318 + 00	11	121	4.7262	128.24	132.97
318 + 50	12	144	5.6246	127.70	133.32
319 + 00	13	169	6.6011	127.16	133.96
319 + 50	14	196	7.6558	126.62	134.28
320 + 00	15	225	8.7886	126.08	134.87
PVT 320 + 50	16	256	9.9993	125.54	135.54

13.9   Data from the plans:

> PVI station = 22 + 00 at elevation 85.00
>
> Back tangent grade $g_1 = -3.0\%$
>
> Forward tangent grade $g_2 = +3.2\%$
>
> Length of curve $L = 500$ ft

Use 50-ft stages. Use the provided data to make the computations to design a sag curve by the stage method. Draw a sketch and set up and fill out a computation table.

Solution

Draw a sketch of the curve (see Fig. 13-14).

$$\text{Number of stages } n = \frac{\text{length of curve}}{\text{length of stage}} = \frac{500}{50} = 10$$

1. Compute the station and elevation of the PVC and PVT.

PVI station =	22 + 00	PVI station =	22 + 00
− L/ 2	− (2 + 50)	+ L/ 2	+ 2 + 50
PVC station =	19 + 50	PVT station =	24 + 50
PVI elevation =	85.00	PVI elevation =	85.00
+0.03(L/2)	+ 7.50	+0.032(L/2)	+ 8.00
PVC elevation =	92.50	PVT elevation =	93.00

2. Compute H. $H = (g_2 - g_1)\left(\dfrac{L}{2}\right) = [0.032 - (-0.030)]\left(\dfrac{500}{2}\right) = 15.50$ ft

3. Compute C. $C = \dfrac{H}{n^2}x^2 = \dfrac{15.50}{100}x^2 = 0.1550x^2$

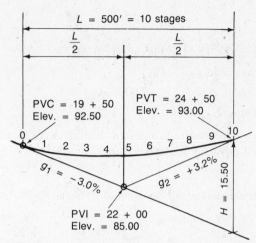

Fig. 13-14 Vertical sag curve solved by the stage method.

4. Compute the values in C column by multiplying each x^2 value by 0.1550.

5. Compute the elevations on the back tangent at the stage points, starting with the elevations at the PVC and subtracting 3% $\times$ 50 ft = 1.5 ft successively.

6. Compute the curve elevations by applying the tangent corrections C to the tangent elevations. See Table 13-13.

Table 13-13   Form for Computing Curve Elevations by the Stage Method (Prob. 13.9)

Station	x	x^2	$C = 0.1550x^2$	Tangent Elevation	Curve Elevation
PVC 19 + 50	0	0	0.0	92.50	92.50
20 + 00	1	1	0.1550	91.00	91.16
20 + 50	2	4	0.620	89.50	90.12
21 + 00	3	9	1.395	88.00	89.40
21 + 50	4	16	2.480	86.50	88.98
22 + 00	5	25	3.875	85.00	88.88
22 + 50	6	36	5.580	83.50	89.08
23 + 00	7	49	7.600	82.00	89.60
23 + 50	8	64	9.920	80.50	90.42
24 + 00	9	81	12.56	79.00	91.56
PVT 24 + 50	10	100	15.50	77.50	93.00

7. Calculate the low point on the curve.

$$a = g_1\left(\frac{L}{g_1 - g_2}\right) = -0.03\left(\frac{500}{-0.03 - 0.032}\right) = 241.94 \text{ ft}$$

$$C = \left(\frac{a}{L}\right)^2 H = \left(\frac{241.94}{500}\right)(15.50) = 3.63 \text{ ft}$$

PVC station =	19 + 50
+a	+ 2 + 92
Low point station =	21 + 92

PVC elevation =	92.50
+(−0.03)(242)	− 7.26
	85.24
+C	+ 3.63
Low point elevation =	88.87

13.10 Given:

    PVI station = 32 + 11.61 at elevation 60.00

    Back tangent grade $g_1 = -4\%$

    Forward tangent grade $g_2 = +7.00\%$

    Length of curve $L = 600$ ft

Compute all elements required to design this sag curve by the direct method. Draw a sketch of the curve and fill out a computation form.

Solution

1. Compute the station and elevation of PVC and PVT.

PVI station =	32 + 11.61	PVI station =	32 + 11.61
$-L/2$	$-$ (3 + 00.00)	$+L/2 =$	+ 3 + 00.00
PVC station =	29 + 11.61	PVT station =	35 + 11.61

PVI elevation =	60.00	PVI elevation =	60.00
$+4\%(L/2)$	+ 12.00	$+7\%(L/2)$	+ 21.00
PVC elevation =	72.00	PVT elevation =	81.00

Sketch the curve; see Fig. 13-15.

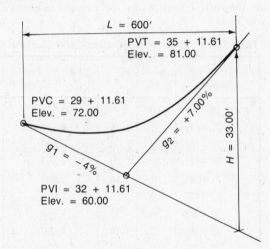

Fig. 13-15 Sag curve solved by the direct method.

2. Compute H.

$$H = (g_2 - g_1)\left(\frac{L}{2}\right) = [90.07 - (-0.04)](300) = 33.00 \text{ ft}$$

3. Compute C.

$$C = \frac{H}{L^2} a^2 = \frac{33}{360\,000} a^2 = 0.00009167 a^2$$

4. Compute values for the C column by multiplying each a^2 value by 0.00009167. See Table 13-14.

5. Compute the elevations on the back tangent at the regular stations. The first step to station 29 + 50 is only 38.39 ft, so add $38.39(-0.04) = -1.54$ ft. The following steps are each 50 ft, so add $50(-0.04) = -2$ ft successively.

6. Compute curve elevations by applying the tangent corrections C to the tangent elevations. Fill out the computation table (see Table 13-14).

7. Calculate the low point.

Table 13-14   Form for Computing Elevations on a Vertical Sag Curve by the Direct Method (Prob. 13.10)

Station	a	a^2	$C = 0.00009167a^2$	Tangent Elevation	Curve Elevation
PVC 29 + 11.61	0	0	0.0	72.00	72.00
29 + 50	38.4	1 500	0.14	70.46	70.60
30 + 00	88.4	7 800	0.72	68.46	69.18
30 + 50	138.4	19 200	1.76	66.46	68.22
31 + 00	188.4	35 500	3.25	64.46	67.71
31 + 50	238.4	56 800	5.20	62.46	67.66
32 + 00	288.4	83 200	7.63	60.46	68.09
32 + 50	338.4	114 500	10.50	58.46	68.96
33 + 00	388.4	150 900	13.83	56.46	70.29
33 + 50	438.4	192 200	17.62	54.46	72.08
34 + 00	488.4	238 500	21.86	52.46	74.32
34 + 50	538.4	289 900	26.58	50.46	77.04
35 + 00	588.4	346 200	31.74	48.46	80.20
PVT 35 + 11.61	600	360 000	33.00	48.00	81.00

$$a = g_1\left(\frac{L}{g_1 - g_2}\right) = -0.04\left(\frac{600}{-0.04 - 0.07}\right) = 218.18 \, \text{ft}$$

$$C = \left(\frac{a}{L}\right)^2 H = \left(\frac{218.18}{600}\right)^2 (33) = 4.36 \, \text{ft}$$

PVC station =	29 + 11.61
$+a$	+ 2 + 18.18
Low point station =	31 + 29.79

PVC elevation =	72.00
$+(-0.04)(218.18)$	− 8.73
	63.27
$+C$	+ 4.36
Low point elevation =	67.63

Supplementary Problems

Complete answers to these problems will not be given. Instead several stage elevations for each problem will be given so that computational procedures may be checked.

STAGE METHOD OF SUMMIT CURVE COMPUTATION

13.11   Using the following information, find the high point of the curve and all curve elevations. Assume that each stage = 100 ft.

PVI = 184 + 00 at elevation 127.24 ft   Forward tangent grade $g_2 = -2.40\%$

Back tangent grade $g_1 = +0.86\%$ Length of curve $L = 1000$ ft

Ans. $H = -16.3$; $C = -0.1630x^2$; elevation at station 186 + 00 = 120.97; high point at station 181 + 63.80; high point elevation = 124.08 ft

13.12 On the basis of the following information find the curve elevation at each stage.

Each stage = 50 ft   Forward tangent grade $g_2 = -2.20\%$

PVI = 175 + 00 at elevation 120.02 ft   Length of curve $L = 600$ ft

Back tangent grade $g_1 = +1.20\%$

Ans.

Station	Curve Elevation, ft
173 + 00	117.34
174 + 50	117.65
176 + 00	116.69
177 + 00	115.34
178 + 00	113.42

13.13 A summit curve is to be designed on a country road. Find the curve elevations and the high point from the following information:

Each stage = 100 ft   Forward tangent grade $g_2 = -1.7\%$

PVI = 200 + 50 at elevation 88.27 ft   Length of curve $L = 800$ ft

Back tangent grade $g_1 = +2.3\%$

Ans.   High point station = 201 + 10; high point elevation = 84.36 ft

Station	Elevation, ft
199 + 50	83.72
201 + 50	84.32
203 + 50	82.92

13.14 Using the following information, find all curve elevations required to design this summit curve, and locate the station and elevation of the high point.

Each stage = 50 ft   Forward tangent grade $g_2 = -3.0\%$

PVI = 301 + 10 at elevation 421.03 ft   Length of curve $L = 700$ ft

Back tangent grade $g_1 = +3.0\%$

Ans.   High point station = 301 + 10; high point elevation = 442.03 ft

Station	Elevation, ft
299 + 10	414.07
300 + 60	415.67
302 + 60	414.82

13.15 Using the following information, find the high point and all curve elevations.

Each stage = 50 ft   Forward tangent grade $g_2 = -3\%$

PVI = 13 + 02.24 at elevation 48.32 ft   Length of curve $L = 650$ ft

Back tangent grade $g_1 = +4\%$

Ans.   High point station = 13 + 48.67; high point elevation = 42.75 ft

Station	Elevation, ft
9 + 77.24	35.32
11 + 27.24	40.11
13 + 77.24	42.70
15 + 77.74	39.94

DIRECT METHOD OF SUMMIT CURVE COMPUTATION

13.16   Using the following information, find the high point and the curve elevations.

PVI = 210 + 00 at elevation 255.00 ft   Forward tangent grade $g_2 = -2.500\%$

Back tangent grade $g_1 = +0.97\%$ Length of curve $L = 800$ ft

Ans.   High point station = 208 + 23.63; high point elevation = 222.02 ft

Station	Elevation, ft
207 + 00	221.87
209 + 50	221.87
212 + 00	219.13
213 + 50	216.20

13.17   On the basis of the following information find all the curve elevations by the direct method.

PVI = 178 + 50 at elevation 123.75 ft   Forward tangent grade $g_2 = -2.34\%$

Back tangent grade $g_1 = +1.34\%$ Length of curve $L = 750$ ft

Ans.

Station	Elevation, ft
175 + 00	119.05
177 + 00	120.50
178 + 50	120.30
180 + 50	118.32
181 + 00	117.52
PVT 182 + 25	114.97

13.18   A summit curve for a country road is to be computed by the direct method. The following information is given:

PVI = 204 + 55 at elevation 95.95 ft   Forward tangent grade $g_2 = -1.6\%$

Back tangent grade $g_1 = +2.4\%$ Length of curve $L = 700$ ft

Find the elevations of the curve and the high point station and elevation.

Ans.   High point station = 205 + 25; high point elevation = 92.59 ft

Station	Elevation, ft
202 + 00	89.57
204 + 50	92.43
206 + 50	92.14
PVT 208 + 05	90.35

13.19   Using the following information about a summit curve, compute the high point station and elevation and the curve elevations. Use the direct method.

PVI = 15 + 75.02 at elevation 49.76 ft   Forward tangent grade $g_2 = -3.18\%$

Back tangent grade $g_1 = +4.30\%$ Length of curve $L = 600$ ft

Ans.   PVT = 18 + 75.02 at elevation 40.22 ft; high point station = 16 + 19.94; high point elevation = 44.28 ft.

Station	Elevation, ft
13 + 00	37.90
14 + 50	42.48
16 + 50	44.98

13.20   The following pertinent facts are given about a road in the Smoky Mountains.

PVI = 18 + 61.67 at elevation 1700 ft   Forward tangent grade $g_2 = -2.87\%$

Back tangent grade $g_1 = +2.45\%$ Length of curve $L = 550$ ft

Compute the curve elevations and high point station and elevation.

Ans.   High point station = 18 + 39.96; high point elevation = 1696.37 ft

Station	Elevation, ft
16 + 00	1693.57
18 + 00	1696.28
20 + 00	1695.12
PVT 21 + 36.67	1692.10

STAGE METHOD OF SAG CURVE COMPUTATION

13.21   Using the following information, find the low point and curve elevations by the stage method.

Each stage = 50 ft   Forward tangent grade $g_2 = +2.8\%$

PVI = 245 + 50 at elevation 201.00 ft   Length of curve $L = 400$ ft

Back tangent grade $g_1 = -2.6\%$

Ans.   Low point station = 245 + 42.59; low point elevation = 203.70 ft

Station	Elevation, ft
243 + 50	206.20
244 + 00	205.07
245 + 00	203.82
246 + 00	203.92
PVT 247 + 50	206.60

13.22 The following data are given for a sag curve near Buckeye Lake in Ohio. Find the curve elevations by the stage method.

PVI = 186 + 00 at elevation 496.04 ft   Forward tangent grade g_2 = +2.6%

Back tangent grade g_1 = −2.1%   Length of curve L = 450 ft

Ans.

Station	Elevation, ft
184 + 25	499.85
185 + 75	498.66
186 + 75	499.17
187 + 75	500.73

13.23 Using the stage method, find all the curve elevations and the low point station and elevation for this short curve.

PVI = 162 + 50 at elevation 181.05 ft   Forward tangent grade g_2 = +2.10%

Back tangent grade g_1 = −1.76%   Length of curve L = 500 ft

Ans. Low point station = 162 + 27.98; low point elevation = 183.44 ft

Station	Elevation, ft
161 + 00	184.08
162 + 50	183.46
164 + 50	185.35
165 + 50	186.30

DIRECT METHOD OF SAG CURVE COMPUTATION

13.24 The following information is given on the road plans for a sag curve in the mountain state of Colorado. Notice the low grade of the tangents. Solve for the elevations and low point.

PVI = 19 + 20.15 at elevation 6201.85 ft   Forward tangent grade g_2 = +0.45%

Back tangent grade g_1 = −0.031%   Curve length L = 500 ft

Ans. Low point station = 18 + 74.10; low point elevation = 6202.31 ft

Station	Elevation, ft
16 + 70.15	6202.62
17 + 50	6202.43
18 + 50	6202.32
20 + 00	6202.43

13.25 Using the direct method, find all the curve elevations and the low point for the following curve. Given:

PVI = 34 + 32.00 at elevation 103.76 ft   Forward tangent grade $g_2 = +1.3\%$

Back tangent grade $g_1 = -1.7\%$ Curve length $L = 550$ ft

Ans. $C = 0.00002727a^2$; low point station = 34 + 68.67; low point elevation = 105.79 ft

Station	Elevation, ft
32 + 00	107.54
33 + 50	106.17
34 + 00	105.91
35 + 00	105.81

13.26 A curve in the Arizona mountains on route 666 has the following features:

PVI = 324 + 10 at elevation 4106.82 ft   Forward tangent grade $g_2 = +1.4\%$

Back tangent grade $g_1 = -3.8\%$ Curve length $L = 450$ ft

Find the curve elevations.   *Ans.* $C = 0.000005778a^2$

Station	Elevation, ft
322 + 00	4114.81
324 + 00	4109.87
325 + 50	4109.20
326	4109.55

13.27 Given the following data for a sag curve, compute by the direct method the low point and the curve elevations.

PVI = 450 + 94 at elevation 60.67 ft   Forward tangent grade $g_2 = +3.5\%$

Back tangent grade $g_1 = -4.0\%$ Curve length $L = 450$ ft

Ans. $H = 16.88$ ft; low point station = 451 + 09.00; low point elevation = 64.87 ft

Station	Elevation, ft
449 + 00	68.51
450 + 50	65.16
451 + 50	65.01

13.28 Using the following information about a sag curve, compute the low point and the curve elevations. Use the direct method.

   PVI = 376 + 64 at elevation 765.78 ft   Forward tangent grade $g_2 = +2.6\%$

   Back tangent grade $g_1 = -2.4\%$ Curve length $L = 1000$ ft

   Ans. $C = 0.000025a^2$; low point station = 376 + 44; low point elevation = 772.02 ft

Station	Elevation, ft
372 + 00	776.95
373 + 00	774.98
374 + 50	772.96
377 + 50	772.30

Chapter 14

Drawing Maps

14.1 INTRODUCTION

Maps have a great impact upon human activities. Today the demand for maps is greater than ever before. They are needed in the fields of engineering, urban and regional planning, environmental management, conservation, construction, agriculture, geology, military operations, and many others.

Maps may show various features such as property boundaries, topography, soil type, vegetation, and transportation routes. Most of us use the road map for transportation purposes. In engineering, project locations, designing of facilities, and estimating the contract quantities are some of the uses of maps. The military services need a constant flow of up-to-date maps and charts; 500 million copies of maps covering 400 000 square miles of the earth's surface were printed during World War II.

14.2 TYPES OF MAPS

Generally speaking, there are two basic types of maps used in surveying operations: area maps and strip maps. Area maps are for the development of areas such as real estate projects, plant layouts, school layouts, and airfields. Strip maps are used for constructing all forms of line transportation such as railroads, streets, pipelines, and, of most importance, highways.

Control (i.e., a framework of fixed reference points) for an area map will usually consist of loop traverses, and sometimes connecting traverses based on triangulation. Also needed is a system of bench marks connected by bench mark leveling. Control for a strip map consists of a long, single traverse and bench marks off this traverse. These traverse and bench mark lines nearly always are located along the centerline of the project.

Maps show the positions and elevations of all topographic features. Buildings, streams, roads, contour lines must all be shown. Horizontal and vertical measurements are made connecting these features to the control system. Whenever the project is large enough, ties are made by analytical photogrammetry (surveying by aerial photography). The use of analytical photogrammetry instead of ground methods can reduce the cost of a large survey by as much as 50 percent. However, aerial mapping requires a system of horizontal and vertical control on the ground which includes ground ties to selected points that appear on the photographs. The basic control system and the ground ties are made by methods described in the following sections. Ties can be rather long, often extending half a mile from the control.

14.3 HORIZONTAL TIES

A complete horizontal tie must always consist of at least two measurements between the control and the point to be located. These measurements consist of the following: (1) an angle and a distance, (2) two distances, (3) two angles. Figure 14-1 shows various combinations. Extra measurements may be made as a check.

Loci

Locus is a set of all points, lines, or surfaces which satisfy a given requirement (*loci* is the plural of locus). Each measurement establishes a line on which the topographic feature must be placed on the map. This line is a locus of the feature. The place where the lines (loci) of two measurements cross is the location of the specific point. Figure 14-1 shows that these loci are always either straight lines or circles.

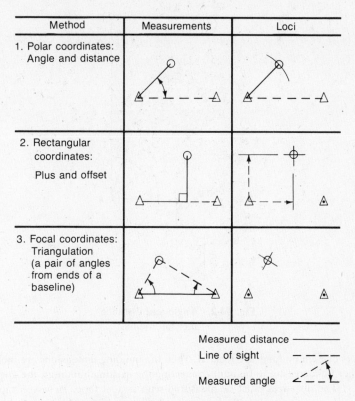

Fig. 14-1 Methods of making horizontal ties.

Creating Loci.   There are four main methods of creating loci:

1. A single distance measured from a point on the control system shows the topographic point is somewhere on a circle whose center is at the control point and whose radius is the distance measured.

2. A distance measurement from a line on the control system shows the topographic feature is somewhere on a straight line parallel to the control line and at the measured distance from it.

3. An angle measurement made at a point on the control shows the topographic feature to be somewhere on a straight line extending from the point where the angle was measured and in the direction indicated by the value of the angle.

4. An angle measurement made at the topographic feature between the directions of the two control points shows the topographic feature is on a circle passing through the two control points and the topographic point.

Strength of Horizontal Ties

Horizontal ties are strongest when their loci intersect at 90°. As the angles depart farther from 90° the tie becomes weaker. If the location on the map is in error more than the error of measurement, then the ties are weak.

The ties in Fig. 14-1 are shown in the order of their importance. Tie method 1 is most useful for an area map. Tie method 2 is used in strip maps. These two methods are used almost exclusively.

Angle and Distance

EXAMPLE 14.1   Use the angle-and-distance method to locate the two buildings shown in Fig. 14-2, and show the field notes.

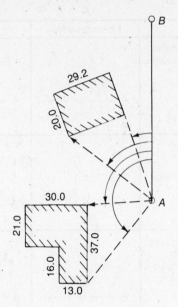

Fig. 14-2 Angle and distance
measurements.

Solution: Two corners of each building are located. Then the building dimensions are measured along the sides of the buildings. If other buildings are to be built connecting the existing buildings, the angles must be measured with extreme accuracy. The distances must be measured with a steel tape. A woven tape may be used if the distances are not great, but a steel tape will give greater accuracy. Figure 14-3 shows how the field notes will appear.

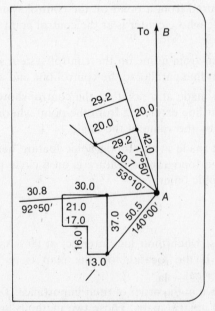

Fig. 14-3

Plus and Offset

EXAMPLE 14.2   Use the plus-and-offset method to locate the building shown in Fig. 14-4, and set up the field notes.

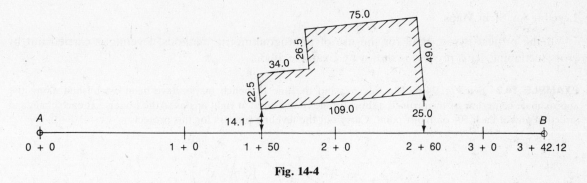

Fig. 14-4

Solution: The traverse line is first marked off in stations which are lined in with the transit. The rear taper holds the zero of the tape first at $1 + 0$. The head taper is on line with the line of marked stations and at the point on the line where he or she estimates a perpendicular from the traverse line would strike the building corner. Next the plus ($+50$) is measured and then the 14.1 ft to the corner of the building. The 14.1 ft is called an offset. In this case the offset is a left offset (LO). For the next corner the rear taper holds at $2 + 0$, and the process is repeated. Finally the dimensions of the building are measured. The field notes are shown in Fig. 14-5.

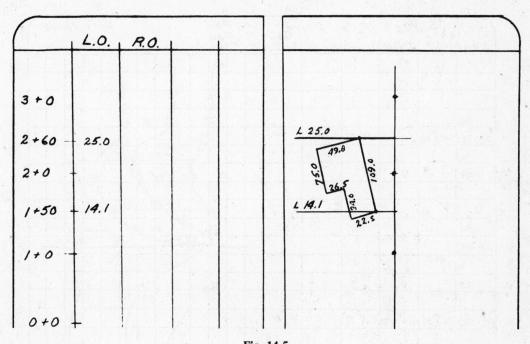

Fig. 14-5

Accuracy

The precision with which horizontal ties are measured depends on the purpose of the survey. If high precision is required, steel tape is used for all measurements. Then the values are placed on the map. Other distances required for the plans are done by mathematical computation. If the distances to be measured for mapping are not too far from the control line, they may be made with a woven tape.

14.4   VERTICAL TIES

Vertical ties for area maps are measured by leveling, as described in Chap. 5. They may also be measured by the stadia method described in Chap. 3. Strip maps are usually done by leveling.

Leveling for Strip Maps

If the project is too small for the use of photogrammetric methods, leveling is carried out by cross sectioning. This process is shown by Example 14.3.

EXAMPLE 14.3 See Fig. 14-6. The traverse and the line of bench marks have been established along the approximate centerline of the project; a short profile is measured at right angles to the traverse at each change in direction and at each 50- or 100-ft point. Carry out the leveling process for this project.

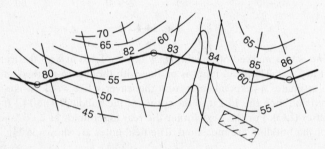

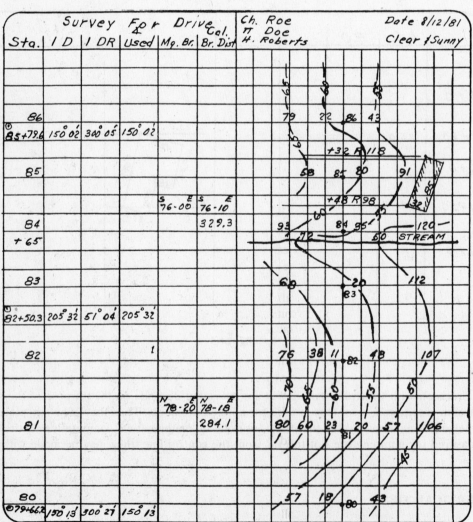

Fig. 14-6 Traverse and line of bench marks in field book.

Solution: Elevations should be established far enough away on each side of the baseline to cover all possible earthwork. Right angles are estimated. Elevations are taken at the centerline, at the breaks, and at the ends of each profile.

Offsets are measured to break in slope, and the pluses and offsets to topographic features are determined. The leveling is carried by ordinary levels from bench mark to bench mark with side shots to the points where elevations are determined. Usually only the centerline elevations are determined with the engineer's level and the contour lines.

The traverse starts at station 79 + 66.2. It extends 284.1 ft at a calculated bearing of N78°18′E, thence turning to a calculated bearing of S76°10′E and running for 329.3 ft. This point is station 85 + 79.6. At 43 ft right of station 80 + 0, the elevation is 45; at 18 ft left the elevation is 50; and at 57 ft left the elevation is 55. At 20 ft right is station 81 + 0 with an elevation of 55; at 57 ft right the elevation is 50; and at 106 ft right the elevation is 45. At 23 ft left the elevation is 60; at 60 ft left the elevation is 65; and at 80 ft left the elevation reads out as 70 ft. Since the plot of the traverse is drawn to scale, the right-hand map in the field book can be sketched in and the elevations called out thereon.

Note: The section beyond 82 + 50.3 does not allow for the change in direction of the centerline.

14.5   MAP SCALE

Choosing a scale for a map is dependent upon the size of the project, the required precision, and the purpose for which the map is designed. Map scales are given in three ways:

1.   Representative fraction or ratio, as 1/2000 or 1 : 2000.
2.   Equation, such as 1 in = 200 ft.
3.   Graphically. Even though the paper on which the map is made changes dimensions, two graphic scales at right angles to each other will permit accurate measurements. It will also provide the proper scale when reduction copies are made.

Map scales are classified as large, medium, or small. A large scale would be 1 in = 100 ft (1 : 1200) or larger. A medium scale would be 1 in = 100 ft to 1000 ft (1 : 1200 to 1 : 12 000). A small scale would be 1 in = 1000 ft (1 : 1200) or smaller.

14.6   MAP DRAFTING

Maps are drafted in two steps. The first step is to prepare a manuscript; the second step is to draft the final map. Generally the manuscript is compiled in pencil on a heavy grade of paper. All features and contours are carefully located in complete detail. Lettering need not be done with extreme care. Careful preparation of the manuscript will ensure a high-quality final map. In addition, carefully prepared maps provide a good graphical check on calculated distances or angles.

The completed version is either drafted in ink or scribed on plastic film. With either drafting or scribing the final map will be traced from the manuscript; thus fastidious care in preparation of the original manuscript is of utmost importance.

In inking, the manuscript is placed on a light table and the features traced on a transparent overlay material. The usual procedure is to letter first. The planimetric features and contours are then traced on the overlay material. In scribing, special tools which can vary line weights and make standard symbols are used to put the information on a transparent, stable-base material which is coated with an opaque emulsion. Manuscript lines are transferred to the coating in a laboratory process. Lines representing features and contours are prepared by cutting and scraping to remove the coating on the *scrib coat*, as it is called. Scribing is usually easier and faster than inking, and its popularity is increasing.

Preparing a Pencil Manuscript

In preparing a pencil manuscript follow these steps:

1.   Plot the control

2. Plot the details

3. Draw the topography and special data

4. Finish the map, including lettering and labeling

14.7 PLOTTING THE CONTROL

Plotting the control depends upon the form in which the control data is available. A traverse control survey can be plotted as a series of angles (using a method described in Sec. 14.8) with distances laid out at the scale you have selected for the map.

Scales normally are 1 in = 10, 20, 40, 50, or 100 ft if you are using the U.S. Customary system. With the metric system the scales are 1 : 100, 1 : 200, 1 : 500, or 1 : 1000 in metric units. Use an engineer's scale in connection with steel scales and dividers to mark control points with an accuracy of 0.02 or 0.01 in or better.

For a traverse in angles and distances, bearings and lengths of the courses should be placed parallel to the lines for easy reading when the user looks at the map from the bottom or the right side. Bearings must be shown in the forward direction and continuously around the traverse. When bearings are read from left to right, but actually run from right to left, an arrow should show the correct bearing direction.

Instead of angles and distances, coordinates can be used for plotting traverses after calculating x and y values for each station. If coordinates are used, the map sheet is first laid out carefully in a grid pattern with unit squares of appropriate size, such as 100, 400, 500, or 1000 ft, and checked by measuring the diagonals. Grid lines are labeled with their coordinate values. Make the origin of the coordinates the most westerly or southerly station of the traverse, even if it is off the map, to assure all positive values.

Control points are plotted by measuring x and y coordinates from the ruled grid lines. Circles $\frac{1}{8}$ in or less in diameter are used to mark the hubs. Errors in plotting may be detected by comparing the scaled distance and bearing of the lines with the length and direction computed or measured in the field. Most topographic maps are drawn with the hub locations either omitted or drawn in light blue ink so they will appear less prominent on the print.

14.8 PLOTTING ANGLES

Three methods are used to plot angles: the tangent method, the chord method, and the protractor method.

Tangent Method

The procedure for plotting an angle by the tangent method, which is used extensively for plotting deflection angles, will be illustrated by Example 14.4.

EXAMPLE 14.4 See Fig. 14-7. Given: A traverse starting at station $0 + 00$ and running in a straight line to station $5 + 20$. It then turns left at an angle of $14°20'$ and runs to station $11 + 64$. At station $11 + 64$ it again turns left at an angle of $16°16'$ and proceeds in a straight line to station $18 + 21$. Find the angles by the tangent method.

Solution: Choose a convenient distance and measure it along the reference line to serve as a base. Following Fig. 14-7, plot a $14°20'$ deflection angle at point A, and mark length AB equal to 10 in on the prolongation of the back line. If 10 in is too large for your paper, select another unit, such as 10 cm; however, keep in mind that the more distance you employ the more accurate will be the angle constructed. Find the natural tangent of $14°20' = 14.3333°$ from a hand calculator or a table of tangents.

$$\tan 14.3333° = 0.25552$$

Since line AB is 10 units long, multiply 0.25552 by 10. This gives 2.55 in, or whatever unit you have chosen. This locates point C and produces the desired angle with line AB.

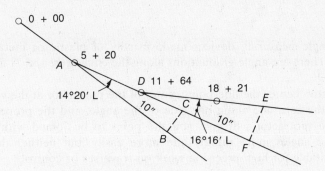

Fig. 14-7 Plotting angles by the tangent method.

Note: Any length of base could be used, but a distance of 10 units requires only moving the decimal place in the natural tangent taken from a calculator or set of tables.

At station 11 + 64 the same procedure is employed. The extension of line *AD* for 10 units beyond *D* with the erection of a perpendicular of 2.92 units will give triangle *DEF*, which gives the desired angle of 16°16′. Find the tangent of 16°16′.

$$\tan 16°16′ = \tan 16.26667° = 0.29179$$

Multiplying 0.29179 by 10 units gives a line *DF* with a distance of 2.92 in (or centimeters or whatever units you have used for your baseline).

Chord Method

To lay off an angle by the chord method, use the following formula:

$$C = 2r \sin \tfrac{1}{2} A$$

where C = the chord for the desired angle
            r = radius of the circle of which the arc subtending the angle is a part
            A = the angle

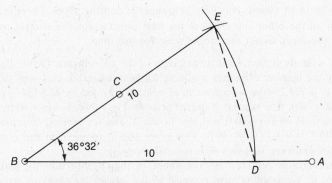

Fig. 14-8 Chord method of laying off an angle.

EXAMPLE 14.5 See Fig. 14-8. Given: An angle of 36°32′ on line *BA*. Set up this angle by the chord method.

Solution: Draw line *BA* as shown in Fig. 14-8. Draw base *BD* a convenient length of 10 units (e.g., inches centimeters, units on a sheet of graph paper). Use *B* as a vertex and swing an arc whose radius is 10 units long. Now with point *D* as the center, and a radius equal to the chord for the desired angle, draw another arc. The intersection of the two arcs locates *E*. The line connecting points *B* and *E* forms the other side of the angle. Use the formula for the chord for a desired angle. Multiply twice the distance *BD* by the sine of one-half the angle.

$$ED = 2BD \sin \tfrac{1}{2}(36°32′) = 2(10) \sin 18°16′ = 20 \sin 18.2666° = 20(0.31344) = 6.2688 \quad \text{or} \quad 6.27 \text{ units}$$

Protractor Method

Protractors are angle-measuring devices made usually of plastic or metal. They come as full circles or semicircles. There are angle graduations along the circumference. A fine point identifies the center of the circle.

To use the protractor, center the fine-point center of the protractor at the vertex of the angle you wish to develop. Put the zero line along one side of the angle, and the proper angle point marked along the edge of the protractor will give you the point to be joined with the vertex. Drafting machines will plot the angles more rapidly and more easily, but neither the protractor nor the drafting machine is suitable for high-precision work on traverses or control.

14.9　TOPOGRAPHIC SYMBOLS

In order to show many details on a single sheet a set of standard symbols has been developed. The symbols represent special topographic features; Fig. 14-9 shows a few of the hundreds of symbols used on topographic maps. The drafter needs considerable practice to draw these symbols well and at a suitable scale. Before the symbols are placed on a map, such items as roads, boundary lines, and buildings are first plotted and inked. The drafter then puts in the symbols by drawing them in or cutting them from standard sheets which have an adhesive on the back and pasting them carefully on the map. The cut-and-paste symbols eliminate differences in drafting quality since often one map will have been prepared by several drafters.

14.10　PLACING THE MAP ON A SHEET

Appearance of the finished map has great bearing on its value. Poorly arranged, carelessly lettered, and rough-looking maps do not inspire confidence. The map should have borderlines which are considerably heavier than all other lines. This greatly improves the appearance of the map.

The first step in map arrangement is to position the control and topography so that the sheet is properly balanced. Figure 14-10 shows a traverse with no topography that has been well placed on the sheet. Before plotting the map, determine the proper scale for a sheet of given size.

EXAMPLES 14.6　See Fig. 14-10. Given: An 18 by 24 in sheet of drafting paper. There is a 1-in border on the left side and $\frac{1}{2}$-in borders on all the other sides. Select the best position for the control traverse so the map will be well proportioned on the sheet. Select the proper scale for this map.

Solution:　A is the most westerly station, so it is the origin of the coordinates. Divide the total departure to the most easterly point C by the number of inches available for plotting in the east-west direction. The maximum scale possible in Fig. 14-10 is 774.25 (the x coordinate of point C) divided by 22.5 (24-in paper width minus $1\frac{1}{2}$-in margin width), giving 1 in ≐ 34 ft. The nearest standard scale to 1 in = 34 ft is 1 in = 40 ft.

Now we must see if a scale of 1 in = 40 ft will also fit the Y axis (or vertical dimension). Divide the total Y coordinate of 225.60 + 405.57 = 631.17 ft by 40 ft. This gives 15.8 in required in the north-south direction. We have 17 in available in this direction, so 1 in = 40 ft is satisfactory.

In Fig. 14-10 the traverse is centered on the sheet in the Y direction by making each m equal to $\frac{1}{2}[17 - (631.17/40)]$, or 0.61 in. Keep the control traverse to the left on the sheet to provide space for the title, notes, and north arrow. So m on the left can be made approximately 1 in.

If the traverse with the topography is to be plotted by any means other than coordinates, a sketch should first be made showing the controlling features. If you make the sketch on tracing paper, you can orient the map for the best appearance on the sheet by rotating and shifting the tracing paper over the top of the sheet on which the final map will be drawn.

14.11　MERIDIAN ARROW

The meridian arrow, commonly called the north arrow, must be placed on every map for orientation purposes. It should be near the top of the sheet and near the right-hand corner. The

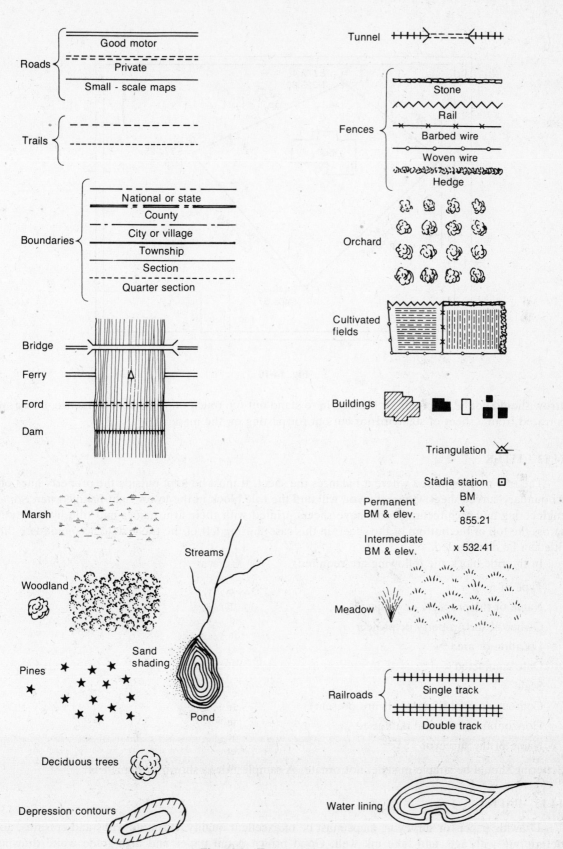

Fig. 14-9 Topographic symbols.

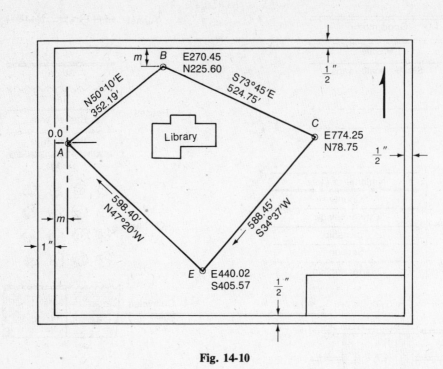

Fig. 14-10

arrow should not be elaborate or blacked in to stand out too much on the map. In practice, an arrow is traced from a sheet of standards or cut out for pasting on the map.

14.12 TITLE

The title may be placed where it balances the sheet. It must be kept outside the property lines on a boundary-survey sheet. Generally you will find the title block in the lower right-hand corner. Some engineering and architectural firms have sheets printed with their firm names and title information across the top or the bottom of the sheet; in this case shifting left of the traverse to accommodate the title can be disregarded.

In the title block the following are required:

Type of map

Name of the property or project

Owner of the property or project

Location or area

Date completed

Scale

Contour interval (if contours are present)

Horizontal and vertical data used

Name of the surveyor

Lettering should be simple in style, not ornate. A sample title is shown in Fig. 14-11.

14.13 PAPER

Drawing paper for surveying maps must be of excellent quality. It should withstand erasures, not deteriorate with age, and take ink well. Good brown detail paper and high-grade white drawing paper are commonly used. Cloth-backed white sheets are desirable for important work.

```
┌─────────────────────────────────────────┐
│         UNITED STATES AIR FORCE          │
│         CIVIL ENGINEERING DEPT.          │
│    Survey of Library     Wright AFB      │
│   Scale: 1 in = 20 ft     Date: 10 Nov. 1983 │
│   Survey by L. Jones, T. Hall    Map by S. James │
└─────────────────────────────────────────┘
```

Fig. 14-11   Title block.

Maps are often drawn or traced on transparent polyester materials. Mylar is the most popular with engineering firms. Polyester sheets, which are dimensionally stable and very durable, are commonly used. Ink can easily be erased from them with water.

14.14 SOURCES OF ERROR IN MAPPING

Following are the most common sources of error in mapping:

1. Not scaling distances when plotting by coordinates
2. Using map sheets varying in dimensions
3. Using too soft a lead on original plot
4. Using a protractor to plot angles when great precision is required

Solved Problems

14.1   Show the notes for the right-hand page of the field book for the topography shown in Fig. 14-12.

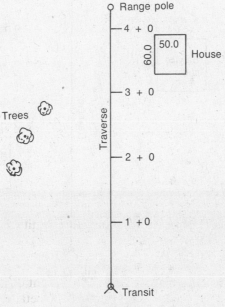

Fig. 14-12

Solution

See Fig. 14-13. Note that in this solution station 1 + 0 is used as the control point on the traverse line to sight in the three trees. Station 2 or 3 could also be used instead of station 1. The angles have been turned to the right as shown by the arrow. To set up the two points required to locate the house, station 3 is used. Angles are turned to the right as shown by the arrow.

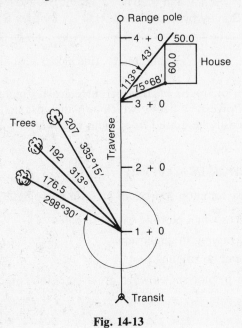

Fig. 14-13

14.2 Show the notes for the right-hand page of the field book for the topography shown in Fig. 14-14.

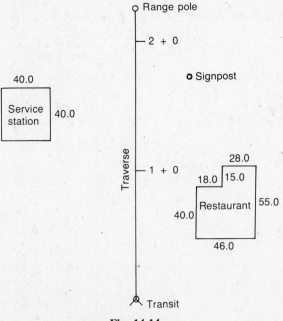

Fig. 14-14

Solution

See Fig. 14-15. Note that station 1 is used as a control point, and angles are turned to the right. The two points needed to locate the restaurant are 50.0 ft at 102° and 71.5 ft at 135°. Station 2 is used to

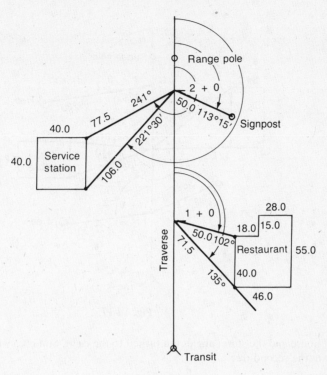

Fig. 14-15

locate the signpost and the service station. All angles are turned to the right; 106.0 ft at 221°30″ and 77.5 ft at 221°30″ give the two points required for locating the service station.

14.3   Show the notes for the right-hand page of the field book for the topography shown in Fig. 14-16.

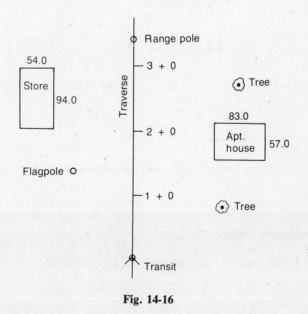

Fig. 14-16

Solution

See Fig. 14-17. In this solution station 1 is used to locate the tree, with the angle turned to the right. Again station 1 is used to locate the flag pole (angle turned to the right). Station 2 is used to locate both

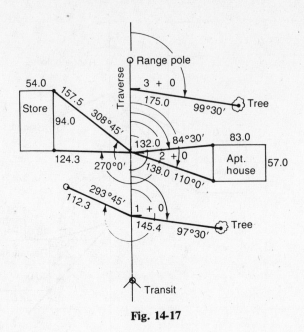

Fig. 14-17

the apartment house and the store; angles are turned to the right. Station 3 with angle turned to the right is used to locate the second tree.

14.4   Show the field notes for both the left- and right-hand pages required for the map shown in Fig. 14-18.

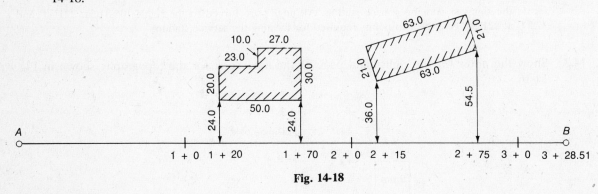

Fig. 14-18

Solution

Figure 14-19 is self-explanatory.

14.5   Show the field notes for both the left- and right-hand pages required for the map shown in Fig. 14-20.

Solution

Figure 14-21 is self-explanatory.

14.6   See Fig. 14-22. Describe the leveling process for this project.

Solution

The traverse, which in this case is just one straight line, starts at station 61 + 80. It runs straight along the road for 501.10 ft at a calculated bearing of N12°29′E. The traverse ends at station 66 + 81.1. Elevations are as follows:

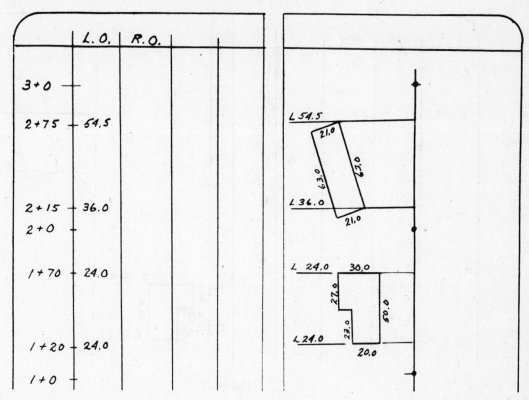

Fig. 14-19

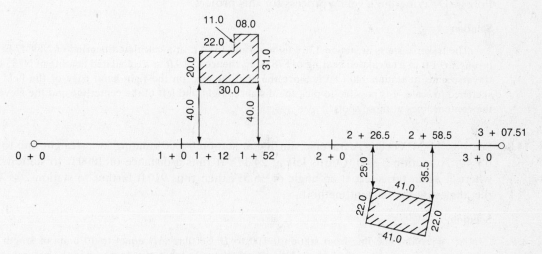

Fig. 14-20

Station 62:   30 ft left, elevation 35

10 ft left, elevation 40

7 ft right, elevation 35

25 ft right, elevation 30

47 ft right, elevation 25

All the other elevations both right and left of centerline may be picked from the contour line sketch on the right-hand page of the field book. The building, a small service station, is 16 ft wide by 68 ft long.

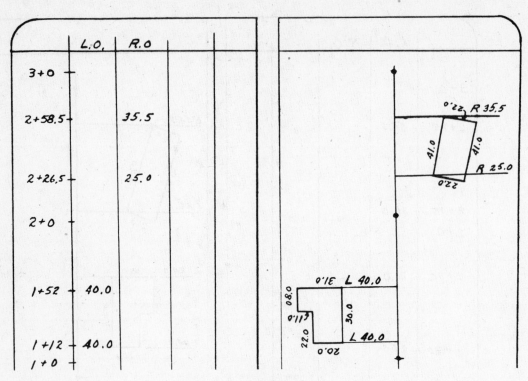

Fig. 14-21

14.7   See Fig. 14-23, which is a survey of a short distance on Tiger Creek Trail in Tiger Creek Forest. Describe the leveling process for this project.

Solution

The traverse starts at station 136 + 00 and runs 151.1 ft at a calculated bearing of N89°47′E. Thence it runs 169.1 ft at a calculated bearing of S78°01′E, thence 159.9 ft at a calculated bearing of N73°00′E. The traverse ends at station 140 + 80.1. Since the contour map on the right-hand page of the field book is sketched to scale, it is possible to pick the distances right and left of the centerline and the elevations of the contour lines at those spots.

14.8   See Fig. 14-24. Given: A traverse starting at station 0 + 00 running in a straight line to station 6 + 40. At station 6 + 40 it turns left at 24°17′. It runs a distance of 1090 ft to station 17 + 30, where it again turns left at an angle of 36°51′, then runs 910 ft farther to station 26 + 40. Plot the angles by the tangent method.

Solution

At *A* prolong the line from station 0 + 00 to *B*. Set line *AB* equal to 10 units of length (inches, centimeters, or any units you choose). Using a calculator or table of tangents, find the natural tangent of 24°17′.

$$\tan 24°17′ = \tan 24.28333° = 0.45117$$

Multiply 0.45117 by 10, as line *AB* is 10 units long: 10(0.45117) = 4.51 units of length. Erect a perpendicular at *B*, which is 4.51 units long; this will be point *D*. *Note*: In this problem the plotted point turned out to be station 17 + 30. This is unusual and does not happen often.

We now have developed the 24°17′ left angle. From station 17 + 30 we turn left at 36°51′ and proceed 910 ft to station 26 + 40. Prolong line *AD* for 10 more units to position *F*. At *F* erect a perpendicular to the left. Find the natural tangent of 36°51′ (= 36.85°) and multiply by 10 units' length (tan 36.85° = 0.74946): 0.74946(10) = 7.49 units of length. So line *FE* = 7.49 units. Connect *D* and *E*, thus producing the angle of 36°51′ between lines *AF* and *DE*. Measure 910 units along line *DE* and stop at the end station, which is 26 + 40.

14.9   See Fig. 14-25. Given: A traverse starting at station $0+00$ and running 830 ft to A, where it turns 48°01′ right. It then runs 870 ft to D, where it turns left 42°12′. It runs thence 1300 ft to station $30+00$, where it ends. Find the angles by the tangent method.

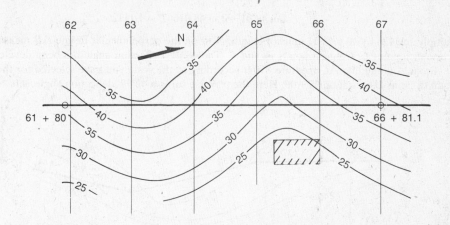

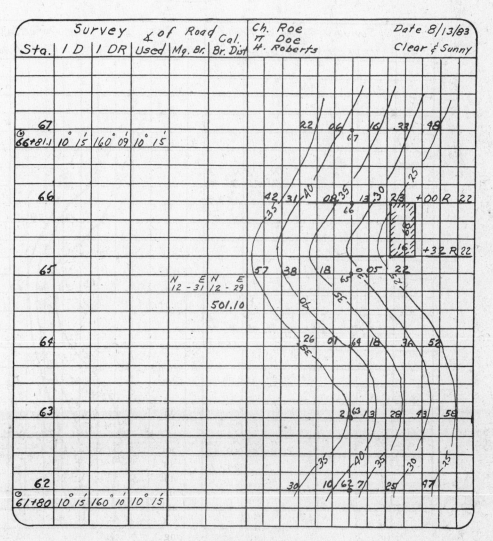

Fig. 14-22

Solution

Prolong the baseline starting at $0+00$ to A. Along the prolonged line measure 10 units from point A to B. The angle is turning right so erect a perpendicular to AB to the right. Compute the natural tangent of 48°01′ in your hand calculator or look it up in a table of tangents.

$$\tan 48°01' = \tan 48.01667° = 1.11126$$

Multiply 1.11126 by $10 = 11.11$ units of length. Now on the perpendicular to line AB measure 11.11 units to establish point C. Connect points A and C. The 48°01′ deflection angle has been developed.

From A along line AC measure 870 ft according to the scale you have selected for the traverse and establish point D as station $17+00$. Here the traverse turns a 42°12′ deflection angle, left. So prolong the

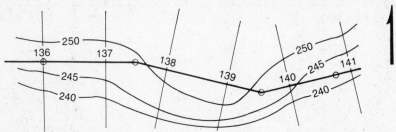

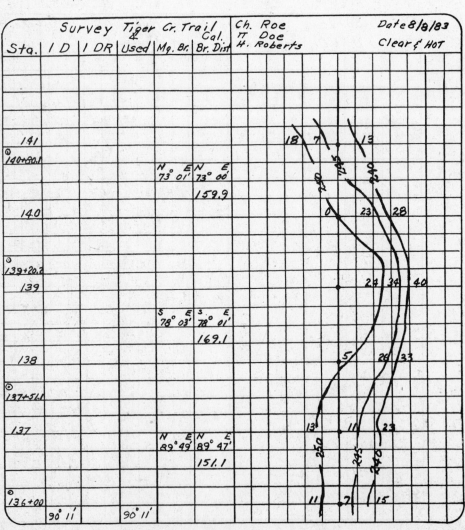

Fig. 14-23

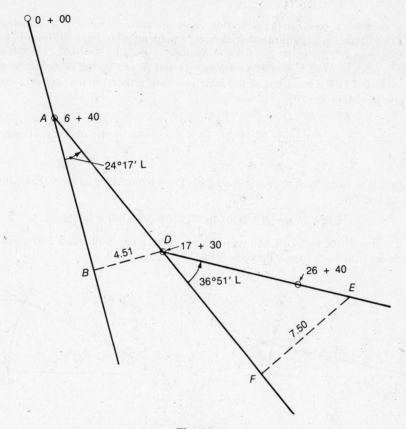

Fig. 14-24

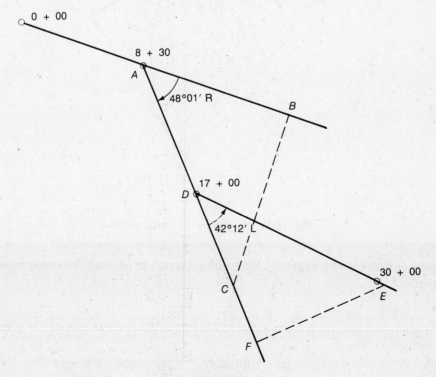

Fig. 14-25

ADC line to establish a base for the deflection angle. Measure 10 units of length starting at *D* and ending at *F*. This deflection angle turned left, so erect a perpendicular going left from *F*. Find the natural tangent of 42°12′: tan 42°12′ = tan 42.2° = 0.90674. Multiply 0.90674 by 10 = 9.07 units. Measure 9.07 units along the perpendicular to find point *E*. Join points *D* and *E*, and the 42°12′ deflection angle has been established. Measure 1300 ft according to the scale you have chosen for your traverse and establish the end of traverse at station 30 + 00.

14.10 See Fig. 14-26a. Given: An angle of 11°30′. Plot the angle by the chord method.

Solution

Lay out a 10-unit base; this is shown as line *MN* in Fig. 14-26b. Swing an arc with this 10-unit base. Use the formula to find *PN*.

$$ED = 2R \sin \tfrac{1}{2}B = 2(10) \sin 5°45′ = 20(0.10019) = 2.00 \text{ units}$$

With *D* as a vertex, strike an arc of 2.00 units' length. Where the 10-unit and 2.00-unit arcs intersect is point *E*. Connect *E* to *B*; the angle between is the 11°30′ angle.

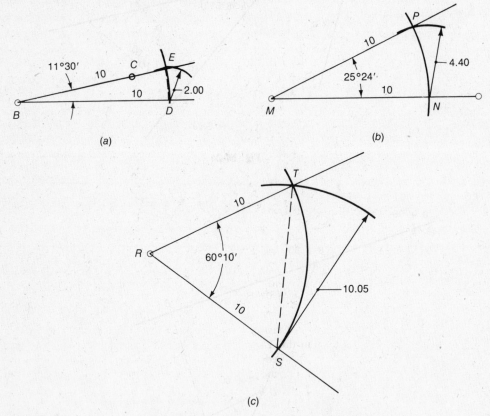

(a)

(b)

(c)

Fig. 14-26 Angles by the chord method.

14.11 See Fig. 14-26b. Given: An angle of 25°24′ with its vertex at *M*. Find the required angle by the chord method.

Solution

Lay out a 10-unit base; this is shown as line *MN* in Fig. 14-26b. Swing an arc with this 10-unit base. Use the formula to find *PN*.

$$PN = 2R \sin \tfrac{1}{2}M = 2(10) \sin 12°42′ = 20(0.21985) = 4.40 \text{ units}$$

Strike an arc of 4.40 units' length from N to P, where the 10-unit arc intersects the 4.40-unit arc. Draw a line from the vertex at M to the arc intersection at P, and the deflection angle of 25°24′ is included between that line and the baseline.

14.12 See Fig. 14-26c. Given: An angle of 60°10′ with its vertex at R. Develop the 60°10′ angle by the chord definition.

Solution

Make the baseline RS 10 units long. Strike an arc from S to T with a 10-unit radius r and vertex at R. Use the formula to find TS.

$$TS = 2r \sin \tfrac{1}{2}R = 2(10) \sin 60°10′ = 20(0.50126) = 10.03 \text{ units}$$

Strike an arc from S of 10.03 units' length. Where this arc and the radius of 10 units intersect at T, draw a line to the vertex R. This is the desired angle of 60°10′.

14.13 See Fig. 14-10. Given: Sheet of drafting paper 18 by 24 in. Control traverse with E station of E440.02, S350.00 and B station of E270.45, N210.00. Locate on the sheet C station of E750.00, N78.76 and give the scale to be used.

Solution

First determine an appropriate scale: 22.5 in is available in the east-west direction. The east-most point on the traverse is 774.25; thus

$$\frac{774.25}{22.5} = 34.41 \text{ ft}$$

The nearest standard scale is 1 in = 40 ft. Check the scale of 1 in = 40 ft in the Y direction. The sum of the Y coordinates is 210.00 + 350.00 = 560 ft. Thus 560/40 = 14 in is required in the north-south direction. We have $16\frac{1}{2}$ in available, so the scale of 1 in = 40 ft is OK.

Center the traverse in the Y direction by making each m distance = $\frac{1}{2}[17 - (560/40)] = 1.5$ in. Use $m = 1.5$ in in both X and Y directions.

14.14 See Fig. 14-10. Given: Sheet of drafting paper 18 by 24 in. Control traverse with coordinates as follows: B station = E210.01, N200.10; E station = E440.00, S325.25; C station = E810.05, N78.00. Find the proper scale and give dimensions and locations of each m.

Solution

The maximum scale possible east and west is 810.05/22.5 = 36.00 ft. The nearest standard scale is 1 in = 40 ft. Check this scale in the Y direction: Sum of Y coordinates = 200.10 + 325.25 = 525.35 and (525.35 ft)/(40 ft/in) = 13.33 in required. Since we have 17 in between margins, the scale of 1 in = 40 ft is OK. Center the sheet in the Y direction:

$$m = \frac{1}{2}\left(17 - \frac{525.35}{40}\right) = 1.9 \text{ in}$$

So the E and B stations will be 1.9 in inside the borders.

C station is E810.05, so 810.05/40 = 20.25 in. There is 22.5 in available between margins, so 22.5 − 20.25 = 2.25 in. If you want to split the distance on the X axis equally, 2.25/2 = 1.12 in for the left-hand margin; $m = 1.12$ in. An even better solution might be to make this starting point 0.5 to 1.0 in from the left margin to provide more space for the north arrow and title block.

14.15 See Fig. 14-10. Given: Sheet of drafting paper 18 by 24 in. Coordinates for B station = E270.00, N310.00; E station = E440.00, S460.00; C station = E950.00, N78.00. Find the proper scale and the m dimensions for X and Y directions.

Solution

The maximum scale possible east and west is $950.00/22.5 = 42.22$ ft. This is more than the standard 1 in = 40 ft and would run the controlling traverse off the paper. The next standard scale is 1 in = 50 ft, which we will use as our scale. It must now be checked to see if it fits in the north-south direction. Total the Y coordinates: $310.00 + 460.00 = 770.00$ ft. Then $(770 \text{ ft})/(50 \text{ ft}) = 15.4$ in required. We have 17 in available between margins, so a scale of 1 in = 50 ft is OK.

Center the sheet in the Y direction:

$$m = \frac{1}{2}\left(17 - \frac{770}{50}\right) = 0.8 \text{ in}$$

So the B and E stations will be 0.8 in from the margins. The C station is E950.00, so $950/50 = 19$ in. There are 22.5 in available in the X direction, so $22.5 - 19 = 3.5$ in total; $3.5/2 = 1.75$ for m on each side of traverse. To allow more room for the north arrow and title block, make the left $m = 1$ in.

14.16 See Fig. 14-10. Given: Sheet of drafting paper 18 by 24 in. Coordinates of stations: B station = E270.00, N115.10; E station = E440.00, S190.50; C station = E410.00, N78.00. Find the best scale to use for the map. Find the m dimension for both the X and Y directions.

Solution

Maximum scale east-west is C station: $410.00/22.5 = 18.22$ ft. This is under the standard scale of 1 in = 20 ft, so our scale will be 1 in = 20 ft. Now check this scale for the north-south direction. Sum the Y coordinates: B station 115.10 + E station 190.50, or $115.10 + 190.50 = 305.60$ ft. Thus $(305.60 \text{ ft})/(20 \text{ ft}) = 15.28$ in required. We have 17 in available, so a scale of 1 in = 20 ft is OK.

Center the sheet in the Y direction:

$$m = \frac{1}{2}\left(17 - \frac{305.60}{20}\right) = 0.86 \text{ in}$$

So the B and E stations will be 0.86 in from the top and bottom margins. C station is E410.00, so $410/20 = 20.5$ in. We have 22.5 in available; $22.5 - 20.5 = 2.00$ in. And $(2 \text{ in})/(2 \text{ spaces}) = 1$ in. So m in the X direction is 1 in on either side.

Index